Mismanagement of Marine Fisheries

ALAN LONGHURST examines the proposition, central to fisheries science, that a fishery creates its own natural resource by the compensatory growth it induces in the fish, and that this is sustainable. His novel analysis of the reproductive ecology of bony fish of cooler seas offers some support for this, but a review of fisheries past and present confirms that sustainability is rarely achieved. The relatively open structure and strong variability of marine ecosystems is discussed in relation to the reliability of resources used by the industrial-level fishing that became globalised during the twentieth century. This was associated with an extraordinary lack of regulation in most seas, and a widespread avoidance of regulation where it did exist. Sustained fisheries can only be expected where social conditions permit strict regulation and where politicians have no personal interest in outcomes despite the current enthusiasm for ecosystem-based approaches or for transferable property rights.

ALAN LONGHURST began his career in fisheries science, but is best known as a biological oceanographer, being the first Director of the Southwest Fisheries Science Center of the US NMFS in La Jolla, California, and later the Director-General of the Bedford Institute of Oceanography in Nova Scotia, Canada. He produced the first estimate of global plant production in the oceans using satellite imagery, and also quantified vertical carbon flux through the planktonic ecosystem. More recently, in reaction to disastrous Canadian management of NW Atlantic cod stocks, he has offered a number of critical reviews of several aspects of fishery management science. He retired in 1995 and now divides his time between South West France and Nova Scotia.

Mismanagement of Marine Fisheries

ALAN LONGHURST

CAMBRIDGE UNIVERSITY PRESS
Cambridge, New York, Melbourne, Madrid, Cape Town, Singapore,
São Paulo, Delhi, Dubai, Tokyo

Cambridge University Press
The Edinburgh Building, Cambridge CB2 8RU, UK

Published in the United States of America by
Cambridge University Press, New York

www.cambridge.org
Information on this title: www.cambridge.org/9780521896726

First published 2010

Printed in the United Kingdom at the University Press, Cambridge

A catalogue record for this publication is available from the British Library

Library of Congress Cataloguing in Publication data

Longhurst, Alan R.
Mismanagement of marine fisheries / Alan Longhurst.
p. cm.
ISBN 978-0-521-89672-6 (Hardback) – ISBN 978-0-521-72150-9 (Paperback)
1. Sustainable fisheries. 2. Fish populations. 3. Marine fisheries–Management. I. Title.
SH329.S87L66 2010
639.2′2–dc22

2010001344

ISBN 978-0-521-89672-6 Hardback
ISBN 978-0-521-72150-9 Paperback

Contents

Preface

A flood of introspection has overtaken fishery science in recent years, and it would be easy to conclude that everything worth writing had already been written. But the recent literature generally avoids criticism of the central and ancient axiom that fisheries are sustained by the density-dependent increase in growth that is provoked by fishing: what has come to be referred to rather inexactly as surplus production. Uncritical acceptance of this axiom then permits the assumption that fishing ought to be sustainable, provided only that appropriate economic and stock management methods are employed. On the other hand, the recent literature has also been characterised by analyses of the present situation of fish stocks that can only be described as alarmist, and which have been widely reported in the press and other media.

Confidence in the theoretical underpinning of fishing has nevertheless been confirmed at recent international meetings of administrators and natural and social scientists, gathered to discuss the crisis in world fisheries. A review of the documents presented at meetings such as the 1995 Rome Consensus on World Fisheries, or the 2004 World Fisheries Congress, reveals few expressions of doubt concerning theoretical sustainability of fisheries. Instead, the participants call for actions 'to eliminate overfishing, rebuild fish stocks, minimise wasteful fishing practices, develop sustainable aquaculture, rehabilitate fish habitats, and develop fisheries for new and alternate species based on principles of scientific sustainability and responsible management'. Note the use of the loaded term 'overfishing' (which I shall largely eschew) and the fact that this text lies at the heart of the Food and Agriculture Organisation (FAO) Code of Conduct for Responsible Fisheries.[1] Similar conclusions concerning the unquestioning adherence to the axiom may be drawn from quite different discussion groups, such as the scientific forum organised by the Royal Society in London in 2004 on 'Fisheries: past, present and future'; there, too, the participants appear not to have seriously questioned the sustainability of fishing, and the theoretical basis of fishery science, provided only that social, economic and some ecological processes could be factored into management.

I learned something of the depth of this conviction among the hierarchy of such organisations when I was invited to lecture at the ICES[2] Centennial meeting in 2001, and proposed the somewhat provocative title '*Fish stock management is an unsolved ecological problem*'. The organisers clearly found that this title reflected an insufficiently positive attitude on my part towards the corpus of fishery science, and agreed to it only conditionally – I was asked to reserve part of the time that had been allocated to my lecture for a formal rebuttal . . .

The arguments of this book continue those of my ICES lecture and suggest that, given the perfectly evident problems that plague the fishing industry everywhere, our response – whether as natural or social scientists – remains insufficiently critical of the central assumption that commercial fishing is sustainable and that catches may be taken indefinitely at levels that are satisfactory for all categories of fishing, from industrial to artisanal. By extension, it is also generally assumed that stocks may be managed sustainably to provide yields either at maximal or at economically optimal yields.

Every book about fishery management that you have ever consulted was based on some version of this conjecture, and most of them assumed that it was so obviously true that no alternative was discussed. Although contrary suggestions are so rare as to be effectively invisible, they do exist: one pillar in the edifice of fishery science, the concept of a maximum sustainable yield (MSY), has recently been described as no more than 'policy camouflaged as science'.[3] So nothing is as simple as it seems to be on the surface!

The assumption of surplus production of fish stocks is usually supported by reference to population models that ignore the interaction of the stock with other biota with which it shares its complex and dynamic natural habitat; Daniel Pauly, however, has suggested that an 'ability to produce a surplus that we can share, year for year, is an emergent property of marine ecosystems, contingent on their continued existence as complex entities'.[4] Because it invites us to think not only about fish, but also about their habitat, this is a much more interesting way of saying the same thing as Pieter Korringa in 1976, who suggested that there is a surplus available to be taken from every fish stock, because catching this surplus is only like 'taking the interest on the capital. On that basis we can manage the fishery. It is a good interest: about

20% increase in fish weight per year. You don't make that everywhere.'[5] Korringa's remarks were made during the period when fisheries scientists had perfect confidence that their simple population models were a sufficient response to the problem of computing the allowable catch for each stock. He went on to express great confidence in the system of regulation and policing that existed in the North Sea at that time. Pauly probably has no such confidence, but nevertheless suggested that fishing should be sustainable at levels of exploitation useful to us: his short statement shall serve as the text for my argument in this book.

It is my intention to examine, as critically as I can, our central assumption concerning surplus production, and the consequences of stock management procedures that are based on that assumption. This will require a rather more wide-ranging examination than has usually been the case in analysis of the sustainability concept, even that of the recent and masterly essay by Sidney Holt.[6] It will be necessary to discuss how fish, their food organisms and their predators respond to the physical forcing of their non-linear environment, and the curiously neglected significance of some aspects of their reproduction, nutrition and growth must also be addressed. As shall become clear, this discussion is intended to suggest that reliance on simple population analysis may lead to erroneous conclusions concerning population growth potential.

However, another thread of my argument is quite different from the analysis of the biology and ecology of fish: it is that we can understand the many failures of fishery management only by understanding the social, economic and political climate in which sustainable fishery management is attempted. Perhaps the problems of understanding the ecology of exploited species, of measuring the state of each stock in near-real time, and of setting appropriate catch limits are all trivial compared with the resolution of political interference in setting quotas, of vote-swapping in international fishery commissions and of cheating on regulations by fishermen at sea; further, as I shall point out, very large regions of the ocean are effectively lawless and here fishery management is, and will probably continue to be in our lifetimes, totally ineffectual. I shall suggest that to concentrate attention only on the scientific issues arising from fishing wild stocks, and to ignore the social, economic and political context within which fishing occurs, is to ensure the loss of natural resources.

Although my intention is to expose the present reduced state of marine fish stocks and the consequences of their management (where there is any), you should not expect any insistence concerning what must be changed in order to repair the obvious damage that industrial fishing has caused to fish stocks and their habitat: the polemics I will leave to the many other people who are less doubtful than I am about the abilities of an exploding human population to manage its future rationally.[7]

This book is one very small outcome of the unexpected collapse of what appeared to be the well-managed cod stocks of the Grand Banks and of other NW Atlantic regions in the late 1980s. Although I was not then working on fisheries problems, I was sufficiently close to those events to be very shocked by them: at that time, I was employed by the same Canadian federal department that had the responsibility to manage the fishery, although I was engaged quite differently – on research into the ecology of the pelagic production system of the open ocean in relation to carbon flux. But the realisation that a defining moment in fisheries science had occurred, and had occurred right on our doorstep, made it inevitable that I should start to think again about fishery management, even after so long a period of involvement in the problems of pelagic ecology.

It is probably appropriate to explain why a biological oceanographer might choose to become involved in the controversies now swirling around the fishery science community. At the start of my career in the early 1950s, I was recruited to one of the six tropical fisheries laboratories that were then operated by the Colonial Office in London: ours was the now long-forgotten West African Fisheries Research Institute, located in Freetown, Sierra Leone. Each of these six laboratories was decently equipped, with access to a sea-going research ship, and staff were encouraged to study the ecology of the near-pristine fish stocks and marine environment that lay off their respective coasts: the whole enterprise was directed by C.H. Hickling, herring biologist, in London. I spent several years investigating and mapping the benthic invertebrate communities and their ecology from the mangroves to the shelf-edge, and from Sierra Leone to the Gambia. Later, I turned to studies of the bionomics of demersal fish, simply because I inherited much unworked data from research on demersal resources off Sierra Leone.

These studies were continued in the laboratories of the Federal Fisheries Service in Lagos, where my Nigerian colleagues and I undertook the gamut of primary research appropriate to trawl-fisheries: mesh-escapement experiments, and investigations of the growth and reproductive biology of the half-dozen species of fish that were dominant in the landings, together with their age-specific diets; we also established a simple system for examining and monitoring the local oceanographic regime. We made proposals to the relevant federal Ministry concerning mesh regulations and minimum landed sizes, as well as for marine protected areas, where we considered that trawling should be banned to protect stocks for artisanal canoe fishermen. We also surveyed the demersal fish and shrimp resources the length of the Nigerian coastline, and ran a line of stations down to the equator in the international EQUALANT oceanographic survey. They were busy days!

Without that grounding in fisheries science during the heady days of the mid-twentieth century when it seemed that all the answers to fisheries management were coming to hand, I would not have had the courage to undertake this work. Without the assistance of many colleagues and friends I certainly could not have completed it: I am especially grateful to John Caddy, John Field, Bob Francis, Sidney Holt, Peter Koeller, Daniel Pauly, Gary Sharp, Michael Sinclair and many others for stimulating discussions about sustainability and management options in the fisheries. My text was improved by the comments of an anonymous reviewer.

ENDNOTES

1. Chuenpagdee, R. and A. Bundy (eds.) (2005) Innovation and outlook in fisheries: papers presented at the 4th WFC. *Research Reports, Fisheries Center, UBC* **(13)** 113 pp.
2. International Council for the Exploration of the Seas, with headquarters in Copenhagen, is the doyen of international fisheries organisations.
3. Finley, C. (2007) The social construction of fishing. Paper presented at the Oregon State University Conference '*Pathways to Resilience*', April, 2007.
4. This was on the occasion of his receiving the well-deserved International Cosmos Prize in Osaka, as reported in the *Sea Around Us Project Newsletter* (**32**) 1–4.
5. Korringa, P. (1976) In *Biology and the Future of Man.* Paris: University of Paris, p. 202.
6. Holt, S (2006) The notion of sustainability. In *Gaining Ground*, Ed. D. Lavigne. Limerick: IFAW, pp. 43–81.
7. An excellent example of this type of review is given by Pauly, D. *et al.* (2002) Towards sustainability in fisheries. *Science* **418**, 689–95.

1

From certainty to doubt in fishery science

'Less fishing is wasteful, for the surplus of fish dies from natural causes without benefit to mankind.'

W. M. Chapman, 1948

Fishery science is usually perceived by its practitioners as being a critical and quantitative activity, deeply dependent on mathematical analysis; indeed, the introduction to a well-known text on fisheries science suggests that only those who are comfortable writing computer programs or playing with numbers should become involved in fisheries management.[1] This blunt statement demonstrates what went wrong with the science: it forgot that it is heavily dependent on two other disciplines – biology and ecology – in which numerical predictions are quite often unsatisfactory. Consequently, there is a fundamental contradiction between the potential capability of fishery science and its stated task of making routine and quantitative predictions concerning the effects of specified levels of fishing on a stock of fish.

Biology is notorious for its lack of predictive theory, and for its high content of inductive and a-priori generalisations that are based on simple observation of nature. As Murray noted, 'the fact that biology lacks ... universal laws and predictive theory ... poses a serious problem for both biologists and philosophers'.[2] Nevertheless, it is possible to deduce simple biological laws and to verify them by the prediction of ecological observables: I shall discuss one such example in Chapter 2. Such laws may be used to falsify theories that have been arrived at by inductive methods.

Ecologists (and I use the term in its original sense) have long recognised that their discipline, being a subset of biology, similarly lacked rigour and predictive ability. Critical reviews and debates about many of its central assumptions characterised the development of ecology during the second half of the twentieth century:[3] trophic levels, succession and stability, energy flow, co-evolution and density-dependent population regulation were all subjects of warm (not to say heated) discussion, although very little consensus was achieved.

Density-dependent population regulation is the central concept in the theory of fish population dynamics, stemming from the logistic function of Verhulst that describes the absolute growth rate of a population that is constrained by a limited resource: in this, the absolute growth rate is maximal at the inflexion between increasing and decreasing growth rates that occurs when population size is exactly half the potential maximum.[4]

The population growth of the entities represented by the logistic function is thus constrained by changing rates of reproduction and mortality, the causes of which are not specified, so that the rate of population increase is a simple function of biomass. In the real world, the entities suffer both parasites and predators, must accommodate to a food supply that changes with time, and their social behaviour may respond to progressive crowding. Thus, the trajectory of the unadorned logistic function resembles nothing in the natural world, and the consequent tension between theory and reality has been at the centre of the great debate concerning environmental population regulation (Andrewartha and Birch) and density-dependent growth (Nicholson) since the 1930s, which is still generating as much heat as light. One of my correspondents suggests that this debate is the nearest thing to scholasticism that he has seen in biology.

Entomologist Berryman remarked in 2002 that population regulation had been recently described as a 'bankrupt paradigm', 'a monumental obstacle to progress' and 'a mind-set, a dogma, a faith'.[5] Although there are semantic problems concerning regulation and density-dependence, he concludes that regulation is not universal, but is merely one of several population behaviours possible in complex ecological systems; a recent study of long-term records of size-at-age and biomass data for 16 populations of marine and freshwater fish appears to confirm his remarks. Significant density dependence was detected in nine populations, while in four others where the relationship was not statistically significant there were point estimates of growth that were consistent with an among-population effect. In three populations, no relationship could be observed, even though other studies had demonstrated density-dependent growth in these species.[6] Despite the difficulty of observing it in every case where it would be expected to occur, regulation remains critical to management theory, and density dependence continues to be quoted as the central tenet of fisheries science,[7] because 'density dependence gives populations the resilience required to sustain elevated mortality from fisheries' – as a recent student text has it.[8]

In fact, the great Andrewartha–Birch–Nicholson debate seems not to have concerned fisheries scientists as much as it did general ecologists and it did not displace the idea of maximal yields, based on the assumption of logistic population growth, that came to take a central place in management theory after the rediscovery of the logistic curve in the 1920s and its exploration by Hjort and Graham. The logistic curve was also the origin of M. B. Schaefer's logistic model of surplus production of tuna populations under non-equilibrium conditions, which was formulated in the mid 1940s while he was working in the US federal fisheries laboratory in Hawaii.[9]

Thus was the formal proposal for management for MSY hatched from Verhulst's logistic, and one might assume that its rapid passage into the heart of fishery science must have occurred because it had received positive peer reviews from the fishery science community. But, in reality, the rapid acceptance of the principle was almost entirely politically motivated, and occurred prior to the eventual publication of Schaefer's model. In writing its epitaph, Larkin evoked the level of enthusiasm and certainty in American fisheries science during what he calls that golden age for the model of maximum sustained yields, when it was the duty of fisheries science to ensure that the seas everywhere were harvested to this maximum.[10]

This singular state of affairs, in the years immediately following the end of the Pacific war, resulted from the fact that the US State Department was then working to enforce a policy of open seas and open skies, intended to minimise the ability of coastal states to restrict the freedom of American naval and fishing vessels to operate everywhere outside the narrow territorial seas. In the latter stages of the war, US fishermen had worked freely over much of the SW Pacific to help feed the armed forces, and the industry wanted that freedom to continue and, indeed, to expand to other oceans. Japanese fishing was forbidden in much of the NE Pacific by a clause in the peace treaty with that country, but Ecuador, Peru and Chile had other ideas, and the upcoming Santiago Declaration of a 200-mile national jurisdiction was already in the cross hairs of a worried State Department. This movement was at least in part a response to the 1945 Truman Doctrine and the unilateral declaration of sovereignty, including the establishment of fishing conservation areas, over the US contiguous continental shelf. There was some dissidence among US fishermen concerning this initiative for, although it was heartily endorsed by the West Coast salmon fleet, it was not at all appreciated by the tuna fleet, which wanted free access everywhere and feared exclusion from foreign 200-mile zones.

Schaefer was among the fishery scientists who advised the State Department in those years and, in 1948, his colleague and friend Wilbert Chapman (to whom I owe a personal debt of gratitude[11]) was appointed under-secretary of state for fisheries. Chapman considered that it was not fit, at this time of stress in the world and concern for food supplies, that the stocks of sea fish should be wasted through lack of fishing: this anthropocentric notion rapidly came to lie at the core of fishery science.[12] Global landings of marine fish and invertebrates were then only about 15 million tons annually and there were quite unreasonable expectations – greatly in excess of catches actually achieved at the end of the century – of the total potential of the oceans to supply proteins for human consumption.

Chapman had a reputation for moving fast, and he quickly crafted a US High Seas Policy which, as Mary Finley has noted,[13] formally specified MSY as the goal of US fishery management policies on the high seas, although the relevant work of Schaefer had until then been published only in an un-refereed house bulletin of the State Department. Within a few weeks of the publication of the US High Seas Policy, in January 1949, a bilateral treaty had been signed with Mexico and multilateral treaties to establish two fishery commissions, for Pacific tuna and for the North Atlantic, respectively, had also been signed. Each of these three treaties formally specified that management should be for MSY, and this fact profoundly influenced the future course of fishery science – even though these treaties had been signed some years prior to the publication of Schaefer's paper of 1954, so the US High Sea Policy was based on the acceptance of work untested by peer review: Finlay suggests that MSY was, at that time, no more than policy disguised as science.

The European nations, and perhaps especially the UK where Graham was a dominant figure of the times, were not at all enamoured of managing for MSY because their theory of fishing, and its application to management, was going in the direction of analytical models based on cohort analysis, and involving assumptions of age-structured estimates of recruitment, growth, fishing mortality and natural death; subsequently, MSY played little part in European negotiations, and management went in rather different directions on each side of the ocean. 'The European side' wrote Sidney Holt[14] recently 'sought to find ways ... to ensure continuity and stability in the face of growing fishery pressures ... the North American side sought optimisation of fishing, especially through setting the target of ... MSY'. Despite this divergence, by the time the UN Conference on the Law of the Sea had established a 200-mile Exclusive Economic Zone, each of the 39 regional fisheries organisations (RFMOs) then in existence had already accepted MSY as the basis for management.[15]

Perhaps it was in part the relative simplicity of Schaefer's model that ensured its rapid acceptance and propagation, at least in North America: it has a much lower information demand than the more sophisticated dynamic pool models being developed in the same years elsewhere in succession to Baranov's yield-per-recruit formulations. Later, of course, the simplistic MSY concept came to be modified by the inclusion of economic considerations, for a maximum economic yield, and by the use of a more cautious approach – as in New Zealand, where a policy of management for maximum constant yield, set at 2/3 of the computed MSY, was adopted during the 1980s.[16] This level of fishing mortality (corresponding to $F_{0.95}$ in a Yield/Recruitment function) was later recommended by Ray Beverton as a universal standard which, he suggested, might have prevented the 1968 crash of the Icelandic summer-spawning herring had it been in use at that time. Despite these modifications of the original simple idea, MSY survived remarkably well after such a strange beginning. Indeed, it is only in recent years and after the recognition that a global fishery management crisis was at hand that MSY has been fundamentally questioned in North American circles: Larkin's 1977 'Epitaph for the MSY' was not at the time taken sufficiently seriously. As shall be discussed in Chapter 11, the manner in which the new paradigm of ecosystem-based fishery management (EBFM) has been taken into the policy of fisheries agencies prior to extensive evaluation and peer review is startlingly reminiscent of the origins of MSY.

It is difficult, at this distance, to understand the confidence felt by fishery scientists in the mid-twentieth century that their numerical solutions to the problem of stock management were valid, and would function well, as soon as some uncertainties, such as the relationship between stock size and recruitment, were resolved by observation. This relationship was, of course, another inductive generalisation that many authors have struggled to define for individual stocks – because, intuitively, it was felt that such a relationship must exist and that it must be rather simple. Yet it has proved very intractable to this day, for reasons to be discussed later.

There was a general sense at this time that the various techniques then being developed – and used to quantify allowable catches – were satisfactory, even if diverse, leaving only the subsidiary tasks associated with managing each individual stock still to be investigated: obtaining data on size-specific escapement through the meshes of cod-ends, on stock-specific age and size at first maturity and on the current status of each fished stock. It was thought that these and other simple matters could safely be left to the biologists, even though this was a period when, as Ray Beverton noted in his posthumous essay,[17] 'biology became subservient to maths, in both staffing and philosophy' in the European fishery laboratories.

During this period, fishery scientists apparently did not listen to J. Z. Young's contemporary Reith lectures broadcast by the BBC: his studies of brain function, as related to speech and thought, led him to suggest how we interpret (and can discuss) the realities around us. He pointed out that the Descartian method of interpreting the human body in mechanistic terms – using the analogy of clockwork – could be extended to our scientific view of the world. Machines, he said, 'are the products of our brains and hands. We therefore understand them thoroughly and can speak conveniently about other things by comparing them with machines. The conception of living bodies as machines, having, as we say, structures and functions, is at the basis of the whole modern development of biology and medicine.' As Professor Young suggested, our brains may use such methods to create a model of the reality around us; unfortunately, it may not be easy to separate the virtual world, so created, from reality.

If we substitute mathematical models for machines, and populations of fish for J. Z. Young's 'other things' we have an explanation for why the development of mathematical models in fishery science led to the comfortable feeling that – because we 'understand them thoroughly' – we also understand the biological system that they represent. Many will disagree with this assessment, for models of ecosystem function and of the dynamics of single populations are now very widely available and widely used in ecological and fishery science; I shall return quite frequently in the following chapters to comment on the limitations of the understanding that we can obtain with such tools.

Although Beverton wrote of those days as being carefree, I prefer to remember the ambient certainty that fishery science was on the right track; there was an air of self-confidence in the Lowestoft laboratory that was palpable, for this was one of the places where age- and sex-structured analytical models of fish populations were being developed from such origins as Russell's formulation; one principal result of this work was published almost simultaneously with Schaefer's dynamic pool model and subsequently became the basis of management in Europe.[18] Here, the principal impetus that led to the development of such techniques was not strategic, as in the United States, but rather the 1946 European Overfishing Convention: in retrospect, it seems remarkable that such activities were undertaken only a year or so after the end of the Second World War, so that, as Beverton put it, 'by 1951, the pioneering age of stock assessment was in full swing, with international trials, research cruises and exercises'.

The certainty that correct solutions to the fishery management problem were to hand persisted, at least in the teaching of fishery science, for several decades; John Gulland's text of 1974 exuded great confidence in the stock management methods that had been brought into general use during the previous 15–20 years.[19] Gulland simply explained how a fishery scientist, to predict stock size in coming seasons, should use a series of mathematical expressions – more or less complex – that represented and predicted the evolution of the target fish population under fishing pressure.

However, by the turn of the century, less than 30 years later, the certainty that simple mathematical models could satisfy the requirements of stock management was starting to evaporate. Nevertheless, it is still evoked as a control rule in setting allowable catches, as in the long-term plan for northern hake management formulated by the EU Commission in 2009 which notes that 'This MSY approach will in the long-term ensure an increased number of fish in the seas, greater yields for fishermen and stable catch limits.'[20]

Although there are many analyses of the progression of fisheries science from confidence to uncertainty, the graphic account by Ray Beverton is especially revealing of what he called the 'gruesome story' of fishery science during the last decades of the twentieth century, dominated by the entirely unexpected stock crashes in the NE Atlantic after 25 years of what was thought to be rational management. He describes the difficult discussions in ICES and elsewhere concerning matters such as virtual population analysis, stock and recruitment relations, the introduction of reference points and the gradual evolution of understanding that year-to-year differences in ocean conditions had important effects on stock development. Already, the use of population models in stock management had progressively declined as their derivatives, biological or target reference points, were progressively incorporated into operational management.

I note in passing that Beverton's retrospective take on reference points was that they caused 'only confusion and gave managers an excuse to do nothing'. Whether managers did nothing because of confusion, or because they had no fishery control plan in effect to specify the actions to be taken under each set of circumstances and previously agreed with industry, is not clear. Nevertheless, resulting from international discussions in 1982 and 1995, target or limit reference points came to be incorporated in the UN Convention on the Law of the Sea and in the UN Agreement on Straddling and Highly Migratory Stocks; signatories of this agreement agreed that 'States shall take measures that when reference points are approached, they shall not be exceeded. In the event that such reference points are exceeded, States shall, without delay, take the additional management and conservation action determined under paragraph 3(B) to restore stock(s).' Further, included in this instrument was a precautionary approach to fishery management that must be 'acknowledged at every step of management from planning through implementation, enforcement and monitoring'.

The specific wording of the UN Agreement reflected the progression of fishery science towards management by reference points: it was very specific, stating that 'fishery management strategies shall ensure that the risk of exceeding reference points is very low' or, in plainer words, that the precautionary principle was to be adopted. As Richards and Maguire commented in 1998, a core component of the precautionary approach is that the absence of adequate scientific information shall not be used as a reason for postponing or failing to take conservation and management measures. Although the same authors described them as new directions for fisheries management science, these admonitions from the UN really brought nothing novel to the scene, because they simply reflected the already-developed philosophy and practice of the more influential fishery scientists who had been involved in the drafting of the instruments. The use of biological reference points had long been central to the work of ICES and other management organisations responsible for advice on the levels of total allowable catch (TAC) to be set annually for stocks under their jurisdiction: what the UN instruments achieved was perhaps to influence the more rapid adoption of modern techniques by nations and agencies that had not yet thought to do so.

Biological reference points are now widely used in providing advice to the management sector and are often formulated as the level of fishing mortality (F) that can safely be imposed on a stock, either as limit or targets; ideally, the chosen level should be sensitive to both management priorities and to observations of stock abundance. A value of F that has been very commonly used in ICES recommendations is the mortality (F_{max}) that would produce the maximum sustainable yield, MSY, as computed by the yield per recruit curve of Beverton and Holt.

However, because MSY must vary according to the level of natural mortality (M), computation of F_{max} is not always satisfactory and a simpler reference point ($F_{0.1}$) is often used. $F_{0.1}$ corresponds to the level of fishing mortality (or effort) at the point on the yield/effort curve where the slope is 1/10 of that at the origin. Other formulations have been used, such as a set of arbitrary reference points (F_{high}, F_{med} and F_{low}), which are based on a scatter diagram of spawning stock biomass against recruitment.

Progressively, it came to be understood that a direct method of initiating management action was required so that if targets were missed this would automatically define what management action should be taken. From this understanding came the suggestion by Mace in 1994 that simple threshold values for stock biomass and for fishing mortality might be established in every fishery, so that when critical values for these were transgressed, action to reduce F would automatically be required. This would, it was thought, be more effective than attempting to maintain an ideal relationship between, for instance, the value of F imposed by the fleet and stock biomass as observed by assessment surveys.

This was, as Caddy remarked, how limit reference points (LRP) were initially envisaged, after which they were rapidly incorporated into the discussions of those who were at that time planning what came to be called a precautionary approach to fishery management. For this to be functional, it was understood that it would be essential for agreement with each respective branch of industry to be reached, prior to fleet operations at sea, that a certain cause of action – usually a restriction on fishing activities – would automatically ensue should the value of any LRP be transgressed. Caddy likened this control plan to the action of a thermostat that controls the rate of burning in a domestic heating furnace.

Although I shall also be returning later to recount the multiple causes of the infamous collapse of the cod stocks of the NW Atlantic,[21] it will be useful here to mention just one aspect of this event. These fish had been decimated by heavy and largely unregulated fishing on the outer shelf and offshore banks by foreign trawlers prior to the unilateral declaration by Canada of a 200-mile extended fishery jurisdiction in 1978. Subsequently, using an approach that was based on an $F_{0.1}$

target reference point, associated with data from commercial catches and their own stock assessment surveys, Department of Fisheries and Oceans (DFO) scientists were confident that cod stocks were rebuilding in a sustained manner: contrary suggestions by inshore fishermen who noted that their stocks were in decline appear to have been ignored. The Canadian DFO biologists in the St. John's laboratory were said to have been stupefied when it became clear that the predicted strong growth of the northern cod stocks had not, in fact, occurred.[22]

Their confidence in simple constructs appears to have been nurtured by assumptions that the universe is mechanistic and governed by a few simple laws – that the marine ecosystem is robust and insensitive to small perturbations, and that its state varies around a natural dynamic equilibrium – although others had already pointed out that the application of $F_{0.1}$ to a situation in which recruitment had been reduced to a very low level was both inappropriate and dangerous. The use of reference points in setting allowable catches requires, in this case as in others, an assumption that the steady state of fish stocks is dominated by a very small number of knowable variables whose effects may be readily modelled. All of which is very far from reality, but criticism – both from formal enquiries and informal discussion – failed to deter the management trajectory in Ottawa. Confidence in their assessment data, and in their mechanism for setting TACs, was unshakeable, until at last it was clear to everybody involved that the stocks had passed the point of no return.

With the collapse of cod stocks on the Grand Banks, it came to be acknowledged, perhaps especially in North America, that the future in fishery science could in no way resemble the past: there were now too many examples of stocks managed with widely accepted techniques, but that nevertheless collapsed, to support the belief that the same techniques would serve on into the new century: people began to talk of the need for Kuhnian paradigm shifts in the science. The search for new ways of managing stocks developed – as I shall discuss in a later chapter – into something resembling a feeding frenzy.

But it is surprising, in retrospect, that it should have taken so long for somebody to point out that fishery management models – whether of surplus production (e.g. Schaefer), or yield per recruit (e.g. Beverton/Holt) or of spawner/recruitment (e.g. Ricker) – all put the cart before the horse, as Bob Francis so wisely commented.[23] By this, he meant that they were used to infer something about nature – rather than the reverse process: this is, of course, a rather common error in scientific methodology. He further suggested that population dynamics models should have been known rather as population statics models, because all operational models used in fishery management assume a static environment. This is a point to which I shall refer very frequently in this book, because it is not only in operational models that this assumption is commonly made; simulation models intended to represent the functioning of the principal elements of a regional ecosystem only too often include no external forcing from a variable physical environment, and therefore have no predictive value.

The conclusion that Francis drew is well worth quoting verbatim, because it so clearly foreshadowed one of the strongest movements in the remaking of fishery science today: 'The practicality of fishery science is that of understanding and managing ... interacting fish populations and their aquatic environment ... systems which have properties and structure of their own which cannot be understood by reductionist study.' He further noted that because the analytic models, being gross abstractions of reality, have long-term equilibrium properties, then so must the populations to which they are applied. He was not, of course, the only perceptive writer in the decade of the 1970s: others were also already discussing the need to preserve diversity in order to prevent irreversible ecological changes in the ecosystem of which the target fish species formed a part.

During the latter part of the twentieth century, and as events in the fisheries unfolded, it became easier for the research community to accept comments that were critical of their models: typical of these was the suggestion by Schnute and Richards that all of the models that were used in management characteristically obscure the truth by narrowing scientific understanding to the realm of quantifiable events.[24] These authors pointed out that fisheries modellers, like Pygmalion, had come to believe that their creations correctly represent reality, echoing the comments of Professor Young noted above. For this reason, mathematicians may fail to appreciate that the output of all of their models must be qualified by three major uncertainties: (i) *process error*, in the case where the mechanism of some natural process in the ocean, such as recruitment to the adult stock, is improperly understood yet is essential for the model; (ii) *measurement error*, as in the incorrect interpretation of assessment survey data, or in the choice of an inappropriate survey design; and (iii) *model uncertainty*, this being an error that is not subject to quantification.

Obviously, the predictive reliability of any fisheries ecosystem model must be inverse to the level of available information concerning environmentally forced changes in the stock sizes of each managed species – and also in the abundance of their prey and predators, if all of these are not included in the simulation. Yet the users of these models often ignore the possibility of natural changes in ecosystem state that occur in response to natural changes in climatic forcing, such as those in the Western Channel in response to environmental warming conditions from 1925 to 1935 that were repeated after 1971.

Fisheries models, like any others, require input data. In this case, values that represent the present state of the stock that is being assessed: ideally, with the use of fishery research vessels, we must determine: (i) the quantitative distribution of the fishable stock, including differences in stock density between different sub-regions of the management area, (ii) the age structure of the entire population, both generally and in different regions and (iii) the relative strengths of recent year-classes that are not yet recruited to the fishery and perhaps cannot yet be caught quantitatively in fishing gear used for stock surveys.

It is rarely simple and sometimes it is impossible to obtain the required data, and it is always extremely difficult to judge the validity of such data as are obtained at sea or at landing places. Some stock survey methods, such as those based on acoustics, may yield data that appear to be more accurate than they actually are: differences in the angle-of-incidence of the acoustic beam on the target fish may yield biomass estimates that differ by at least an order of magnitude, even when the size-specific target strength of individuals of each species likely to be encountered is well known. It is also easy to under-estimate the difficulty of obtaining data on population age structure from fish sampled either at sea or the dockside because reliable reading of otoliths or scales requires great skill and much patience and the final data set may easily be contaminated by one unskilled reader. More generally, as shall be discussed in Chapter 3, even oceanographers find it very difficult to grasp or accept the extent of our ignorance concerning the structure of marine ecosystems and their dynamic response to atmospheric forcing. The understanding of general principles is very different from the quantitative description of the temporal sequence of individual processes at specific locations.

Information must also be obtained directly from the fishing industry on the age- or size-specific catches that have been made as the season progresses, and this is even more complex because of the extent of illegal, unreported and unregulated (IUU) landings and also because of the extent of wastage at sea during fishing operations. The financial pressure on fishing enterprises at all levels is sufficiently high that both owners and skippers routinely cut corners and cheat on regulations to survive. I shall be returning to this almost global problem in a later chapter, but here it will be useful to note that such practices are perhaps the principal obstacle to obtaining correct data on F required by stock assessment scientists. The observation of F, a model parameter that is critical to the setting of catch limits for future fishing seasons, is beset with problems; even if techniques have been developed, as for the Revised Management Plan of the International Whaling Commission (IWC), to simulate real total fishing mortality,[25] it would be an optimist who would claim that in most instances what is used is any more than a gross approximation.

Nevertheless, using such inexact information on the real value of F imposed by a fishery, managers set quotas for the coming season; these are computed to produce precisely the level of fishing effort required by the population model. It is hard to be sure, but I have the impression that this problem, and the general unreliability of stock assessment data, was only progressively understood by fishery science during the latter half of the twentieth century: here, again, there appears to have been a progression from certainty – or, at least, from great self-confidence – to doubt.

* * * * * * * * * * *

It has not been my intention here to develop a formal critique of the evolution of scientific management of fish stocks, but simply to point out that it has clearly come a long way since the days when Wilbert Chapman suggested that although he knew 'fairly clearly for several stocks the level that corresponds ... to the MSY ... and the probable error of that estimate', he suggested that 'fishery science, management procedures and ocean science in general are not yet in good enough shape ... to undertake rational management of ... living resources of the sea'.[26]

Nevertheless, there is some sense today that the present widespread depletion of fish stocks represents the result of a failure of fisheries science to bring forward satisfactory approaches to management in the 50 years that have passed since Chapman's remarks. But, I shall suggest that the responsibility for the depletion lies elsewhere, and I reject Peters' implication that fisheries science, which is a subset of ecology, must be considered to have failed if problems of the fisheries remain unsolved: his argument that 'unattainable solutions are no solutions at all' and that science must provide 'feasible solutions' ignores the relationship and tensions between scientific advice, politics, commerce and society. It also ignores the fact that funding levels are only very partially under the control of scientists and are usually derisory in comparison with real needs.

Failures in the scientific management of fish stocks there have certainly been, and I shall discuss several of these, which were the unfortunate but perhaps inevitable consequences of the learning curve of fisheries science during the twentieth century. I shall suggest that although science has produced apparently satisfactory solutions to the fishing problem, these may be insufficiently sensitive to the largely undescribed complexity of ecological interactions in the ocean, and to the high degree of variability in the natural forcing of marine ecosystems. I shall also emphasise the fundamental inadequacy of the ecological data that are habitually available to those responsible for fish stock management and the difficulty and expense of obtaining what is routinely required.

In the end, I shall have to admit my pessimism for the notion of sustainable fishing except in a few favoured regions of the oceans. The evidence suggests that the commercial interests of the fishery industry, with all the associated economic, social and political pressures that influence management decisions, are such as to render rational management illusory in most regions of the oceans. The problems of the fisheries can be resolved (if at all) by social and political, rather than scientific, solutions.

ENDNOTES

1. Hilborn, R. and H.J. Walters (1992) *Quantitative Fish Stock Assessment*. Berlin: Springer.
2. Murray, B.G. Jr. (2001) Are ecological and evolutionary theories scientific? *Biol. Rev.* **76**, 255–89.
3. Peters, RH (1991) *A Critique for Ecology*. Cambridge: Cambridge University Press.
4. Verhulst, P.F. (1838) Notice sur la loi que la population poursuit dans son acroissement. *Corr. Math. Phys.* **10**, 113–21.
5. Barryman, A.A. (2002) Population regulation, emergent properties and a requiem for density dependence. *Oikos*, **99**, 600–06.
6. Lorenzen, K. and K. Enberg (2001) Density-dependent growth as a key mechanism in the regulation of poplations: evidence from among-population comparisons. *Proc. R. Soc. Lond. B*, **269**, 49–54.
7. Rosenberg, A.A., *et al.* (1993) Achieving sustainable use of renewable resources. *Science*, **262**, 828–9.
8. Jennings, S., *et al.* (2001) *Marine Fisheries Ecology*. Oxford: Blackwell Science.
9. Schaefer, M.B. (1954) Some aspects of the dynamics of populations important to the management of commercial fish populations. *Fish. Bull. US Dept Int.*, **52**, 191–203.
10. Larkin, P.A. (1977) An epitaph of the concept of maximum sustained yield. *Trans. Am. Fish. Soc.* **106**, 1–11.
11. In 1963, Wib Chapman was directly responsible for my improbable move from the Nigerian Federal Fisheries Service to the Institute of Marine

Resources, then under the direction of Milner B. Schaefer, located at the Scripps Institution of Oceanography in California.

12. Scheiber, H.N. (1986) Chapman and the Pacific fisheries. *Ecol. Law Quart.*, **13**, 383–534.
13. Finley, C. (2007) The social construction of fishing, 1949. Paper presented at Oregon State University Conference 'Pathways to Resilience', April, 2007.
14. Holt, S. (2006) The notion of sustainability. In *Gaining Ground: In Pursuit of Ecological Sustainability*, ed. D.M. Lavigne. Limerick: IFAW and University of Limerick, pp. 43–82.
15. Clark, C.W. (1981) Bioeconomics. In *Theoretical Ecology: Principles and Applications*, Ed. R.H. May Oxford: Blackwell, pp. 387–418.
16. Francis, R.C. (1992) Use of risk analysis to assess fishery management strategies: a case study using Orange Roughy (*Hoplostethus atlanticus*) on the Chatham Rise. *Can. J. Fish. Aquat. Sci.*, **49**, 922–30.
17. Beverton, R.J.H. (1998) Fish, fact and fantasy: a long view. *Rev. Fish Biol. Fish.*, **8**, 229–49.
18. Beverton, R.J.H. and S.J. Holt. (1957) On the dynamics of exploited fish populations. *Fish. Invest. London*, ser 2, 19.
19. Gulland, J. (1974) *The Management of Marine Fisheries*. Bristol: Scientechnica, pp. 198.
20. EU Fisheries Commission press release IP/09/408 of 19.3.09.
21. It cannot be sufficiently emphasised that this was a disaster that had multiple causes, although many analysts have tended to look for a single culprit.
22. Finlayson, A.C. (1994) *Fishing for Truth*. Newfoundland: ISER Press, Memorial University, 176 pp.
23. Francis, R.C. (1980) Fisheries science now and in the future: a personal view. *NZ J. Mar Freshw. Res.*, **14**, 95–100.
24. Schnute, J.T. and L.J. Richards. (2001) Use and abuse of fishery models. *Can. J. Fish. Aquat. Sci.*, **58**, 10–7.
25. S.J. Holt, pers. comm.
26. Scheiber, op. cit.

2

The ecological consequences of the exceptional fecundity of teleosts

'The codfish lays ten thousand eggs, the homely hen but one
But the codfish doesn't cackle to tell us what she's done . . .'

anon

It will be useful to begin this investigation of the sustainability of fisheries and the concept of surplus production by enquiring whether the organisms that comprise the bulk of global catches have any special biological characteristics relevant to the argument. Teleost fish dominate the vertebrates of interest to fisheries, and so discussion in the following chapters shall be concentrated on this group, although not to the exclusion of the selachians and marine mammals.

Teleosts have three biological characteristics that strongly differentiate them from all terrestrial mammals and birds:

- their anomalously high fecundity, which is orders of magnitude greater than that of any other vertebrate and is habitually attributed to the rigours of the transient passage of the growing larvae through the planktonic ecosystem, with all that that involves;
- their pattern of life-long growth, which has been almost totally neglected by fishery science, but which has important implications for it; and
- the characteristic variability of their annual reproductive success, which appears to be higher than is typical of terrestrial vertebrates.

The significance of these characteristics of the biology of marine teleosts for the supposed surplus production of their populations and for the management of fishing will be critically examined in this and the following chapters. It will be suggested that the inferences usually drawn are readily open to criticism: all may not be what it seems to be ...

Parenthetically, note that I have used here the term surplus production and shall do so again, but only because it has become almost impossible to discuss the management of fishing without using it: yet, it is self-evident that there is no surplus production in natural ecosystems. What we do not catch, as Chapman pointed out, '*dies from natural causes*' where this almost exclusively involves predation and the nourishment of another element in the food web. The surplus that we take, therefore, is obtained only by a modification of the food web and of the reproductive biology of the target species – a process to which I shall frequently return in what follows.

THE ANOMALOUS FECUNDITY OF TELEOSTS

It is not surprising that fecundity quickly became a focus of early fisheries science, especially because cod are a paradigm for fecundity and because the first formal investigations in fisheries science were those commissioned by the Norwegian government in the mid-nineteenth century to investigate the troubling variation in the abundance of the cod stocks. The great Norwegian fishery scientist Johann Hjort came to the conclusion that a connection must exist between annual fecundity and subsequent year-class strength, thus opening the long debate on this evasive relationship.

The phenomenon of the fecundity of fish came early to public attention, for the *New York Times* of 4 February 1883 carried a report that pointed out that 1000 herring could theoretically produce 25 million new individuals in the following year, except that 'the waste of eggs ... is really enormous'. This report doubtless echoed Thomas Huxley's lectures at the time of the International Fisheries Exhibition in South Kensington, because he suggested that the fecundity of fish was so great that it should enable their populations to sustain fishing at useful levels.

The reproduction of teleosts had already intrigued the naturalists of earlier generations: both Lamarck and Darwin had mused on the significance of their great fecundity. Lamarck suggested that the rapid multiplication of bony fish would ensure that their populations would survive fishing, while Darwin wondered why different species of fish, having widely different fecundity, should be equally successful in maintaining populations – a much more interesting observation. But even if Darwin had access to the same data that we do, he might still be asking the same question; recent studies of the consequences of varying levels of fecundity for recruitment success and for annual reproductive rate have given somewhat equivocal results. Lamarck was probably unaware of the often very rapid population growth of at least some teleost fish, otherwise he might have suggested that their high fecundity might serve them well in some natural situations, as shall be discussed below.

However, it is necessary to emphasise that it is not simple to quantify fecundity: the problem stems from the variety of spawning strategies employed by teleosts. In batch spawners, like cod, most of the eggs that are spawned each year are accumulated in the ovary prior to the annual spawning event, so that at this time the ovary comprises a large proportion of the body mass of a female cod and occludes the body cavity sufficiently as to reduce food intake significantly. This process generally explains the annual cycle of feeding rate in seasonal spawners, and has been observed even in tropical seas.[1]

Nevertheless, despite the appearances of great fecundity of batch spawners, the annual production of eggs by serial spawners may be very much greater. This is especially the case in warm seas where batches of eggs may be released by some pelagic species almost daily wherever and whenever water temperatures are suitable. A recent study (to be discussed below) obtained *minimal* annual reproductive outputs for individuals of 49 species, all at low population density: for large *Gadus morhua* (cod) the annual production was found to be around half a million ova, while for large *Thunnus thynnus* (bluefin tuna) it was 20 times greater! It is, of course, much simpler to compute an estimate of annual fecundity for batch than for serial spawners.

One simple fact remains quite clear, even if it cannot be stated as a quantitative relationship; the high fecundity of teleosts is clearly a factor in the ability of the populations of at least some species to increase very rapidly. The rate of increase achieved by some teleost species that experience natural fluctuations of their population size is extraordinary: the case of the Peruvian anchovy will be discussed in the later chapters, but similar comments would be appropriate to many other recorded rates of increase of population size. Whether or not there is a relationship between fecundity levels observed to be characteristic of species and their natural rates of population increase is yet to be elucidated.

Some relation between fecundity and reproductive success in bony fish has recently been demonstrated by Rickman and others; although it does not tell us how rapidly, under pristine conditions, a population could expand, one important measure is the variability of recruitment of a population between years over an extended period.

One might, therefore, expect fecundity to be related to this variability in some way but a simple cross-species approach, comparing highly fecund with less fecund species ($N = 13$), suggested that fecundity and recruitment variability were unrelated. When the comparison was repeated between pairs of regional stocks of each of the same species (and so somewhat analogous to a paired *t*-test), it was found that the more fecund population had a higher recruitment variability than the other in almost every pair.[2] However, the magnitude of the effect was unrelated to the magnitude of the difference in fecundity, so it is clear that there remains much uncertainty concerning the real significance of this result.

Also unclear are the results of a study by Froese and Luna of the reproductive biology of 49 representative teleost species, selected for their wide range of minimal annual fecundity:[3] the variance of maximal annual reproductive rate (defined as the number of replacement mature spawners per spawner) was found to be smaller than that of fecundity, which spanned five orders of magnitude. This result appears to confirm earlier suggestions that the maximum reproductive rate of fish at low population size is a fundamental parameter that has very little variance, either within or between species.[4,5,6] However, it should be noted that these studies were based on between-species comparisons, rather than paired within-species population comparisons that should, in principle, be more sensitive. Such results suggested to Froese and Luna that 'there is no basis for assuming that high fecundity confers high resilience to exploitation' and also that 'there was no significant relationship between fecundity and recruitment variability'.

These authors suggest that high fecundity evolved in teleosts 'primarily to counter-balance pre-adult mortality typically suffered by offspring ... fecundity is high enough to ensure replacement of spawners and small enough to avoid over-investment in reproductive effort'. But this is a circular argument, which says no more than that there is a balance between recruitment and adult mortality (which we assume must be the case in the long term) and also simply revisits the well-known observation that planktonic organisms suffer high mortality rates.

But more than that, such theories lack any explanation of how evolutionary pressures could have obtained such a result: none explains why female fish with exceptional fecundity should be 'fitter', and more likely to survive to reproduce again, than less fecund fish of the same age. In fact, one might predict the opposite effect because, long ago, Hickling noted that older female herring may be sufficiently stressed by their relatively high fecundity as to significantly shorten their lives in relation to the survival of male fish of the same body length.[7]

So, perhaps a short diversion into the development of deductive theory in answering such questions might be in order: it is possible that ichthyologists have entered the same dead-end as ornithologists, who long believed in David Lack's assertion that fecundity in birds had evolved towards clutch sizes that represented the maximum number of young that could be fed daily or, more generally, that could be produced annually. The original argument was based on the observation that clutch size tended to increase with latitude, and hence with summer day length – even though it didn't take long for someone to point out that the same pattern was followed by nocturnal birds!

Murray suggested that the problem could be generalised by proposing three simple evolutionary laws:[8] (i) genotypes with higher values of the Malthusian parameter increase more rapidly than those having lower values,[9] (ii) in the absence of change in selection forces, a population will reach and maintain an evolutionary steady state, and (iii) selection will favour those females which lay as few eggs or bear as few young as are consistent with replacement because they will have the highest probability of surviving to breed again, their young have the greatest probability of surviving, or both. A simple model for clutch size in birds was based on these laws: *Clutch size at age N = (male/female ratio) + 1/(number of reproductive years) (probability of survival to age N) (mean number of broods)*. This model successfully predicts clutch size in several species of passerines.

I do not intend to explore whether or not a similar model might predict fecundity in teleost fish, and I refer to Murray's approach only to emphasise that it may be helpful to seek potential selection mechanisms whenever a conclusion such as that of Froese and Luna is discussed: in this case, none was proposed. Nor does their study suggest why comparable annual reproductive rates should not equally well be obtained from the higher rate of survival to be expected from a smaller number of young that are given parental care. Nor yet does it explain why so many teleosts should have evolved to spend part of their lives as members of the zooplankton because, although this option has been abandoned several times by bony fish in favour of livebearing or of producing a few, large eggs, the species that have abandoned the planktonic larval option remain in a very small minority.

The general distribution of types of reproduction in fish is indicated by a quick scan of 2764 species in Myers' data tables: these comprised 60.1% pelagic spawners, 20.2% that lay demersal eggs,

7.6% (all Scorpaenidae) that are partly viviparous, 0.5% (all Ariidae) that incubate their eggs in their mouths and 5.5% (all selachians) that are viviparous or lay just a few large eggs. For the remainder, the reproductive mode remains undescribed. Of those that lay demersal eggs, only Clupeidae and Osmeridae are pelagic; the remaining demersal egg-layers (dominated by Blennidae, Cottidae, Gobidae, Liparidae, Pomacentridae and Stichidae) are mostly small fish associated with rocky terrain or reefs. There is, as would be expected, at least a weak relationship between fecundity and egg type, and in a much smaller sample of only 121 species, the fecundity of those depositing demersal eggs ranged from very low numbers up to 2×10^6 eggs deposited annually: almost half of those with planktonic eggs release higher numbers than this maximum, and none release the very small numbers characteristic of some demersal spawners.

THE ECOLOGICAL ROLE OF PLANKTONIC LARVAE

One might suppose that if rapid population growth was the key factor in the evolution of high fecundity in teleosts, then this could be achieved best through recourse to planktonic larvae capable of nourishing themselves. But it is still strange that general ichthyologists should have concerned themselves so little with trying to confirm that is indeed why so many teleosts should have taken this route rather than brooding, or giving birth to, just a few, large young.

This is especially odd since all elasmobranchs and even some teleosts, the livebearers and nest-builders, succeed very well with such a strategy. Information concerning the life history parameters of selachians is not so readily available as for teleosts, but some generalisations may be made: compared with teleosts, species of selachians tend to have high longevity (10–75 years) and mature relatively late (2–30 years), while fecundity is very low (<100 young, and often only 2–10) and the stock–recruitment relationship is rather tight. It is not clear how to balance the rapid population growth that potentially accrues from the high fecundity of teleosts against the uncertainty of their annual recruitment from planktonic larvae. Nevertheless, it is to be noted that the population doubling time of selachians tends to be relatively longer than for teleosts.

Of course, other possibilities come to mind when thinking about the high fecundity and planktonic larvae of teleosts; the most obvious alternative explanation is the simple analogy with the broadcast seeds of terrestrial plants, in this case their essential means of distribution.

However, there appears to be no correlation between the problems of dispersion that individual species would be expected to encounter and the type of reproduction that each has evolved: it would be difficult to support a suggestion that the cartilaginous fishes have significantly different problems of dispersion or retention than bony fish!

It is even more pertinent to observe that the propagules of terrestrial plants are passively distributed by a wide variety of mechanisms while, as Andrew Bakun remarked 'Fish larvae are not drift-bottles.'[10] Because they have very limited swimming capabilities, and because inertia is only a minor consequence of motion for small planktonic organisms, it is only by the using the differential transport of the upper and lower layers of the superficial water column that fish larvae can move significant distances; this enables individuals to be retained in, or reach, locations where they may metamorphose and enter the post-larval population. Bakun notes that the use of restricted spawning and nursery areas by pelagic fish in the California Current constrains their populations to smaller sizes than the carrying capacity of the adult habitat; to avoid wind-induced turbulence and offshore transport, the adult populations of sardines and anchovies perform long migrations to spawn in a rather restricted area in the Southern California Bight. It is relevant to note that species of copepods in the California Current (likewise having no capability of horizontal migration over long distances) have evolved a very complex stage- and season-specific pattern of vertical migrations that avoids their being swept seawards within the near-surface offshore flow of upwelled water.[11] That planktonic larvae can, at once, renew parent populations and act as dispersal agents has been shown by studies of naturally and artificially tagged coral reef fish larvae.[12]

The passive migration of fish larvae from spawning areas to nursery areas may be quite complex and it is on wide continental shelves, such as the North Sea, that these long-distance drifts of larvae have been most extensively studied: here the individual populations of extensively distributed species may have individual routes from spawning to nursery areas. The drift of larval plaice from spawning areas in the southern North Sea, around Texel and so in to the Dutch coastal nursery grounds is very complex; it requires specific responses to tidal streams and to the boundary between stratified and well-mixed coastal water masses.[13] The extent of the dispersal of larvae of populations of sedentary demersal, littoral or reef species, and therefore the genetic connectedness between their neighbouring populations, is not easy to generalise.

But it must be emphasised that the first problem faced by planktonic larvae is not their retention in a hospitable environment after larval life is completed: most crucially, they must find food – and find it very quickly – despite the fact that the feeding success of newly hatched larvae may be very low indeed. When a novel prey-type is encountered, the larvae undergo a learning period before feeding strikes become reasonably successful so that, all in all, things can go rapidly and fatally wrong for newly hatched planktonic fish larvae:[14] the planktonic larvae of a variety of species, from soles to anchovies, have a characteristic critical period that varies from 3 to 8 days, after which they reach a 'point-of-no-return' if suitable food has not been found at suitable concentrations. On the contrary for herring, which produce relatively large benthic eggs, the critical period to their point-of-no-return is about three weeks.

There is a rich harvest of ecological studies of these processes, so that we now understand quite well how and why fish larvae are preferentially distributed in three dimensions in the water column in relation to the distribution of micro-crustaceans and other members of the zooplankton: the aggregations of such biota that lie within the biomass-rich layers of the thermocline are especially important for the survival of fish larvae. None of these studies appears to support a seed-dispersal analogy for planktonic fish larvae, and although dispersal of individuals to previously unoccupied regions must occur – in order for each species to survive in a dynamically changing ocean – it appears rather to be strays from pre-adult and adult migration routes that support this function: of sea-going salmonids that succeed in returning to a spawning river, something like 1% stray to a river other than their natal stream. Genetic evidence suggests that the replacement of the west Greenland cod stock, after a period of absence, was accomplished by strays from the seasonal migrations of the Icelandic stocks.

But if the evolution of planktonic larvae by bony fish is not primarily related to problems of dispersal, why did their reproductive habits evolve in that direction? There appears to have been little concerted enquiry concerning the evolution of this reproductive strategy rather than one similar to that of elasmobranchs. Indeed, general ichthyologists seem to have been more concerned with the range of different reproductive strategies of teleosts than with asking why these fish should have evolved a reproductive process similar to that of tunicates. Exactly when the evolution of planktonic larvae occurred in vertebrates is perhaps not easily answered, although both external eggs and internal development of embryos probably occurred in fish very early: Silurian Placoderms had claspers like dogfish today, although modern Agnatha produce external eggs.

One suggestion concerning the evolution of planktonic larvae by teleosts is already on the table, and although to date it has been very largely neglected, I believe that it may be at least a partial answer to the conundrum and will certainly repay closer examination. The argument may be somewhat teleological, but it suggests that the extraordinary fecundity of marine teleosts *coupled with their recourse to planktonic larvae* may be a partial solution to the general problem of the nourishment of large predators in an ecosystem dominated by micron-scale plants and millimetre-scale herbivores: fecundity levels may not respond only to the problems of population growth but also – at least under some circumstances – to the nutrition of adult fish.

To address this possibility, I take as my point of departure the remarkable – but equally remarkably ignored – contribution of Walter Nellen, who suggested that there is a connection between the piscivory that is so characteristic of teleost fish and their recourse to planktonic larvae.[15] Because very large numbers of very small young fish occur in the same habitat as their piscivorous parent generation, Nellen suggested that 'bony fish may have adapted to be nourished to some good extent by their children rather than vice versa as it is with other vertebrates'. I think that Nellen should have added the observation that 'their very small children also have to look after themselves, rather than being nourished by their mothers' as are the children of other groups of vertebrates, including the elasmobranch fish that inhabit the same environment as teleosts.

It has been suggested to me by a correspondent that Nellen's hypothesis has suffered relative neglect because it appears to raise the question of group evolution, which is one of the more difficult concepts in evolutionary theory to confront. I do not propose to open that Pandora's Box here, and will only point out that the simple definition of the concept is not appropriate because it 'leads to the maintenance of traits favourable to the population, but selectively disadvantageous to genetic carriers within populations', says Odum.[16] In this case, the entire adult population may draw an energetic profit from the phenomenon, as I hope to demonstrate; if this is so, then those who have ignored Nellen's hypothesis because it seems to require group selection have done so for no good reason. In any case, it is more important to investigate the ecological and population consequences of reproduction by planktonic larvae in today's oceans, than to try to understand how it might have arisen by natural evolution in the deep past.

Nellen pointed that a rather simple food chain runs directly from phytoplankton cells, through small crustacean herbivores to small fish, starting with first-feeding larvae. Further, the same food chain may be traced through to progressively larger size-classes of teleost fish via cannibalism. If this analysis is correct, then reproduction in teleosts may not involve an energy cost: on the contrary, at least some happy parents may experience an energy gain through reproduction.

This apparently ridiculous suggestion may be evaluated by comparing the energetics of dogfish (*Mustelus canis*) and cod (*Gadus morhua*). Large individuals of each species are about 1 m in length, but the annual reproductive output of their females is strikingly different: the viviparous dogfish produces 10–15 young in the range 20–25 cm in length while the oviparous cod releases 5–10 million ova, each <2.0 mm in diameter, batch-spawned over a relatively short season of less than 2 months. If even a tiny fraction of these survive to reach the same life stage as the newborn dogfish, the initial energy expenditure of the female cod has been enormously more effective in terms of potential population increase than that of the female dogfish.

But more than that, the tiny cod larvae nourished themselves, rather than depending on a significant maternal contribution to the ovum from which each individual has developed, as did the young dogfish. Subsequently, many small cod contribute nourishment to the adult cod by cannibalism, thus returning to their parent cohorts at least a part of their original maternal energy contribution and probably more than that: individuals that are consumed in this way are <3 years old, after which they appear to become exempt. Cannibalism in cod increases progressively with age, becoming especially characteristic of individuals >100 cm in length.

Can these ideas be generalised? Perhaps they can, and perhaps the clue lies in the particle-size spectrum of the oceanic biosphere to be discussed in Chapter 4. For present purposes, recall that the herbivore trophic level in the ocean is largely represented by millimetre-scale invertebrates and that the food-chain transfer thence to large fish is performed almost entirely by carnivory: this can be energetically successful only in the presence of a chain of progressively larger organisms, all themselves carnivores, of suitable size to be captured and eaten by individuals of the next larger size-class in the chain. If every link is to be filled by a different species, such a chain cannot be found in every oceanic habitat; where this condition cannot be fulfilled, as in cold seas, then cannibalism is the only means by which fish may achieve large size.

Cannibalism – including filial cannibalism – is certainly widespread among teleosts of many habitats and is especially characteristic in some extreme places. Consider the case of habitat that is newly available to fish, as occurred after the introduction of a predatory cichlid into two Central American lakes. In one lake, population expansion was almost entirely built on cannibalism, with only the youngest individuals feeding on the few, small species of fish that were already present in the lake.[17] In the second lake, with a more diverse native population of other species of fish already present, the introduced cichlid exhibited very much lower levels of cannibalism during the expansion of its population although, as discussed below, this may not represent a steady-state situation. This observation is a model for the difference in the spectrum of potential prey species that may be available to large predatory teleosts. It is also a model for the irruption of teleost species into natural habitat newly available to them: although, at the time scales that we can personally observe, this occurs only very rarely, it is something that happens frequently at the scale of natural climate changes and enables the rapid invasion of new habitat.

Reviewing cannibalism in teleosts, Smith and Reay found positive evidence in 31 families, but suggested that 'it is considered to be more widespread than this. Finding examples of cannibalism is not difficult, and it may be more interesting to look for taxa in which the behaviour does not take place.'[18] Unfortunately, it is difficult to ascertain to what extent the role of cannibalism differs in different marine habitats, at different depths or in different latitudes. If, as Smith and Reay conclude, 'the main proximate advantage conferred by cannibalism is assumed to be nutritional', then it would be reasonable to suppose that the utility of cannibalism would be greater in the low-diversity ecosystems of cold seas. In tropical seas, a wide range of fish species occupies the entire size spectrum from centimetre to metre scale, enabling an efficient transfer of energy from very small to very large species by predation.

In cold seas, gaps occur in this size spectrum, and these may effectively be filled only by young individuals of the dominant large species; such a situation represents the ideal condition for what has been described by community ecologists as a lifeboat mechanism, in which energy is obtained from conspecific individuals by cannibalism under conditions lacking any alternative food source.[19] Here, cannibalism may permit a population to persist under conditions in which '*a non-cannibalistic but otherwise identical population would go extinct*'. The essential condition for the lifeboat effect is that the victims should obtain external energy for growth between their birth and the time they are eaten by a conspecific individual.

The lifeboat effect may result in a bistable population that switches between the two states: stunted, in which the largest fish are rather small, or piscivorous in which the adults reach large maximum sizes: this so-called Hansel and Gretel effect has been observed in freshwater perch populations that switch between these two states at decadal time scales, the switch-over being forced by the relative survival of recruits, itself determined by the availability of their planktonic food. Other theoretical studies of cannibalism by community ecologists have concentrated not on the trophic consequences of the phenomenon, but rather on the effect it has on population stability and population size structures – these are numerous and complex.

What is known of cannibalism in gadoid populations is not inconsistent with such theory, although the evidence for the extent of the habit is somewhat contradictory: individual cod may be consumed by their conspecifics (and even by larger members of their year-class) until their third year of growth, after which they are apparently exempt. The relative proportion of the diet obtained by cannibalism in cod increases progressively with length, becoming especially high in individuals >100 cm in length.

Some investigators have suggested that a relative absence of small species that are usually dominant in the diet may induce a relative increase in cannibalism, although Bogstad *et al.* believe that that the extent of cannibalism by cod depends principally on the relative abundance of small cod in the regional population. In the case of a large incoming year-class, many 1–2-year-olds will be consumed, as occurred on Flemish Cap in 1991 when a very strong class of 1-year-old recruits was associated with a rapid increase in cannibalism by older fish. It had earlier been suggested that events such as this may contribute to a natural, density-dependent mechanism for population regulation, as in the case of the cyclical alternation in relative abundance of adults and young fish in the Arcto-Norwegian cod stock. A similar situation has been observed in the eastern Baltic, where stomach content analysis of more than 60 000 cod collected over a period of 17 years showed that 25–38% of each incoming 0-group were consumed by adult cod, together with 11–17% of the subsequent 1-group: thus, as Neuenfeldt and Koester explain, 24% of the post-larval fish of the initial cohort is consumed before it reaches the age of 2 years. The rates of cannibalism are highest in years when adult stocks are large and the density of young fish in the water column is especially dense.

It should perhaps be noted here that uncovering what might have been the behaviour of cod in a pristine ecosystem has now become very difficult: as Link and Garrison remark, during the second half of the twentieth century, the diet of the cod population of the NW Atlantic changed. Piscivory became relatively reduced in the individual diet, and the role of cod in the predation of other species was globally reduced; such a process is, of course, what we should expect as the age and size structure of the cod stocks changed in favour of smaller, younger fish, and we must consequently mistrust any stomach content analyses performed during these decades if we are trying to understand the performance of a pristine population.

Cannibalism in gadoids – the transfer of energy along a food chain from small to large individuals – is an example of what occurred to Walter Nellen. In our example of cod, we might very well conclude that the function of very large incoming year-classes was not to enable subsequent rapid population growth, but rather to contribute to the nourishment of the entire adult population by transforming food particles too small to be useful to fully adult cod (even smaller fish, euphausiids, hyperiids, etc.) into items of food (themselves) large enough to be consumed profitably. In the North Sea, 0-group cod can form as much as 20% of the diet of older fish and, in their review of cannibalism in fish, Smith and Reay note that around 80% by number of prey items were smaller individuals of the same species in *Micromesistius poutassou* on Porcupine Bank.

Although cannibalism probably does not exceed 10% of the annual food intake of even the largest cod in various North Atlantic regions, in other gadoids the overall contribution may be much greater: Juanes[20] records levels of <40% in silver hake (*Merluccius bilinearis*), <49% in walleye pollock (*Theragra chalcogramma*), and >70% in Cape hake (*M. capensis*). Small and large hake are very often distributed differently, with young age groups inshore in shallower water and older fish deeper and farther offshore; younger cohorts of the Cape hake occupy specific inshore nursery grounds, and it is suggested that this leads to cannibalism and to the selection of faster-growing individuals.[21] In this species, levels of cannibalism are determined neither by the overall density of small hake nor that of alternative suitable prey – rather, large hake have a 'dietary preference for small conspecifics'[22]: this finding recalls the comments of Bogstad *et al.*, discussed above, concerning the dietary preferences of older cod. One of the most striking cases of cannibalism yet known occurs in another high-latitude gadoid, the population of pollock in the eastern Bering Sea,[23] because the dominant predator of juveniles are adults of the same species. Rates of cannibalism vary from year to year and can reach levels of 60–70% by weight of the food of adults, as in 1981. The mechanism that determines this variability is the co-occurrence in the same water mass of small fish and adults; larvae and juveniles are normally isolated in superficial water masses from deeper-living adults that prefer warmer off-shelf water but, as stratification breaks down seasonally, the smaller fish come to be accessible and are preyed on heavily by their parents.

The larger larvae of North Sea mackerel (*Scomber scombrus*) feed actively on smaller individuals, and these may comprise as much as 70% of their food intake. Spent herring *Clupea harengus*, returning from spawning migrations in the southern North Sea, consume herring larvae spawned by the eastern Channel stock that have been advected across their route. Unusually large individuals (mostly spent males) may be important consumers, towards the end of the spawning season, of unhatched benthic egg masses; most herring, of course, habitually leave the spawning beds immediately after spawning.

I do not treat here the filial cannibalism that occurs commonly among demersal nest-building teleosts, except to note its existence, and the assumption that it is adaptive because the growing young (having secured their own food for growth) provide an alternative food source for parents and so an investment in future reproductive success.[24] Partial filial cannibalism by the male, nest-guarding clingfish (*Diademichthys lineatus*) of the Indo-West Pacific induces growth to large and robust form, so enhancing the ability of males to maintain their territoriality in habitats where suitable nests are limited in number.[25] Nevertheless, it seems unlikely that this behaviour could have been a precursor to the cannibalism observed in gadoids and other demersal fish of the open-water habitat.

It is no surprise to find that motile organisms other than fish have evolved feeding habits that include cannibalism: John Caddy has analysed cannibalism in the pelagic cephalopod *Illex*, suggesting that this may be a strategy to bridge gaps in the pelagic particle-size spectrum so that the species as a whole is able to exploit food particles too small to be usable by larger, maturing squid.[26] Squid seem generally to be very dependent on this process, and a very brief review of the literature at once reveals that cannibalism in this group is very common: the giant Humbolt squid of the eastern Pacific, *Loligo pealii* and *Illex illecibrosus* of the NW Atlantic, *L. duvauceli* of the Indian Ocean, *Notodarus gouldii* of Australian seas, *Loligo vulgaris* off South Africa and the Gonatidae (armhook squid) of northern boreal Pacific.

Finally, we should take some note of the apparently very odd reproductive behaviour of some clupeids, especially sardines and anchovies, among which egg cannibalism may be intensive, although each episode is necessarily brief. Having demonstrated experimentally that northern anchovy (*Engraulis mordax*) recognise a patch of their own eggs as food, and begin filtering intensively within the patch, Hunter (1981) found that the stomach contents of anchovies off California indicated a daily ration of 86 anchovy eggs per adult individual, equivalent to about 32% of the daily mortality of eggs; rates of egg cannibalism increase exponentially with the density of eggs in the sea. Further, Hunter suggested that because larvae are digested totally within 30 minutes while ova are more resistant to digestion, it is extremely probable that larvae, especially yolk-sac larvae, suffer equivalent mortality from parental cannibalism. Apparently, cannibalism and intra-guild predation is common among clupeids, and has been noted among the populations of *Engraulis capensis* and *Sardinops ocellatus* off southern Africa. Stomach contents of both species contain anchovy eggs (mean numbers per stomach being 24 in anchovies and >2000 in sardines): this represents 6% and 56%, respectively, of total anchovy egg mortality.[27] It should perhaps be no surprise, then, that it has been shown that egg cannibalism is one of the most important regulatory mechanisms that determine recruitment in anchovies off California and Peru.[28]

At the other extreme of ecological conditions, those of the tropical open ocean where seasonal environmental changes are relatively weak compared with the North Atlantic, reproductive patterns of teleost fish are characteristically different. Tuna mature relatively early in relation to their total life span and, once mature, produce batches of eggs almost continuously at intervals of only a few days wherever and whenever water temperatures are suitable: in the case of yellowfin (*Thunnus albacares*) and skipjack (*Katsuwonus alalunga*), this occurs when temperatures exceed 25°C. Some regions of the ocean are thus continually supplied with new cohorts of very small tuna, so that small fish are continually present and available as food for adults. In the case of surface-feeding skipjack, up to 13% of the total intake of adults may be due to cannibalism of 0-group individuals, although yellowfin, which feed principally at thermocline depth, take young individuals of their own species less frequently.

Naturally, I do not wish to exaggerate the potential importance of cannibalism in the ecology of teleost fish or to present it as the sole or principal reason for the adherence of these fish to reproduction by means of planktonic larvae: there are, as Walter Nellen has reminded me, many contrary examples of species that are highly prolific, that produce planktonic larvae – and hence many small juvenile individuals – but for which there is neither evidence nor expectation that cannibalism occurs. Such is the pelagic, oceanic oddity *Mola mola*: this large species has one of the highest counts of ova of any batch spawner, yet gelatinous organisms famously dominate its diet, with no evidence of cannibalism. It is not difficult to find other similar examples.

Apart from such exceptions, it is not yet certain to what extent we can generalise the hypothesis that cannibalism serves to permit the growth of populations of large predatory fish in regions where the entire particle-size spectrum is not fully inhabited or where, in other words, there is an insufficiency of smaller species suitable for their food. Some doubt is cast on the thesis by a study of the predator–prey size relationship in the diet of populations of habitually piscivorous gadoids; this comforts rather the alternative thesis that the benefit derived from cannibalism is to slow down the progressive decline in trophic niche breadth experienced by most large predatory fish as they grow. There is evidence for selection of conspecific prey that are smaller than other species consumed in only one of the species studied (*Merluccius bilinearis*). In one other species (*Theragra chalcogramma*), conspecific prey items tend to be larger than others, while in the remaining three species studied there is little distinction.[29] These observations do not, of course, contradict the original Nellen thesis that reproduction for some fish is, in part, a nutritional investment.

Whether this thesis is correct or not, cannibalism must be included in any evaluation of the natural dynamics of teleost populations because it has, or may have, important population effects. The extreme example of this is perhaps that of the eastern Baltic cod stock, in which an average of 25–38% of the individuals of the 0-group and 11–17% of the 1-group is consumed by older adults, so that each year-class may lose as much as *c*. 45% of its members to cannibalism, resulting in a trophodynamic control on recruitment success.[30]

Crucial as such observations may be for understanding the ecology of teleosts, they are not the key to an understanding of what might contribute to the claimed resistance of teleost fish to the pressures of a modern fishing industry; yet one consequence of the extraordinary fecundity of teleosts is the potential ability of a population to increase its total numbers very much more rapidly than could – for instance – a population of elasmobranchs or mammals. At least theoretically, a small change in the mortality rate of larval and post-larval cohorts can result in an anomalously large incoming cohort of recruits to the adult population. Such a mechanism has the potential to return a depressed population to full occupancy of its habitat, and the history of fished populations shows quite clearly that this can happen.

The circumstances leading to the occurrence of strongly depressed populations of teleosts in the absence of a fishery shall be discussed in Chapter 5, because it is to deal with these circumstances that the potential for rapid population growth has perhaps evolved. Obviously, such an ability is clearly of critical importance in the potential of teleost populations to rebuild after a fishery-induced collapse; the role of this ability in the rehabilitation of stocks that have been reduced to a small remnant by overfishing will be discussed in Chapter 8. In this context, we are close to the proposition that if a population is maintained by a fishery at smaller stock biomass than the pristine level, a useful surplus production results: the problems resulting from this ecologically simplistic concept shall be discussed in the following chapter.

We may well conclude, then, that the mode of reproduction of teleost fish, unique among vertebrates and otherwise shared essentially only with insects, does give some support to Pauly's axiom and to the assumption that the ocean may be a more sustainable source of protein from vertebrates than the land.

ENDNOTES

1. My Nigerian colleague Vincent Sagua described this process in *Brachydeuterus typus* off Lagos in 1965 (*Nigerian Fed. Fish. Serv. Res. Rpt.* **2**, 61–3).
2. Rickman, J., *et al.* (2000) Recruitment variability related to fecundity in marine fishes. *Can. J. Fish. Aquat. Sci.*, **57**, 116–24.
3. Froese, R. and S. Luna. (2004) No relationship between fecundity and annual reproductive rate in bony fish. *Acta Ichthy. Pescat.*, **34**, 11–20.
4. Myers, R.A., *et al.* (1999) Maximum reproductive rate of fish at low population sizes. *Can. J. Fish. Aquat. Sci.*, **56**, 2404–19.
5. Mertz, G. and R.A. Myers. (1996) Influence of fecundity on recruitment variability of marine fish. *Can. J. Fish. Aquat. Sci.*, **53**, 1618–25.
6. Denney, N.H., *et al.* (2002) Life history correlates of maximum population growth rates in fishes. *Proc. R. Soc. Lond.*, **269**, 2229–37.
7. Hickling, C.F. (1940) The fecundity of herring in the southern North Sea. *J. Mar. Biol. Ass. UK*, **24**, 61–63.
8. Murray, B.G. (2001) Are ecological and evolutionary theories scientific? *Biol. Rev.*, **76**, 255–89.
9. Fisher used this term to denote the rate of change of the genotype.
10. In *Patterns in the Ocean* (1996, California Sea Grant, p. 323).
11. Peterson, W.T., *et al.* (1979) Zonation and maintenance of copepods in the Oregon upwelling zone. *Deep-Sea Res.*, **26**, 467–94.
12. Palumbi, S.R. (1999) The prodigal fish. *Nature*, **402**, 733–4.
13. Cushing usefully discusses this and other larval migrations in his 1975 *Marine Ecology and Fisheries*, Cambridge: Cambridge University Press.

14. Hunter, J.R. (1981) Feeding ecology and predation of marine fish larvae. In '*Marine Fish Larvae: morphology, ecology, and relation to fisheries*, Ed. R. Lasker. Washington: University of Washington Press, pp. 33–79.
15. Nellen, W. (1986) A hypothesis on the fecundity of bony fish. *Meeresforsch. Hamburg* **31**(11), 75–89.
16. Odum, E.P. (1971) *Fundamentals of Ecology*. Toronto, London, Philadelphia: W.B. Saunders, 3rd edition, p. 274.
17. Bedart, A.T., *et al.* (2001) Initial six-year expansion of an introduced piscivorous fish in a tropical central American lake. *Biol. Invas.*, **3**, 391–404.
18. Smith, C. and P. Reay (1991) Cannibalism in teleost fish. *Rev. Fish. Biol. Fish.*, **1**, 41–64.
19. Claessen, D., *et al.* (2003) Population dynamic theory of size-dependent cannibalism. *Proc. R. Soc. Lond. B*, **271**, 333–40.
20. Juanes, F. (2003) The allometry of cannibalism in piscivorous fishes. *Can. J. Fish. Aquat. Sci.*, **60**, 594–602.
21. John Field, University of Cape Town, pers. comm.
22. Macpherson, E. and A. Gordoa. (1994) Effect of prey densities on cannibalism in Cape hake off Namibia. *Mar. Biol.*, **119**, 145–9.
23. Francis, R.C. and K.M. Bailey. (1983) Factors affecting recruitment of selected gadoids in the NE Pacific and East Bering Sea. In *From Year to Year*, Ed. W.S. Wooster. Seattle: University of Seattle Press, pp. 35–60.
24. Manica, A. (2002) Filial cannibalism in teleost fish. *Biol. Rev.*, **77**, 261–77.
25. Gomagano, D. and M. Kohda. (2008) Partial filial cannibalism enhances initial body condition and size in paternal care fish with strong male–male competition. *Ann. Zool. Fennici*, **45**, 55–65.
26. Caddy, J.F. (1983) The cephalopods: factors relevant to their population dynamics. *FAO Fish. Tech. Pap.*, **231**, 1–452.
27. Szeinfeld, E.V. (1991) Cannibalism and intraguild predation in clupeoids. *Mar. Ecol. Progr. Ser.*, **79**, 17–26.
28. MacCall, A.D. (1980) The consequences of cannibalism in the stock–recruitment relationship of planktivorous pelagic fishes such as *Engraulis*. *UNESCO/IOC Workshop Rep.* **28**, 201–19; see also Csirke, J. (1980) Recruitment in the Peruvian anchovy and its dependence on the adult population. *Rapp. Réun. Cons. Int. Explor. Mer.*, **177**, 307–13.
29. Juanes, F., op. cit. (see note 20 above).
30. Neuenfeldt, S. and F.W. Koester. (2000) Trophodynamic control on recruitment success in Baltic cod: the influence of cannibalism. *J. Cons.*, **57**, 300–09.

3

Indeterminate growth, negative senescence and longevity

'Youth comes with age . . .'

Vaupel *et al.*, 2004

The second characteristic of marine fish that requires some preliminary discussion is their great longevity and indeterminate growth, especially when compared to that of terrestrial mammals; the continued growth of fish, almost throughout life, proves to have major consequences for their reproductive pattern that are critical to understand when contemplating the ability of teleosts to serve as a resource base for industrial fisheries. This aspect of their biology (like many others) was almost entirely neglected by classical fish stock management techniques. Consequently, although it has been known for more than 50 years that the eggs of older female teleosts are liable to be larger, to contain more yolk and to produce more viable larvae than those of younger fish, these observations were ignored until very recently, perhaps because they were made on freshwater fish and were published in Russian.[1] This is unfortunate, because they could have led to an earlier understanding that reliance on the surplus production axiom may contribute significantly to stock failure; this problem is the focus of this chapter.

Studies of the reproductive ecology of long-lived Pacific rockfish recently introduced a new concept into contemporary fishery science – that of the BOFF (or big old fecund female); these, it is now suggested, should be protected from fishing mortality because of their critical importance to the reproductive success of the population. Still, I have the impression there is not yet a proper appreciation of the fact that the existence of individuals with these characteristics in pristine populations of teleosts is the general result of the progressive partition of energy between growth and reproduction in all organisms having indeterminate growth trajectories. This observation suggests that a discussion of the comparative ecology of organisms that have indeterminate and determinate growth, respectively, may be in order. It will be suggested in the course of this discussion that the indeterminate growth pattern of fishes has several important consequences for fisheries management – both positive and negative – other than that of the progressive efficiency of reproduction during the period of continued growth of adult fish after maturity.

NEGATIVE SENESCENCE: A CONSEQUENCE OF INDETERMINATE GROWTH

Indeterminate growth, or growth that continues past maturation to the end of life, has consequences that have been little appreciated by fisheries science, even though this is the growth pattern of almost every fish species. There has been little understanding of the fundamentally different patterns of ageing in organisms that grow in this way from those – like you and me – that have a determinate growth trajectory that flattens at a certain size, usually near the size at maturity. It has been suggested recently that indeterminate growth usually evolves as a trade-off between high fecundity and risk-avoidance in stochastic environments.[2] The authors of this study suggest that while determinate growth has evolved among 'many mammals and birds', the optimal strategy for 'organisms in a number of other taxa' is indeterminate growth.

I suspect that this statement would not cause significant comment among fishery scientists, but it is really very misleading: it would be closer to the truth to remark that indeterminate growth is the fundamental growth pattern of multicellular animals and that only exceptional groups exhibit determinate growth. Indeterminate growth is also the general rule among plants, the principal exceptions being annual herbs and grasses which die after flowering.

Those exceptional animals that exhibit determinate growth are dominated by species that weigh more than a few grams and which stand erect on legs or fly (thus, all terrestrial mammals and birds). In these groups, the evolution of optimal adult form and dimension for each way of life – running, climbing, digging or flying – is dominated by the effects of gravity: birds having the physical dimensions of ostriches or moas have had to forego the ability to fly.[3]

Metamorphosis between stages of the life history also requires determinate growth, as in the case of the higher insects, which are a special case: while insects such as Diptera and Lepidoptera continue to grow throughout their larval stages, growth must cease after metamorphosis into the imago. The same is true for Copepoda, which exhibit stronger metamorphosis towards an adult form than some other crustaceans, such as Cladocera or Malacostraca. Some interesting exceptions to this otherwise general rule occur in exceptional environments, in which steady state is induced by very low energy flow, as shall be discussed below.

Thus, most animals which take their full weight on their legs while walking grow rapidly to a maximum adult size appropriate to their sex, after which growth ceases; reproductive activity usually begins only once adult size has been reached and energy is subsequently diverted from growth into reproduction. It is the few exceptions that demonstrate the validity of this as of many other rules: in just a few mammals, males continue to grow very slowly almost throughout life. These are species that live in small social groups that are dominated by a single alpha-male, such as some kangaroos, some ungulates, lions, elephants and apes. Among reptiles and amphibia, only a few isolated species (like Blanding's turtle) show determinate growth, while the great majority grow continually towards an asymptote.

Organisms that swim are generally not constrained to evolve a determinate growth pattern, having some degree of buoyancy in their fluid habitat. So fish growth, it goes without saying, follows an indeterminate trajectory, as does that of marine mammals, from large whales to seals: Steller sea lions (*Eumetopius jubatus*) mature at around 6 years of age, after which asymptotic growth continues until death at 15–25 years.[4] The larger males approach their asymptotic length somewhat later than females. A similar pattern of indeterminate growth occurs in the South American sea lion (*Otaria flavescens*), in which maturity is attained prior to the growth asymptote being approached.

The consequences of longevity in organisms with determinate growth have attracted the attention of biologists studying human senescence, because our own pattern of determinate growth is associated with age-related changes that adversely affect vitality, fertility and other functions and, most importantly, cause an increase in our mortality rate.[5] These characteristics are common to all organisms with deterministic growth, which suffer a progressively increasing individual mortality rate throughout their lives after maturity. Because this is our own growth pattern, we humans do not easily appreciate the consequences of life-long growth in fish and other organisms that have a totally dissimilar growth pattern from our own; I suggest, however, that we need to think about indeterminate growth very carefully, because it holds both promise and danger for industrial fishing.

Early debates concerning the consequences of longevity in animals culminated in an understanding that senescence is inevitable and universal, a thesis that was considered to have been proved by Hamilton in 1966[6]: 'senescence cannot be avoided by any organism', he stated. At this time, the only outstanding problem appeared to be why senescence had not been eliminated by selection, since it is accompanied by reductions in fecundity and survival. Enquiry into the mechanisms of senescence led from early and blind-alley analogies concerning the mechanical wearing-out of machines towards more satisfactory analysis of cellular senescence and the accumulation of DNA damage, including shortened telomeres, with increasing age. Thus, it was found that normal diploid, differentiated cells lose their ability to divide after about 50 divisions.

Such studies ignored the fact that as long ago as the 1930s, the potential immortality of organisms like fish that grow throughout their entire lives, and experience negligible senescence, had been remarked on by Bidder.[7] The fact that fish also have high telomerase activity in all their organs, unlike mammals in which it is restricted to stem cell derivatives,[8] was unknown to Bidder – but would have significantly strengthened interest in his thesis when he presented it; unfortunately, it appears to have been largely neglected until quite recently.

Progressively, it came to the attention of gerontologists that the death rates of some, indeed of many, organisms rose only very slowly, if at all, with increasing age so that these species – of both animals and plants – experience negligible senescence, a term coined by Finch in 1990 as the end-member of the series rapid, gradual and negligible senescence.[9] Rapid senescence occurs in annual herbs and grasses and many fish such as Pacific salmon, eels, congers, some shads, some smelts, silversides and capelin. Gradual senescence is what you and I can expect if we are lucky, while the organisms that experience negligible senescence, said Finch, are those in which 'dysfunctional changes have not yet been detected' and whose reproductive adult phase may span decades or even centuries. Perhaps because, on the side, he discussed the effect of extrinsic causes (e.g. diet) in human senescence, his book rapidly became central to gerontological and more popular debate: the apparent relevance of negligible senescence to biomedical gerontology has been used by some biologists to attract interest to their studies.[10]

A case was made recently by James Vaupel and his colleagues that Finch's series should be continued beyond negligible senescence;[11] they pointed out that if senescence is defined by age-related changes that adversely affect the vitality and functions of an organism, then in those many organisms in which death rates decrease and relative fecundity increases throughout life, the term negligible senescence is not really appropriate and we must speak rather of *negative* senescence. They pointed out that in many organisms, once maturity has been achieved, mortality declines even as relative fertility increases almost throughout life. This pattern of fertility and mortality is characteristic of organisms having indeterminate growth, while gradual senescence is generally associated with deterministic growth.

Since senescence is not the fundamental, universal and intrinsic property of living organisms that had previously been widely assumed, Vaupel and his colleagues suggested that the chronological age of an organism does not correspond to its biological age very well, and perhaps this 'may be better captured by the average age of an individual – i.e. by some appropriate measure of the *average age of the organs, body parts or cells* of an individual – than by chronological age'. The average age, thus defined, of an individual can be much lower than the chronological age; moreover, average age in organisms experiencing indeterminate growth declines over time as each individual grows and its component parts are renewed by cell division. In this sense a centennial rockfish is greatly younger than a centennial human whose somatic cells have far outlived their potential for renewal by division and have accumulated potentially deleterious mutations and DNA damage.

Indeterminate growth is so widely expressed in both animal and plant kingdoms that we must expect that negative senescence would occur widely, and so it does. Finch discussed a number of cases that have been reported for modular plants and animals, such as long-lived clonal coral colonies (*Muricea*, *Goniastrea*, *Platygyra*) in which mortality is reduced as the colonies grow and age.[12] Several candidates for negative senescence have been reported for plants, from brown algae to clonal groves of aspens (*Populus*) that may be very ancient, together with many large forest trees; macroscopic fungi are perhaps a special case because they are immortal, in the sense that their propagules are clonal. Clear evidence for negative senescence is observed in several phyla of invertebrates, such as marine gastropods (*Umbonium*, *Littorina*) and bivalves (*Yoldia*); the echinoid *Echinometra lucunter* is long-lived and telomerase activity has been detected in its tissues, the cells of which show a lack of size-related telomere shortening as would be expected in an organism with indeterminate growth.[13]

Many reptiles are very long-lived and adults appear to have declining mortality, as in the lizard *Scoleporus undulates*;[14] the reproductive success of sand lizards (*Lacerta agilis*) is strongly and progressively related to adult size.[15] In the following section, I shall discuss the extent and significance of negative senescence in fish, among which there is widespread evidence for its existence. I have no doubt that it is yet to be observed in very many other vertebrate and invertebrate taxa, and will prove to be a widespread phenomenon and characteristic of those organisms that exhibit indeterminate growth. The expression of the Vaupel model is highly variable in the 30 000 odd species of fish recorded in FishBase,[16] ranging in size from $<$8 mm (a newly found *Paedocypris* of Sumatran peat swamps) to $>$12 m (*Rhincodon*, the whale shark of the open ocean), but I suggest that it is entirely appropriate for those species of marine fish in which the fisheries are interested, and that it has consequences for their management that have not so far been evaluated.

In such a diverse group as the fishes, one would not expect indeterminate growth to be expressed similarly in all species: in some exceptional groups, growth ceases or slows significantly after maturity and well before their characteristic life expectancy is approached. This effect is a response to environmental conditions, and is characteristic of those species that occupy highly structured habitat within which individuals occupy and hold physical niches that it would be counterproductive to outgrow: territory-holding species of coralline reefs, whether in the shallow-water tropics or at the shelf-edge in cooler seas, characteristically take this growth trajectory. Thus, after a normal growth trajectory to age 10–12 years, the pomacentrid *Parma microlepis* essentially ceases linear growth and puts energy preferentially into reproduction. Not all deep hard-ground species perform similarly, for the lutjanids and serranids that are ubiquitous on this habitat in warm seas do not hold territory and continue to grow throughout life.

Similarly, the very long-lived species that inhabit oceanic banks and the continental slope to depths of a kilometre or more have growth curves that flatten somewhat after maturity is achieved and may experience very long intervals between successful recruitment: *Sebastes*, *Hoplostethus*, *Macrourus* and *Allocyttus* and others all grow in this way, and many are very long-lived indeed. In their case, the flattening out of the growth curve at maturity appears to be a response to the steady-state ecosystems that they inhabit. The same growth pattern occurs in Arctic charr that inhabit very cold lakes having very low energy flow that leads to steady state, a high degree of uniformity of the size of adults of all ages and very low levels of recruitment. It is likely that this response to a steady-state habitat and relatively cold temperatures accounts for the occurrence of a similar pattern of growth, uniformity of adult size and low recruitment of *Sebastes* and other teleosts that inhabit deep oceanic banks and the continental slope, but do not live in physical niches on rough ground.[17]

BOFFs: GROWTH TRAJECTORIES, SELECTIVE ENERGY ALLOCATION AND REPRODUCTIVE STRATEGY

Of the consequences that flow from the indeterminate and lifelong growth of fishes, perhaps the most important is the length of the reproductive period. In most reviews of fish physiology and reproduction, emphasis is placed on the relation between length at maturity and asymptotic length to suggest that fish habitually mature at 65–80% of their maximum length (L_{inf}), but this conceals the importance of the great length of life after maturity, a period when the individual rate of natural mortality progressively declines.

An overview of the length of reproductive life has been obtained from a file of data[18] concerning the ages at first maturity (T_{mat}) and of maximum age (T_{max}) of 161 individual populations of 50 species, of all groups of fish of interest to fisheries and at all latitudes, the mean value for T_{mat} was 0.29 of T_{max} recorded for each stock: consequently, the mean length of the reproductive period was about 14 years. For a few stocks, the indicated age at maturity was very much higher ($T_{mat}/T_{max} = 0.5–0.8$) and it is thought that these represented stocks that had been heavily fished and lacked many older age groups; it is very likely that the pristine mean ratio for the kinds of fish in which we are interested is significantly lower than indicated by this data set.

So, associated with the indeterminate growth trajectory there is a long period in the life of a fish when the potential to modify the allocation of energy between growth, reproduction and maintenance gives flexibility of response to changing circumstances and for the evolution of individual reproductive strategies. This response during the reproductive period is not available to organisms having determinate growth trajectories.

Perhaps in consequence, teleosts have evolved two different and contrasting reproductive strategies which serve to reduce the level of uncertainty of the recruitment process: females may either direct their reproductive effort into maximising the number of eggs produced annually through exceptional fecundity (as discussed in the previous chapter), or they may increase the probability of the survival of each larvae by increasing the maternal bestowal of energy stored in fewer and larger eggs; the concept that a *fecundity/bestowal* continuum exists in teleost reproductive strategies was suggested by Miller.[19] Two different processes appear to be involved in the transition from a fecundity to a bestowal reproductive strategy: (i) a progressive increase during ageing in the size of their ova that probably occurs in all female iteroparous fish, and (ii) the selection for larger, fewer eggs that occurs principally in those species that occupy habitat where recruitment is uncertain. Although these two processes are intimately related, it may be useful to treat them as somewhat distinct.

This approach recalls the r- and K-selection continuum that is so much part of the corpus of classical ecology but which was intended to refer only to species whose abundance is determined by fluctuating or stable adult mortality, rather than responding to fluctuations in reproductive success.[20] It is therefore not appropriate as a model for life history strategies in teleosts whose larvae commonly undergo mortality rates of 25–75% each day during the first weeks of life, during which the strength of the subsequent recruitment may be determined. K-selected species having long lives and few offspring are supposed to be characterised by relatively stable population sizes but, as will be discussed below, within species-groups those species having the greatest longevity are those whose recruitment to the adult stock is the most variable: this appears to be contrary to expectation from r–K theory.[21]

The approach of Miller to the problem of understanding relative egg size in teleosts also differs somewhat from the classical ecological investigations of the optimal relations between maternal and propagule sizes in animals. The ongoing debate among ecologists concerning evolution of the multivariate correlations between maternal age, size, fecundity and per-propagule investment can be resolved only if the ecological context of both maternal and offspring fitness are explicitly considered. Propagule size is unlikely to be selected for on the expectation of offspring fitness alone, as mother and offspring form a single evolutionary unit, so 'it is necessary to consider explicitly the ecological context in which a mother is producing eggs, not just that into which offspring will enter'.[22] Although the propagule size debate among theoretical ecologists has centred principally on energy allocation and maternal fitness, rather than – as in fisheries science – around recruitment success, the central conclusions are not dissimilar.

How the allocation of energy between growth, maintenance and reproduction evolves in fish is too well known to require any elaboration here: it is sufficient to recall that during growth, weight increases more rapidly than length, so that $W = a^{*}L^{b}$ where the exponent b usually lies between 3.1 and 3.5.[23] Since fecundity scales with gonad weight and since gonad weight is usually proportional to body weight, then absolute fecundity increases progressively during life after maturity: this is an important consequence of indeterminate growth. However, this increase is greater than linear, so the relationship of absolute fecundity with body length must be curvilinear and is expressed as relative fecundity, or fecundity per unit fish mass, which in many species increases significantly during adult life.

This effect can be very important in some species: during the growth from 12.5 to 20.5 cm of *Boops boops*, a Mediterranean sparid, absolute fecundity increases by an order of magnitude, and this increase is associated with a change in relative fecundity from about 300 to about 2500 ova/g.[24] However, it is probable that an increase by a factor of 2 or 3 is more usual: relative fecundity in *Sebastes melanops* rises from 374 to 549 ova/g between the ages of 6 and 16, in *Melanogrammus aeglefinus* from 274 at age 2 to 493 'in older year classes', and in *Micromesistius australis* from 200 to 500 from Lt 40 to 85 cm.[25] The general increase of relative fecundity with age must contribute to the exceptional fecundity of teleosts that was discussed in the previous chapter and that dominated studies of reproduction and recruitment in teleosts in the past.

The bestowal (or investment) reproductive strategy (*sens.* Miller) has attracted little attention until recently, but its importance is now increasingly recognised as central to understanding the significance of the low relative fecundity that is associated in some fish with a large egg size. This relationship has been investigated in 75 teleosts,[26] most of which were from freshwater or shallow marine habitats, and these have relative fecundities approximating 10^{-4} eggs per gram, with ova volume ranging from 0.1 to 1.0 mm^3; the remaining 20 species were from deep marine habitats and these were found to have relative fecundities lower by 1–2 orders of magnitude associated with larger eggs, from 2.0 to 20.0 mm^3. These data on the mean size of ova in fish are similar to those obtained from a different data set representing 309 North Atlantic species,[27] which indicated a median diameter of 1.1 mm (mean 1.6 mm, with a tail of the distribution extending to 16 mm); these data demonstrated both a direct relationship between the dimension of eggs and of newly hatched larvae, and also indicated a significant variability of 2–20% in ova diameter between individual conspecific females.

It seems likely that the observed relationship between fecundity and egg volume, varying with the degree of maternal investment, is general in teleosts, and supports Miller's suggestion. While clear demarcation cannot be made between fecundity and bestowal strategies, the species in which the bestowal strategy is most strongly developed are almost exclusively deep-water species. This strategy is surely a response to uncertainty of reproduction, and is typical of rockfish (*Sebastes*, *Hoplostethus*) living at shelf-edge depths and whose recruitment is episodic and uncertain: good year-classes may be produced only every 20–25 years by these stocks: such species have very low fecundity and produce unusually large eggs.

There are now sufficient observations to suggest that we may generalise the increase in size of ova during maternal growth in the kinds of teleosts in which fisheries are interested, for it has now been observed in many species: North Atlantic herring, capelin, winter flounder, turbot, cod, haddock and striped bass; South Atlantic Argentine hake; New Zealand orange roughy; North Pacific rockfish of several species; and, in Russia, several species of freshwater fish.[28] One of the latter, *Coregonus peli*, is reported as having a terminal decline in egg size in the oldest (7+) age group.

The consequences of changing character of the ova produced by females as they age was investigated experimentally in the black rockfish (*Sebastes melanops*) obtained from the central Oregon coast by the late Steve Berkeley and his associates. The results were unequivocal: 15-year-old females produce larger eggs that contain larger oil globules than fish only 5–6 years old, and their extruded larvae grow more rapidly and starve only after a longer period without food. The differences are very significant: the oil globule volume in eggs of older fish is 0.012 compared with 0.003 mm^3, growth rate of larvae post-extrusion is 0.06 compared with 0.02 mm/day and their starvation time to 50% mortality is 13 compared with 6 days.

We can now be reasonably confident of the generality (at least among those teleosts of significant interest to fisheries) of the mechanisms involved in these processes, in which older females:

- have higher fecundity,
- spawn more frequently, and start earlier in the season,
- spawn larger and more buoyant eggs,
- having higher fertilisation and hatching rates, and
- which produce larvae that swim faster.

This pattern has now been observed in a wide range of commercial teleosts, very largely in the North Atlantic: cod, haddock, winter flounder, turbot, striped bass, herring and capelin.[29] In haddock (*Melanogrammus aeglefinus*) of British seas, egg size is positively correlated with fish length, although with a change in slope of the relationship between very small reproductive fish (mostly 2-year-olds) and older year-classes that have a higher relative fecundity by a factor of about 2; since in some years very young spawners form a large proportion of the spawning stock, the effect on recruitment may be very significant.[30] In Argentine hake (*Merluccius hubbsi*), young spawners ($\leq$5 years old) have greater variability in egg size and weight than older females.[31] In species such as haddock and hake in which the maternal age effect is limited to the very young age groups, the lifetime egg production of a stock may not be so sensitive to removal of the older year-classes by fishing as it would be in species in which maternal age effects are progressive throughout life.[32]

In Norwegian coastal cod, egg volume increases by 25% between the first and second spawning season and further increases occur progressively, so that females of 75 cm produce eggs that are 38% greater in volume than those produced by fish only 56 cm in length.[33] The quality of the eggs also progressively improves, as measured by hatching success: only 40% success at first spawning, but 70% by the third season. However, these observations are not of universal application, for in some observations (such as those made by Buckley and associates on American winter flounder) no relationship has been found between egg size and hatching success.[34] Compensatory, or catch-up, growth by those young fish that have had a maternally influenced slow start of originally undersized larvae has also been observed on rainbow trout.

Recruitment success may also be influenced by maternal size in other ways than quality of ova. Older, larger females characteristically begin to spawn earlier in the season than younger fish, and continue spawning over a longer period: this has been observed in sufficient individual species to accept that it is probably a general phenomenon: cod (three stocks), herring, plaice, winter flounder, small- and large-mouth bass, rock bass and white sucker.[35] The best-worked example of the potential consequences of these maternal age effects must be that of the Vestfjord cod of northern Norway: early season eggs are larger (a general effect of first batches in a batch spawner like cod, plus the effect of older fish spawning earliest), and these eggs appear to have a high diversity in their buoyancy so that all vertical habitats in the highly stratified water masses of the fjord shall be occupied.[36]

A similar situation has been observed in the Baltic, where cod spawning occurs in the deep basins having strong density stratification above a low-oxygen bottom water mass; it is thus critical that egg buoyancy should be such as to retain them in suspension in the shallow, oxygenated surface layer and prevent sinking into the anoxic layer: the eggs of older females have this characteristic. Consequently, recruitment levels at age 2 are positively related to the percentage of females ≥ 5 years of age in the maternal population, and hence to the fraction of the egg production that originated from older fish.[37] Similar relationships between strong recruitment and the relative presence of older fish in the population has been observed in Norwegian herring and Icelandic cod: these effects appear to be at least as important as total stock size. Despite these observations, the mortality of larvae from Baltic cod has also been shown to be size-dependent. Those hatching at or below mean length for the population die after about 12–13 days, while those surviving beyond 20 days are those that hatch at larger than mean length for their cohort: these effects are habitat-independent.[38]

WHAT HAPPENS IF WE CURTAIL THE DURATION OF THE REPRODUCTIVE PERIOD?

Strikingly, the effect of these maternal age effects on spawning patterns and egg quality question a basic assumption of fisheries science (as Trippel and his associates put it), 'that spawning stock biomass is proportional to reproductive potential remains one of the greatest untested assumptions in fisheries science'. This comment raises the question of the modification of population age structure that appears to be an inevitable result of fishing – whether artisanal, industrial or recreational – in relation to the pattern of decreasing natural mortality rate and increasing relative fecundity during life that is characteristic of teleost growth.

Although it is very difficult to model the characteristics of pristine populations with any confidence, it is very probable that each annual cohort within the population would maintain an output of ova that would not reflect directly their numerical reduction with the passage of years. The inevitable and progressive reduction of the older fish once fishing mortality is imposed, and catches are selectively taken from older individuals, may have unexpected consequences for total stock fecundity. This has been little investigated, although it has been suggested that the economic value of older fish is significantly less than their reproductive value and that the natural economic capital represented by uncaught fish can only be assessed by age-based accounting.[39]

An attempt was made some years ago to evaluate this effect (in ecological but not economic terms) for a few species considered to be typical of those that are important in the industrial fisheries.[40] Species-specific models of the lifetime evolution of numbers of individuals, and their collective fecundity, within a year-class cohort were drafted so as to represent a pristine population of each species in the absence of fishing mortality. As far as possible, the values used were derived from published observations, and consultation with experts familiar with the species concerned, and search of the early literature was given preference over standard relationships derived from the von Bertalanffy equation.[41] It is more difficult to obtain satisfactory values representing natural mortality (M) in pristine populations than for growth in weight or for fecundity, for which observations made on populations early in the twentieth century are usually available – as in the case of the herring fecundity assumptions, based on observations made by Hickling before 1940. Published suggestions for representative values of M are almost always expressed as single values that are supposed to be representative of all age-classes, and suggestions for the derivation of M from other parameters or observables (e.g. K, L_{inf}, $T_{ambient}$, and so on) are equally unrepresentative.

In the cohort fecundity models considered here (Figure 3.1), values proposed either in FishBase archives or in the late Ram Myers' database were used, but were adjusted so that mortality decreased during early adult life from values representative of pre-reproductive years and then increased again in the few years prior to the death of the last survivor of the cohort. It is thought that this better represents the real course of natural mortality in unfished stocks, especially in long-lived fish like cod, than single-value estimates of M. The problem is that certainty cannot be reached because we have such incomplete knowledge of the population characteristics of stocks prior to any fishing. Archaeological evidence from the frequency distribution of bones from large and small individuals found in mediaeval middens is often cited to support suggestions of pristine stock structure,[42] but even in such cases the stock, by definition, was not pristine: it was already fished, or there would have been no bones in the midden! What is quite sure is that the survival of old, large fish was such that the population structure of northern cod stocks 500 years ago was entirely different from what it has been in the recent past, when a physiological competence for a life span of 25–30 years, achieving lengths of almost 200 cm, has rarely been realised.

The models suggest that the production of eggs by each cohort does not decline at a rate simply determined by the number of survivors after natural mortality is accounted for, but significantly more slowly: in the case of a cod stock of the NW Atlantic, the model suggests that the annual egg production of the single dwindling cohort is substantially sustained over a period of 15 years, when the survivors are

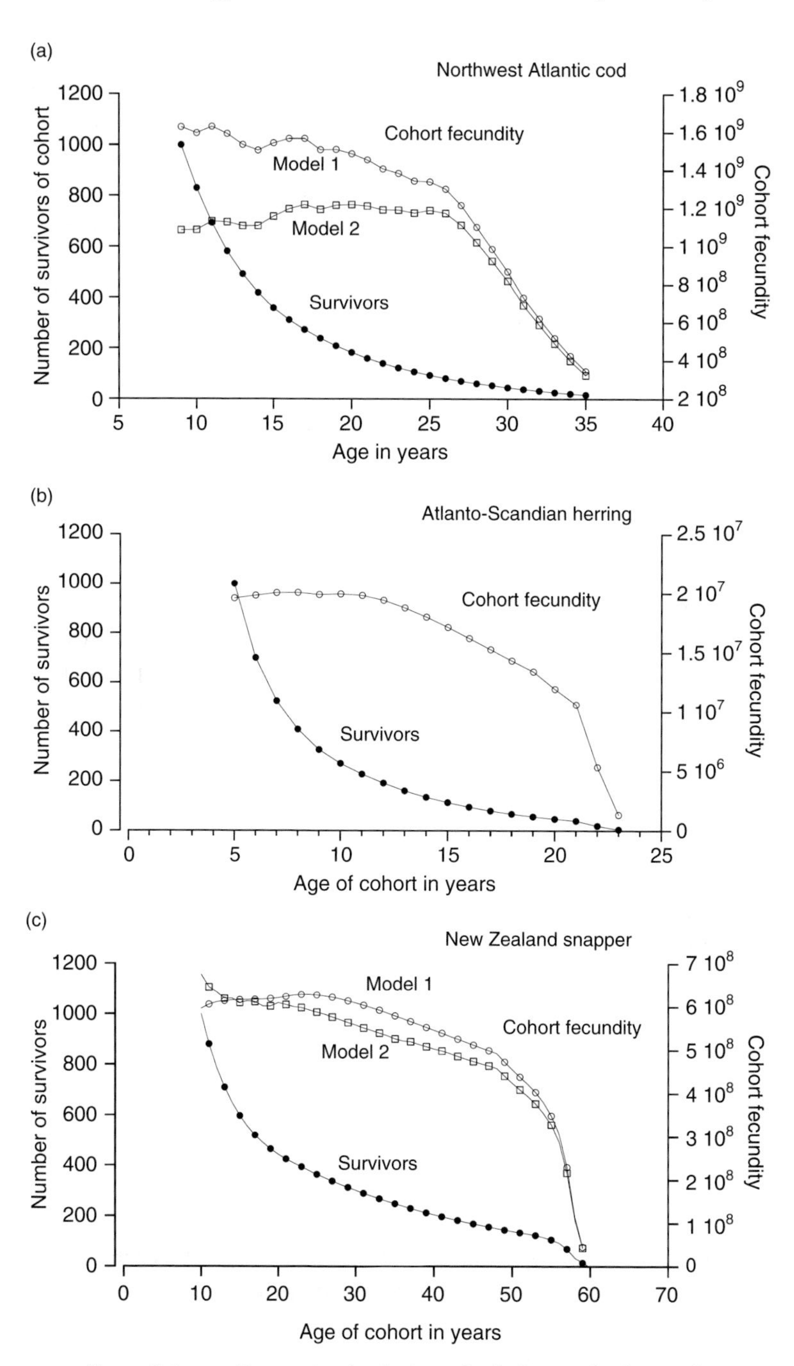

Figure 3.1a–c. Egg production by hypothetical annual cohorts of 1000 females at first maturity: number of survivors and the summed annual egg production of each cohort.

approaching 25 years of age and it is only in the last period of the life of each year-class that cohort fecundity decreases substantially. Such a pattern of cohort fecundity would permit a stock to maintain sufficient reproductive potential to produce a good year-class even after a very long period of unfavourable oceanographic conditions, and even were the entire population to be dominated by very few widely separated year-classes.

The models representing Atlanto-Scandian herring and *Chrysophrys auratus* (a sea bream of the continental shelf targeted by trawl fisheries[43]) off New Zealand both suggest that this pattern is not a unique attribute of high-latitude gadoids: the fecundity of these stocks appears to be disproportionally sensitive to loss of older fish. The *Chrysophrys* model suggests that 73 fish at 26 years of age have the same reproductive potential as 810 fish at 10 years of age – each group having the potential to produce about one billion eggs. Additionally, of course, we may also assume that the eggs of the older fish would have a higher individual chance of successful recruitment.

The ability to maintain a high level of cohort fecundity over the entire reproductive life of each cohort appears to be a critical factor in the life history strategies evolved by recent teleost fish, and the expression of this ability appears to respond to the exigencies of the environment (both living and non-living) of each species and, especially, to the probability of successful annual recruitment: as a generality, species that experience variable recruitment have evolved longer reproductive periods than those which experience more reliable recruitment success. This effect was first observed about 40 years ago in a small group of five species of clupeids, among which those with the most unreliable recruitment pattern had the greatest longevity.[44] These observations by Garth Murphy suggest that fish populations having the characteristic longevity of a K-selected stock have competitive advantage in a variable environment over r-selected stocks; he thus turned the classic notion of r- and K-selection strategies on its head.

The r-selection, K-selection model is usually interpreted by the extreme cases. Thus, r-selection, where r is the intrinsic rate of population increase, is held to represent the ideal response of species in which high density-independent mortality requires repeated re-colonisation of habitat, and is associated with early maturity, high fecundity and a brief life span. At the other extreme, density-dependent mortality does not require constant re-occupation of habitat, and is associated with late maturity, low fecundity and great longevity: these are the characteristics of K-selected organisms, where K defines habitat carrying capacity. One of the many problems with the general theory, as stated, is that it is irrelevant to typical life histories of fish, because it doesn't respond to uncertainty in their reproductive success.

Because of the current ease of access to major data archives of the critical characteristics of marine fish, it is now relatively trivial to explore the relationship explored by Murphy rather widely; this has been done, and the results confirm the original thesis, although with some interesting constraints.[45] As we would expect, the relationship is not valid when species from different families of fish are compared, because of divergence in their life history pattern, which is one of the consequences of the design constraint barriers in the ecology of species discussed by Stearns.[46] However, when species within groups are compared, the relationship proves to be rather general, except for a single anomalous group that suggests a major distinction between fish that inhabit the physically structured benthic habitat and those that swim freely in the water column, either as demersal or pelagic species.

This is perhaps most clearly illustrated by analysis of 117 individual stocks of clupeids for which both recruitment variability (cvR) and longevity are available. This suggests that neither longevity nor recruitment respond directly to latitude as we might expect them to do. Rather, in each natural oceanographic region they take values apparently appropriate to its characteristic environmental variability. *Clupea* is a good example of this response: oceanographic variability in both Atlantic and Pacific Oceans is exceptionally strong where subtropical western boundary currents encounter cold sub-polar gyral water. In these places, we find the highest values of cvR for *Clupea* (197% off Newfoundland and 183% off Hokkaido, together with greatest longevity, reaching 23–25 years). But the Baltic is protected from basin-scale oceanographic forcing, and two stocks of Baltic herring have a cvR of only 34% and a T_{max} of only 10 years.

Consequently, both collectively, and within each genus for which we have sufficient data, clupeids exhibit a more-or-less clear relationship between longevity and the uncertainty of recruitment. For herrings (*Clupea*) and sardines (*Sardina, Sardinella*) the relationship is unequivocal so that T_{max} increases progressively with increasing recruitment variability; the range of values is greater in the longer-lived herrings than sardines (Figure 3.2a). For the very short-lived anchovies (*Engraulis, Cetengraulis*), for which we have data from only eight stocks, the relationship is unreliable and is forced by the data from a single stock: in this genus, whose life-history characteristics are constrained by their very brief lives, mostly of 3–4 years, the possibility of extending

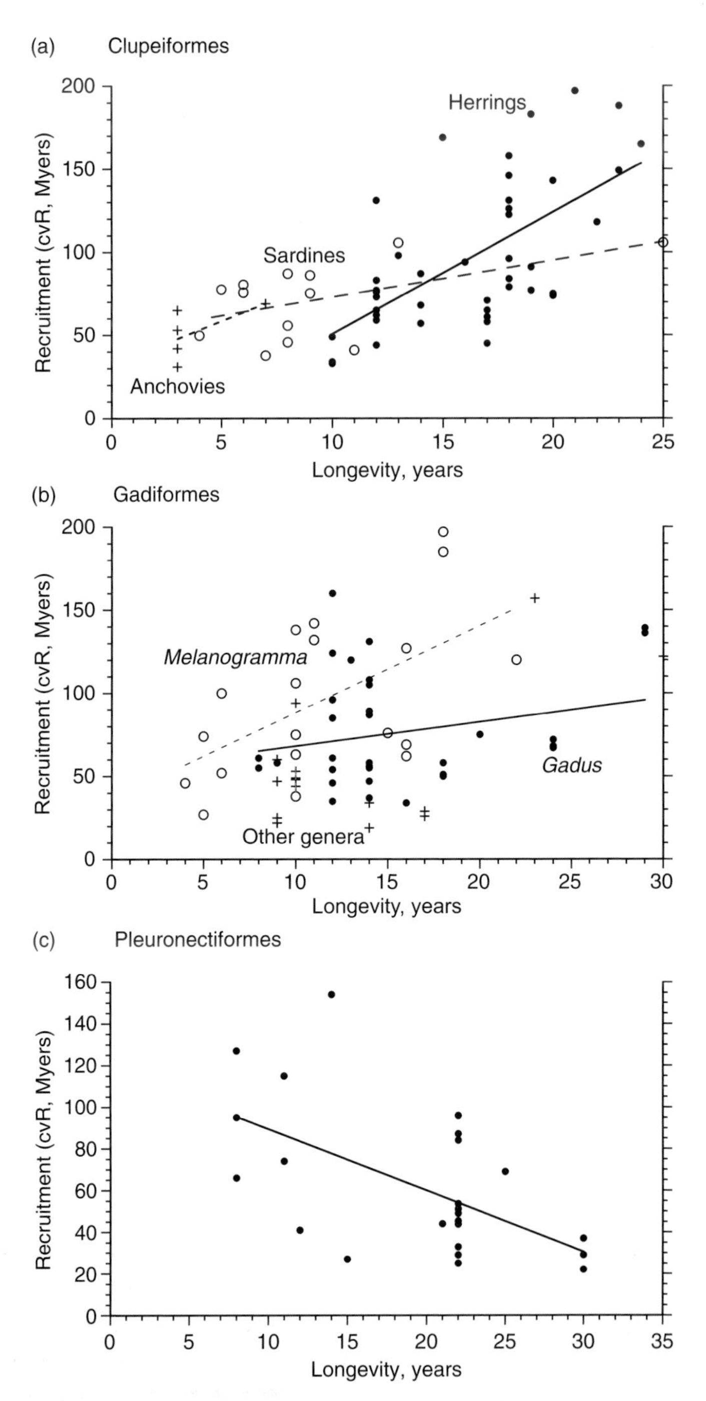

Figure 3.2a–c. 'Murphy plots' illustrating the relationship between longevity and the uncertainty of recruitment in clupeiods, gadoids and flatfish.

longevity may not exist: a clear distinction is necessary between anchovy and sardine species for them to occupy the same pelagic habitat. Somewhat similarly, the very long-lived rockfish of deep habitats appear to have a similar ecological design constraint that restricts the range of longevity to between 60 and 90 years in the species for which data are available; but, even across this relatively narrow range and within a rather small group of stocks ($N = 10$), those with the greatest longevity do have the most uncertain recruitment.

The relationship between T_{max} and cvR is also expressed clearly in gadoids, both collectively and individually within the genera (*Melanogramma* and *Gadus*) for which there are sufficient data for analysis (Figure 3.2b); this occurs across a group of fish with a rather wide range of natural longevity although, ecologically, they are rather homogeneous. Unfortunately, I have been unable to find satisfactory data to investigate the relationship within the warmer-water genera associated with hard grounds. Such a data set might prove very interesting, because the only clear exception to the general relationship discussed here are the pleuronectiform flatfish: although the data set is not large ($N = 25$ stocks), the result appears to be unequivocal – stocks with the greatest longevity have the most reliable recruitment (Figure 3.2c). The three stocks having longevity of 30 years are Greenland halibut, American plaice off Newfoundland and Alaska plaice in the eastern Bering Sea, while the three stocks having the smallest longevity are of *Solea vulgaris* around Britain.

How to explain this apparently strange result? Perhaps the relatively low mean cvR for flatfish, of only 49%, compared with about 75% among the gadoids, is relevant as is their form of recruitment to the bottom community by a unique pattern of larval metamorphosis which is 'abrupt, systematic and apparent even to casual observation'.[47] Associated with this is the suggestion that the limited areas of suitable sediment for larval settlement act as a density-independent, two-dimensional spatial filter that regulates year-class strength within relatively narrow limits. In the transformation of the larvae of the demersal fish of the water column, the limiting factors change from year to year. Such a constraint on the range of recruitment success may perhaps explain the observations and this possibility makes it all the more unfortunate that it has not been possible to locate a suitable set of data for hard-ground demersal fish like lutjanids and pomadasyds, among which the same effect might be observed. Among reef fish (like *Parma*, whose termination of growth in middle age was discussed above) we may expect to find the same situation – but, once again, suitable observations are not to hand.

The most direct explanation of the observed relationship is the evolution of specific longevity as a response to the level of uncertainty of recruitment experienced by each stock as it encounters novel ecological conditions, either by the invasion of novel habitat or by changing conditions within habitat previously occupied. It must be admitted frankly that such arguments, based simply on observed correlation and logical explanation, are too frequently advanced in ecology, and also that fisheries ecology freely participates in arguments of this kind that are logically inadmissible; yet, the reality is that without their use, we should make little progress. In this case, the practical implications that are suggested are too important to be ignored.

The ability to sustain growth and fecundity throughout adult life should enable the remnant adult individuals of a stock that has been significantly reduced in numbers to respond with a density-dependent increase in their individual growth rates and fecundity levels at any age, right up to the very last period of life. Such an ability must enhance the rapidity with which a numerically depressed population can rebuild its numbers compared with any animal, terrestrial or marine, that has a determinate growth pattern: in these, only the reproductive potential of surviving adults can respond to the density-dependent increase of food. And this, of course, is an ability that progressively decreases with the age of the individual, unlike the opposite case in fish. This potential for rapid population increase of fish stocks lies at the very core of classical fisheries management techniques.

Like fish, those marine mammals having an indeterminate growth pattern may also exhibit a density-dependent response to population size. Etnier has described this process in northern fur seals at the Pribilofs from archived sub-fossil mandibles collected prior to the start of the fur seal fishery: the size of individuals was found to be negatively correlated with population abundance.[48] There is also some evidence that North Pacific sperm whales responded to relatively light industrial exploitation in the 40-year period after 1950; growth rates, especially of males, increased significantly.

Because of their typically deterministic growth pattern, we should not look for the same individual growth responses in terrestrial vertebrates; the most that we should expect is a relationship between survival and initial growth of young animals. A probably typical example of what occurs to adults under extreme conditions of population density is that of populations of feral donkeys in Australia: pregnancy rate, juvenile condition and mortality and juvenile growth rate all respond as expected to relative population density. Such observations would support an expectation that populations of marine vertebrates, dominated numerically by teleost fish, are more likely than terrestrial mammals to survive the imposition of hunting or fishing mortality on the natural mortality that is appropriate to their pristine natural environments. These reflections serve to support a suggestion that marine vertebrates should be better able than terrestrial forms to support imposed mortality, such as that from fishing.

Unfortunately, there are several counter-indications to this comfortable suggestion that all stem from the structural difference between exploited and pristine populations and the increasing reproductive potential of older fish: a progressive reduction in the relative numbers and biomass of the older fish in an exploited population appears to be an inevitable consequence of almost all fisheries. I personally know of only one example in which this is not the case – that of the small-boat fishery for hake on the French Mediterranean coast, because this targets the smaller individuals that are more readily available, while the larger, older fish live offshore and much deeper, and so escape capture.

Indeed, it seems to be almost inevitable that a fishery should reduce the reproductive potential of a stock not only by reducing the relative numbers of the most fecund individuals within the population, but also by reducing the number of years over which each annual cohort is able to maintain an adequate output of ova. The fact that each fished stock has, after several years of exploitation, a set of life-history traits that no longer resembles those naturally evolved in the habitat appears to have been of no significance to fisheries managers, and (as shall be discussed in a later chapter) I can see little evidence that the significance of this inevitable result has been understood widely.

ENDNOTES

1. References to the works of Martishev, Anisimova and Severtsov are to be found in Nikolsky's 1969 book *Theory of Fish Population Dynamics*, Edinburgh: Oliver and Boyd.
2. Katsukawa, Y., *et al.* (2004) Indeterminate growth is selected for by a trade-off between high fecundity and risk avoidance in stochastic environments. *Popn. Ecol.* **44**, 265–72.
3. Although it has to be admitted that some pterodactyls had approximately the same dimensions as giraffes – and yet managed to fly very well! How they got themselves off the ground or sea and into the air is still not resolved.
4. Winship, A.J., *et al.* (2001) Growth of body size in Steller's sealion. *J. Mammal.* **82**, 500–19.
5. Williams, P.D., *et al.* (2006) The shaping of senescence in the wild. *Trends Ecol. Evol.* **21**, 458–64.
6. Hamilton, W. (1966) The moulding of senescence by natural selection. *J. Theor. Biol.* **12**, 12–45.

7. Bidder, G.P. (1932) Senescence. *Br. Med. J.* **193**, 583–5.
8. Klapper, W., *et al.* (1998) Telomerase activity in immortal fish. *FEBS Lett.* **434**, 409–12.
9. Finch, C. (1990) *Longevity, Senescence and the Genome*. Chicago, IL: University of Chicago Press.
10. See, for example, the home page at www.agelessanimals.org.
11. Vaupel, J., *et al.* (2004) The case for negative senescence. *Theoret. Popn. Biol.* **65**, 339–51.
12. Babcook, R. (1991) Comparative demography of three species of scleractinian corals. *Ecol. Monogr.* **61**, 225–44.
13. Ebet, T.A., *et al.* (2008) Growth survival and longevity in the rock-boring sea urchin *Echinometra lucunter*. *Bull. Mar. Sci.* **82**, 381–403.
14. Tinkle, D., *et al.* (1993) Life history and demographc varations in the lizard *Sceloporus graciosus*. *Ecology* **74**, 2413–29.
15. Olsson, M. and R. Shine. (1996) Does reproductive success increase with age or with size in species with indeterminate growth: the case of sand lizards. *Oecologia* **105**, 175–8.
16. www.fishbase.org.
17. Gauldie, R.W., *et al.* (1989) K-selection characteristics of orange roughy (*Hoplostethus atlanticus*) stocks in New Zealand waters. *J. Appl. Ichthyol.* **5**, 127–40.
18. Thanks again to the late Ram Myers.
19. Miller, P.J. (1979) Adaptiveness and implications of small size in teleosts. *Proc. Zool. Soc. Lond.* **44**, 263–306.
20. Rochet, M.J. (2000) A comparative approach to life history strategies and tactics among four orders of teleost fish. *J. Mar. Sci.* **57**, 228–72.
21. This has been discussed by Jennings, S., *et al.* (2001) *Marine Fisheries Ecology*. Oxford: Blackwell.
22. Barnado, J. (1996) The particular maternal effect of propagule size, especially egg size. *Am. Zool.* **36**, 216–36.
23. Ware, D.M. (1978) Bioenergetics of pelagic fish. *J. Fish. Res. Bd. Canada* **35**, 220–8.
24. El-Agamy, A., *et al.* (2004) Reproductive biology of *Boops boops* in the Mediterranean environment. *Egypt. J. Aq. Res.* **30**, 241–54.
25. These data are, respectively, reported by Bobko and Berkely (2004), Hislop (1996) and Trella (1998).
26. Pankhurst, N.W. and A.M. Conroy. (1987) Size–fecundity relationships in the orange roughy *Hoplostethus atlantica*. *N.Z. J. Freshw. Mar. Res.* **21**, 295–300.
27. Chambers, R.C. and W.C. Leggett. (1996) Maternal influences on variation in egg sizes in temperate marine fishes. *Am. Zool.* **36**, 180–96.
28. These observations are obtained from Trippel *et al.* (1997) *Early Life History and Recruitment of Fish Populations* (London: Chapman and Hall); Pankhurst and Conroy, *N.Z. J. Freshw. Mar. Sci.* **21**, 295–300; Macchi *et al.*, (2008) *Fish. Res.* **80**, 345–9; Berkeley *et al., Ecology* **85**, 1258–64; and Nikolsky's *Fish Population Dynamics* (Edinburgh: Oliver and Boyd, 1969).
29. Trippel, E.A., *et al.* (1997) Effects of adult age and size structure on reproductive output in marine fishes. In *Early Life History and Recruitment in Fish Populations*, Eds. Chambers and Tripple. New York: Chapman and Hall, pp. 31–62.
30. Hislop, J.R.G. (1988) The influence of maternal length and age on size and weight of eggs and the relative fecundity of the haddock in British waters. *J. Fish. Biol.* **32**, 923–30.
31. Macchi, G.J., *et al.* (2006) Influence of size, age and maternal condition on the oocyte dry weight of Argentine hake (*Merluccius hubbsi*). *Fish. Res.* **80**, 345–9.

32. O'Farrell, R.O. and L.W. Botsford. (2006) The fishery management implications of maternal-age dependent larval survival. *Can. J. Fish. Aquat. Sci.* **63**, 2249–58.
33. Trippel, *et al.*, *op. cit.*
34. Buckley, L.J., *et al.* (1991) Winter flounder reproductive success. *Mar. Ecol. Progr. Ser.* **74**, 125–35.
35. Tabulated by Trippel, *op. cit.*
36. Kjesbu, O.S., *et al.* (1992) Buoyancy variations in eggs of the Atlantic cod *Gadus morhua*. *Can. J. Fish. Aquat. Sci.* **53**, 610–20.
37. Vallin, L., *et al.* (2000) Maternal effects on egg size and egg buoyancy of Baltic cod. Implications for stock structure and recruitment. *Fish. Res.* **49**, 21–37.
38. Grønkjaer, P. and M. Schytte. (1999) Non-random mortality of Baltic cod larvae inferred from otolith hatch-check sizes. *Mar. Ecol. Progr. Ser.* **181**, 53–9.
39. Caddy, J.F. and J.C. Seijo. (2002) Reproductive contributions foregone with harvesting: a conceptual framework. *Fish. Res.* **59**, 17–30.
40. Longhurst, A. (2002) Murphy's law revisited: longevity as a factor in recruitment to fish populations. *Fish. Res.* **56**, 125–31.
41. Three models are discussed. In the *cod model*, M takes lower values (0.125) in the reproductive years than previously (<0.2), increasing again after age 25. Fecundity from observations of Paul Fanning (progressive) and Art May (fixed). Weights-for-age from Fanning and from 2J3KL observations. Maturity in pristine population is estimated at 9–10 years. For the *herring model*, growth is computed from $Lt = L_{inf}(1-e^{-kt})$ with parameter values from FishBase, fecundity is based on Hickling (1940). For the *NZ snapper model*, length-for-age was digitised from Horn (1986), weight for age is obtained from $W = 0.02L^{3.021}$, M is from Myers' data (progressed as for the cod model), relative fecundity is assumed similar to Australian surf bream. Some data sources incorrectly state a maximum age of only 35 years for this species: Horn and other NZ fishery publications do not concur.
42. Amorosi, T., *et al.* (1994) Bioarchaeology and cod fisheries: a new source of evidence. *ICES Mar. Sci. Sympos.* **198**, 31–48.
43. Chosen for modelling for purely sentimental reasons, because I once worked on its stock structure in the Bay of Islands and the Hauraki Gulf.
44. Murphy, G. (1967) Vital statistics of the California sardine and the population consequences. *Ecology* **48**, 731–5.
45. Longhurst, A. (2002), *op. cit.*
46. Streans, S.C. (1977) The evolution of life history traits. *Ann. Rev. Ecol. Syst.* **8**, 145–71.
47. Chambers, R.C. and W.C. Leggett. (1992) Variation in flatfish metamorphic traits. *Neth. J. Sea Res.* **29**, 7–24.
48. Etnier, M.A. (2004) Reevaluating evidence of density-dependent growth in northern fur seals, based on measurements of archived skeletal specimens. *Can. J. Fish. Aquat. Sci.* **61**, 1616–26.

4

Marine ecosystems: their structure and simulation

'As we know, there are known knowns ... the things we know we know and there are known unknowns ... the things that we know we don't know. But there are also unknown unknowns, the ones we don't know that we don't know.'

Donald Rumsfeld, 2002

If we believe that marine ecosystems really can produce a surplus of biomass of some of their component vertebrates beyond what is required to sustain their natural populations, surely we should ask what distinguishes them from terrestrial ecosystems, which do not appear to have the same capacity. If the supposed surplus production of some marine vertebrates is real, and useful to us, then we need to understand what characteristics of marine ecosystems are essential for its production. We need to understand whether surplus production of fish biomass is a consequence of special characteristics of the fish themselves, such as those discussed in the previous two chapters, or of marine ecosystems in a wider sense. Discussion of these issues occupies this and the following chapter and will focus on the very different trophic structure of terrestrial and marine ecosystems, and also on the very different characteristic scales of variability of the two habitats.

At least during the formative period of fishery science, such issues appear to have been of little concern to those who had responsibility for devising – and applying – methods for fish stock management, and it has been suggested that a lack of understanding of the complexity and variability of the habitat of marine fish has been responsible for the failure of fishery science to evolve significantly from its 'quantitative and homogeneous' beginnings until very recently.[1] Indeed, if agricultural scientists had as little knowledge and interest in the relationship between crop success, local geology, hydrology and soil structure as fishery scientists have had (at least in the past) concerning the habitat in which their fish swim, then modern agriculture could not have developed as it has done.

An attempt is made in this chapter to identify those characteristics of the marine habitat that may, on the one hand, support a surplus of fish production and, on the other, may confuse our management models; such a synthesis is perhaps still needed, because both fishery scientists and biological oceanographers have been rather prone to accept concepts without a full understanding of the extent to which they lack the support of verifiable evidence. This has led both groups to indulge in what has been called mythic thinking, 'which is to place special emphasis on a selective conjecture ... thereafter given privileged status over alternative suggestions'.[2] This comment was made in reference to problems in geomorphology, but mythic thinking is a general problem of science, and is certainly to be looked for in reports concerning those environmental problems that have recently captured public concern.

TERRESTRIAL ECOSYSTEMS: NO SURPLUS PRODUCTION OF BIOMASS USABLE BY INDUSTRY

To discover what might be the special characteristics of marine ecosystems that enable them to support industrial fishing, it may be useful to remind ourselves why vertebrates in terrestrial ecosystems could not support industrial hunting, an observation made by others not so long ago.[3] Perhaps a comparison of the natural traits of marine and terrestrial vertebrates may reveal something about the reality of surplus production in marine ecosystems.

Because human anatomy is structured primarily for a diet of non-fibrous plant material, animal protein probably did not figure largely in the diet of early humans, living in family groups like contemporary anthropoid apes. The earliest human societies probably incorporated insects and small vertebrates, such as amphibians, reptiles and fish, into their diet, but the routine capture of mammals must have awaited the progressive development of hunting techniques.

That human societies at the hunter–gatherer stage of social development could and did obtain a part of their diet from large vertebrates is shown by archaeological material testifying to the killing of equids with wooden javelins by pre-*sapiens* hominids in Europe as early as 400 000 BP. However, there is good historical and archaeological evidence to show that the supply of food from terrestrial vertebrates could not be sustained once human numbers began to increase and their technology evolved accordingly: after this evolution, the larger terrestrial herbivores rapidly became extinct or very reduced in numbers.

Although there has been much discussion concerning the relative importance of post-glacial climate-induced habitat change on populations of megafaunal mammals and marsupials, it is now generally agreed that human hunting was, for many species, the proximate cause of extinction. The authors of a recent critical review of the controversy note that there is 'no testable climate change theory for the extinction' of the megafauna, and further comment that the Younger Dryas cooling (12 900 BP), evoked by some for the North American megafaunal extinctions, would have been significantly less stressful than earlier, deeper glaciations that the fauna had survived very successfully.[4] In fact, the progressive extinctions of very large mammals (ground sloths, equids, giant elk, marsupials and other megaherbivores) appears to have occurred very soon after the arrival on each continent of human populations dispersing from Africa: 46 000 years ago in Australia, in America about 13 000 years ago and in Oceania and New Zealand more recently.

The demonstrated ability of groups of 15–20 hunters in the African savannas to bring down and kill adult elephants in the very recent past, using only their broad-bladed throwing spears, shows that the great majority of large herbivorous mammals – including very dangerous species – were vulnerable to hunting even prior to the development of firearms. It has been suggested that the relative survival of the African megafauna may have been due to their progressive adaptation to the evolution of human societies in the continent in which these first evolved: elsewhere, such animals would have been naïve and more vulnerable to newly arrived groups of hunters.

In pristine terrestrial ecosystems, the population numbers of megaherbivores (that became the choice target of hunters) were controlled by large felids and ursids; the additional human-induced mortality probably caused the collapse of the entire ecosystem, leading to massive growth of forest vegetation, and the disappearance of the large predators.[5] Australia may have been a special case, for there the collapse of the marsupial megafauna around 40 000 BP may have resulted in part from 'fire-stick' bush clearance for shifting cultivation.[6]

Hunting groups of palaeoamericans first appeared in the Great Plains regions of North America about 12 000 BP, and they used well-developed weapons (initially characterised by stone tips of the Clovis culture) to kill large mammals, and they managed to exterminate several tens of species within a remarkably short period of less than 500 years: the consequences have been described as a wave of extinction. While bison abundance also declined at this time, these animals continued to maintain large populations in the face of continued hunting by the Native Americans, right up to the era of the European incursion into the region.

This event, of course, led to what is perhaps the ideal modern example of the consequences of the evolution of hunting methods for terrestrial mammals, and which brings to mind the effects of industrialisation of the fisheries: the rapid near-extinction of the bison of the Southern Plains by Comanche and other tribes after their acquisition of horses which were brought to North America by the Europeans. The governor of New Mexico, de Anza, pacified the region around 1770 and then became convinced that the now-swelling Comanche population would rapidly come to exterminate the bison.[7] Accordingly, he built a pueblo for them on the Arkansas River, and tried to induce them to adopt a settled, agricultural way of life: of course, they knew better and refused. But, just as de Anza predicted, they faced starvation within 50 years, due to increasingly critical shortages of bison on the grasslands: it was the change from hunting afoot to hunting on horseback, and their refusal to accept the probable consequences of that evolution, that brought the Comanche to disaster.

This history is a model for what drove the move from hunting towards agriculture and urbanisation that occurred in every society as its population increased and its technology advanced. It is also a confirmation for the truth of the assertion that terrestrial vertebrates – even when their habitat is relatively undisturbed – cannot support anything approaching the level of exploitation that we generally believe to be possible when marine vertebrates are our quarry. Mammals, of course, differ strikingly from teleost fish in their pattern of natural mortality and reproduction: they are characterised by high and invariant adult survival and a low rate of recruitment.

The vertebrate population in settled lands everywhere, except within the confines of areas of special protection, is now a low-diversity remnant of the smaller mammals that were originally present in the pre-human fauna of each region. This remnant includes species whose characteristics evolved in a different ecosystem, now incomplete: the pronghorn antelope evolved high running speeds to avoid a predator that no longer exists, and the giant condor evolved to feed on carcasses of a size that also no longer exist – and so on.

Only when natural predators have been removed can populations of mammals support sustained hunting: in this case, hunting mortality can theoretically reach levels equivalent to predation mortality in the mediaeval forests. Such is the case in the forests of western Europe, now lacking wolves and bears but supporting abundant roe deer and wild boar, and in the case of the herded reindeer culture of the Lapps; perhaps it should be noted that these cases are in no way parallel to the condition of an exploited fish stock, where fishing mortality is additive to natural mortality.

The susceptibility of terrestrial vertebrates to human hunting pressure except under such exceptional conditions clearly suggests that if industrial sea fishing is indeed sustainable, even at a modest level, then this must be due to some very fundamental differences between the biology of oceanic and continental-shelf vertebrates and those which inhabited the plains and forests ashore. Several potential candidates will be identified in the following section, and will be discussed at length in following chapters.

TROPHIC RELATIONS AND ENERGY FLOW IN MARINE ECOSYSTEMS

When an increase in their population forces hunter–gatherer societies to turn to agriculture and develop the axe and the plough, forests are cleared and grasslands ploughed. However, the ocean is relatively untouched by such processes, because fishing requires neither plough nor axe and modifies only very indirectly the patterns of plant growth in the sea. Nor does fishing directly affect the principal marine herbivores that transfer organic material to those higher trophic level organisms in which the fisheries are interested. This fundamental difference between the effects of agriculture and fishing on the ecosystem that each exploits has been little discussed, and requires some examination here.

Comparisons between the single-celled plants of the ocean and the herbs and trees with which we are familiar ashore are two-a-penny, and I do not propose to add significantly to their number. However, none of the comparative accounts that I have seen takes the next step: it is not remarked that – in consequence – the relationship between herbivores and carnivores is totally different in the two environments. On land, herbivores come in all dimensions, from caterpillars to elephants, with carnivores in due proportion, but in the ocean the great majority of large organisms – whether vertebrate or invertebrate – are carnivores, while marine herbivores are almost all very small: an understanding of the consequences of this difference would seem to be vital to an understanding of the ecology of the predatory bony fishes that are the preferred target of our fisheries.

But before we get to that, it will nevertheless be useful to take a short detour through the ecology of oceanic plants because, despite its consumption by herbivores of a type totally foreign to us land-dwellers, primary production in the ocean is still the origin of the food we obtain by fishing. In particular, I want to emphasise the ephemeral nature of plant production in the sea, and its variability in space and time, compared with plant growth ashore. I also want to emphasise the relative lengths of the food chains that support cod or halibut compared with those that support, say, rabbits or deer.

Because sunlight penetrates usefully only a few tens of metres into the ocean, photosynthesis of organic material by the autotrophic phytoplankton can occur only within this zone, where it is performed almost entirely by single-celled phytoplankton – both prokaryotes and eukaryotes – and only an insignificant fraction is contributed by the macroalgae of the littoral fringe or by floating macrophytes of the open ocean.

Although the dominant consumption of plant material in the ocean occurs by filtration of phytoplankton cells from the ambient seawater by a great variety of invertebrates and just a handful of fish, many marine ecologists have emphasised marine herbivory where it most closely approaches the terrestrial equivalent – in the grazing of macroalgae by echinoderms in littoral kelp forests, or of microalgae by several families of teleost fish on coral reefs. Unfortunately for the contribution of ichthyologists and ecologists to fishery science, studies done in the coastal zone dominate the ichthyological and ecological literature for reasons not far to seek: wrasse, for instance, are much easier to study *in situ* than haddock, and the damage done by fishing and other activities is so easily seen in the near-shore, and understood for what it is, by the general public. For such reasons, the importance of the ecosystems of the littoral, often only a few hundreds of metres wide and a few tens of metres deep, has been greatly overemphasised in discussions of the problems of fisheries.

Compared with the terrestrial flora, phytoplankton are restricted taxonomically and comprise only a few thousand species, which are usually considered in three groups: (a) a very small pico-fraction (0.5–2.0 mm) of photosynthetic bacteria and some eukaryotes, mostly small flagellates, (b) a nano-fraction (2–20 mm) of larger flagellates and small diatoms, and (c) a net fraction (>20 mm) of larger diatoms and dinoflagellates. The ecology of each group is singular, but the three groups are complementary in contributing to production of organic material in different regions of the oceans. Of these organisms, the quite recently discovered cyanobacterial prokaryotes of the picoplankton are numerically dominant: these extremely small and simple marine cells are the precursors of the chloroplasts that are present in the cells of the green leaves of terrestrial plants; as Victor Smetacek suggested, trees really exist only to keep their cyanobacteria up in the sunlight.[8]

An essential quality of phytoplankton is the ephemeral nature of their biomass, a fact that was well understood quite early in the history of plankton investigations,[9] and which is a critical factor in the strong spatial and temporal variability of the marine ecosystem compared with terrestrial habitats. The ratio between the standing stock of phytoplankton and its rate of production in characteristic regions of the ocean lies between 0.02 and 0.1: this is equivalent to 1–3 days' production of plant material in polar seas, and about 0.5 days' production in tropical seas.[10]

Consequently, the marine ecosystem (global biomass approximately 3×10^{15} g) lacks any equivalent of the inertial effects of the accumulated living biomass (about 550×10^{15} g) of forests, grasslands or tundra.[11] This enables marine ecosystems to shift rapidly between different states, each of which may be equivalent to an individual terrestrial biome, in response to changes in environmental forcing. Such changes will be discussed in the next chapter.

It is, of course, the consumption of phytoplankton cells by small herbivorous invertebrates and the consequent cycling of essential elements within the euphotic zone that enables the rapid turnover of photosynthetic biomass. Although this process is usually treated in models in a very simple way, the trophic structure of the small pelagic invertebrates that inhabit the phytoplankton-dominated habitat is very complex: an analysis of more than 1500 zooplankton samples from all oceans revealed the broad outlines of this fauna in polar, temperate and tropical oceans and coastal seas.[12] Overall, biomass of planktonic herbivores comprised 65% of total zooplankton biomass, predators 31% and omnivores just 2%; the proportion of herbivores is strongly correlated with latitude, forming 56% in polar seas but only about 20–25% in midlatitudes and in the tropics. Zooplankton, like the oceanic plants that they consume, form a complex and diverse group of organisms, difficult to generalise both in distribution and function. Benthic invertebrates are also significant consumers of phytoplankton biomass, although this is very largely composed of senescent cells that have sunk to the sea floor.

Although planktonic organisms may be represented as a flat spectrum of particle sizes, with equivalent biomass in each logarithmic size interval,[13] this abstraction (much used in modelling) conceals the real complexity of the relations between producers and consumers. Cohen remarks on the paradox that although marine taxa are significantly less diverse than terrestrial organisms, each organism typically interacts with a larger number of other taxa than do terrestrial organisms: marine food chains are typically longer and more open.[14] It has been suggested that the relative rapidity of response of marine food webs to external forcing at all scales is also a function of their relative openness compared with terrestrial ecosystems.[15]

Because the larvae of many species of teleost fish pass their developmental stages as members of the plankton, it is no surprise that some are at least partly herbivorous at this stage. Herbivory is especially prevalent among clupeid larvae at first feeding, so that their stomachs may be crammed with green cells; experimentally, dinoflagellates have been found to support the growth of anchovy larvae.[16] Some of the larger clupeids also feed on diatoms, by swimming open-mouthed in order to catch diatoms on their fine gill rakers. Of these fish, the most important are the menhaden (*Brevoortia tyrannus*) and coastal populations of the bonga (*Ethmalosa fimbriata*) of the tropical Atlantic and of the oil sardine (*Sardinella longiceps*) of the northern Indian Ocean. Until recently (see below) I would have added anchovies, and especially those off Peru, to this list – but that seems to have been a myth waiting to be exposed.

Since macroscopic plants are characteristic of the littoral zone in both warm and cold seas, the importance of herbivorous fish is greatest in shoal water. Here, in warm seas, these fish are dominated by the perciform families Acanthuridae, Kyphosidae, Pomacentridae and Scaridae and are important in the grazing of macroscopic plants on hard substrates where algal turf dominates the flora of rocky reefs and coral flats, and also in the equatorward parts of the range of kelp.[17] In cold seas, large brown algae are grazed primarily by echinoids and (prior to their extermination) by some very large marine mammals.

However, the great majority of teleost fish, including those that dominate fishery landings, are carnivorous and occupy significantly higher trophic levels than the browsing and grazing terrestrial mammals that have been utilised as a source of human food. So fishing has only a minor impact on the natural productivity of the seascape and – perhaps more importantly – does not directly modify the abundance of oceanic herbivores.

Consequently, it has been easy to suppose that trophic cascades of the kind observed on land or in lakes after the removal of top predators would not often occur in the open sea, because the base of the trophic pyramid that supports the fish in which we are interested is relatively unmodified by fishing.[18] This conclusion was supported by analysis of the network structure of marine food webs that suggested that link density (links per species), connectance (links per species^2), mean chain length and relative trophic composition each take relatively high values in marine food webs: the authors of the study concluded that marine food webs may, in consequence, be relatively robust to the loss of even highly connected species.[19]

Since the NE continental shelf of the USA was one of the marine ecosystems included in this study, and since the cod (*Gadus atlantica*) was identified as its most highly connected taxon, I wonder whether the authors understood the extent of the devastation of the pristine ecosystem structure that occurred after cod biomass had essentially been fished out? This phenomenon will be discussed later (see Chapter 9), because we can no longer be quite so complacent today that trophic cascades cannot occur in the open sea, now that we have observed the ecological consequences of the removal by fishing of a dominant large predatory fish species. This study is noted here only to suggest that theoretical food web modelling may be a less-than-satisfactory art, despite the confidence that many place in it; this problem is also discussed more extensively later.

In recent years, much attention has been given to the role that may be played in marine ecosystems by species that dominate the middle trophic levels of a regional ecosystem, directly consuming small herbivores – or even large phytoplankton – and themselves supporting a larger number of larger predators: the classical examples of such wasp-waisted ecosystems are the coastal upwelling systems, where biomass is dominated by one or two species of microphagous clupeids, sardines or anchovies. Strong fluctuations in relative and absolute population size of these species is characteristic of such regions, and have important consequences for both higher and lower trophic levels. One might quite properly equate such situations with the fluctuations of keystone species in terrestrial ecosystems – such as the snowshoe hares of North America, or Asiatic lemmings – and, as with this classical example, care must be taken in interpreting the causes of their fluctuations, which lead to the starvation or dispersal of predators.

CHARACTERISTIC STRUCTURE AND SCALE OF MARINE HABITATS

In the oceans, ecological niche structure differs strongly from the terrestrial habitat, where quasi-permanent, microscale niches are ubiquitous. The specialised microenvironments that are provided by terrestrial plants, from lichens to trees – living, dying or dead – have no equivalent in the oceans except along the littoral zone, in coralline habitats and in some abyssal locations: although these loom large in the public imagination, they occupy only a trivial percentage of ocean space.

This difference in relative richness in niche structure is reflected in the relative numbers of animal species between the two environments, noted in the previous section: a much greater number of species has evolved on land than in the ocean. Nevertheless, terrestrial organisms are more homogeneous taxonomically, simply because some of the major invertebrate phyla – coelenterates, echinoderms, tunicates, and so on – have never made it ashore. If the average oceanic animal interacts with more other species than the average terrestrial animal, this is perhaps because the latter tend to be confined to a spatially definable microhabitat, unlike those whose oceanic habitat is in constant state of flux and change. A similar distinction can be made between the more diverse littoral fish fauna, many of which are associated with individual microhabitats, and that of the open ocean at similar depths.

Yet such arguments can be carried too far. The trawling grounds of the North Sea or the Scotia Shelf, if they could be exposed to our view, would resemble featureless stony or sandy plains on which most of the invertebrate organisms lie buried below the surface, and we might conclude that very little ecological structure existed. Nevertheless, the benthic organisms (like individual species of trees) are really very sensitive to minor differences in the nature of the substrate, so that readily distinguishable communities inhabit different grades: rocks, gravel, shelly sand, sand, muddy sand and mud being the principal benthic facies of the open shelf. The recognition of characteristic benthic communities, globally associated with each bottom-type, was the now-rather-unfashionable revelation of the Danish benthic ecologists in the mid-twentieth century.[20] These communities are locked to their substrates, and can respond to changing external forcing only by geographic species replacement or by changes in relative species abundance.

But it is likewise easy to overemphasise the lack of permanent structure in the pelagic habitat. In discussing its characteristic spatial variability, it is easy to forget that even the pelagic habitat has physical and ecological structure in the vertical plane to which it will return if perturbed and which is critical for pelagic organisms. The first few metres immediately below the surface, and the few tens of metres centred on the seasonal thermocline of higher latitudes or on the permanent feature of tropical seas, both comprise structured and specialised habitats within the pelagos. At the thermocline, where maximum shear in the upper water column occurs and where Brunt–Vaissala frequencies are high, a phytoplankton biomass maximum of very restricted depth develops under stable environmental conditions, responding to the optimal conjunction of illumination and nutrient supply from the sub-pycnocline reservoir. With this is associated a layer of maximal biomass of all herbivorous zooplankters, and here also the specific diversity of the pelagic ecosystem takes maximal values. Layers of omnivores and detritivores preferentially lie deeper.[21] Small mesopelagic fish migrate at night from deeper in the water column into the layer of maximal biomass at the thermocline, there to feed in the layers of high zooplankton biomass.

This is a case in which the dependence of biological diversity on physical structure is remarkably well demonstrated. John Woods has suggested recently that Hutchinson's paradox of the plankton (why are there so many species in an apparently unstructured space?) will only be answered by prognostic forecasting techniques applied to biodiversity;[22] but really a large part of the answer is simply that the pelagos is a far less uniform habitat than imagined by Hutchinson, that great limnologist, and by others who have not observed first-hand its vertical structure in the ocean.[23] That this structure differs in permanence from niche structure in terrestrial or benthic habitats does not negate its reality, because when disturbed by a transient wind event, oceanic niches that are associated with discontinuities in the gradient of density of seawater are very rapidly re-established and the planktonic organisms re-order themselves within the habitual structure.

In what Halley called a seminal paper,[24] John Steele made some very pertinent observations on the differences in scale of the responses to environmental forcing of terrestrial and marine habitats.[25] He pointed out that while the variance of environmental forcing on land is rather consistent over long periods, the variability in the ocean increases with the period of observation. So, growth of variance is dependent not on the number of observations in a series, but on the length of the period that they represent: Steele suggested that marine systems exhibit a 'reddened' environmental variability spectrum compared with that of terrestrial habitats. He later suggested that this difference accounted for the fact that terrestrial ecologists were primarily concerned with interactions between organisms while marine ecologists were (or should be!) more interested in the consequences of environmental variability on the organisms that they study.

Steele illustrated his concept with a sea-surface temperature image of the Gulf Stream eddies obtained from an early satellite, but now that we have simple access to a wide range of satellite products, we can visualise the seascape with much greater precision and greater information content than can be obtained from surface temperature alone. The motion of eddies is routinely obtained from sea-surface elevation, which also enables us to deduce the slope of the thermocline; with this we may associate sea-surface chlorophyll, from which the instantaneous rate of primary production is computed. Although it is not yet available routinely, the chlorophyll data may be dissociated into information on the nature of the dominant phytoplankton taxon at each pixel: diatoms, *Prochlorococcus*, *Synechococcus* or haptophytes – with, of course, a final category of *incertae sedis*.

The pattern that we observe in satellite imagery is imposed by the asymmetry of ocean basins on the general circulation pattern of geostrophic currents: on an Earth without continents, flow would be essentially linear towards the east. The pattern of eddying, and major departures from simple flow, are predictable in their essential form, although not in daily observations; persistent pattern is associated with major features of coastal topography, with eddying in major current systems and along the equator.

This eddying flow follows similar patterns to those that we observe in high- and low-pressure systems in the atmosphere so that, scaled differently, forecasting should be possible in the same manner as for weather systems. But no such forecasting system is in place or is envisaged: the computing power required for a weather forecast is so expensive that there is no foreseeable future in which we shall have similar products for the mesoscale systems of the ocean.

So, although the dynamic response of marine ecosystems to physical forcing must follow rules that we can aspire to understand, these responses are chaotic in the sense that we are quite unable to predict their future state. It is only after an event, and to the extent that we have observed and described it correctly, that our general knowledge of the biology of individual species may be sufficient to allow us to suggest what has probably taken place – although, even then, we can rarely be sure that we have got it right.

TAXONOMIC CHARACTERISATION OF REGIONAL ECOSYSTEMS

Before discussing the functioning of what we have come unthinkingly to call ecosystems, it will be useful to remember that the concept of the ecosystem itself is liable to mislead us: this undefinable term[26] is now widely used, often in such a way as to imply that ecosystems are entities having natural boundaries and an ideal or pristine state within which a natural balance has been achieved by co-evolution. Of course, nothing could be further from the truth. The mix of species that existed in the pre-human ecosystems both on land and in the sea were the result of evolution of organisms *in situ*, together with the random movement of organisms between regions, as this was enabled by the geological evolution of the continental masses and – more recently – by the opening and closing of oceanic passages during glacial periods.

Consider the North Atlantic: if, in the geologically recent past, the coastal plains that now constitute the continental shelf of the Bering Sea had been just a few tens of metres higher in relation to sea level, then trans-Arctic transport of Pacific organisms into the North Atlantic would not have occurred and the fisheries that developed there would have been quite different. In this case, the Pacific taxon now represented by capelin (*Mallotus*) would not be present in the Atlantic and it is unlikely that the perfect pairing of a small forage fish and a large predator would have given rise to the great cod stocks of the North Atlantic. Conversely, the North Pacific would entirely lack gadoids, and hence the great pollock fishery.

The complexity and intermittence of oceanic dispersal is demonstrated by the highly diverse genetic composition of capelin at the present time: *Mallotus villosus* comprises a complex of at least four distinct clades, inhabiting: (i) the NW Atlantic, including Hudson Bay, (ii) from West Greenland to the Barents Sea, (iii) over the Arctic shelf, and (iv) in the northeast Pacific. The relative divergences of the mtDNA haplotypes of these clades show that dispersals into the Atlantic occurred twice; perhaps the Arctic clade is presently poised for a third Atlantic dispersal, since it is present in the northern Labrador Current.[27]

It is not, of course, only fish that are thus offered access to new habitat by passive dispersal, for trans-Arctic transport also modifies the pelagic ecosystem; the Pacific diatom *Neodenticula seminae* has recently been recorded in the southern Labrador Current off Newfoundland in routine plankton recorder surveys that enabled its dispersal to be timed with some precision; a pulse of Pacific water appears to have occurred in 1998–9 via the Canadian Arctic archipelago. Even if the species had already been present, but undetected in Arctic waters, its return into the Atlantic probably occurred at this time: previously, it is known from this ocean only in deep-sea sediment core material dated no younger than the Pleistocene (1.2–0.8 Ma). Since *Neodenticula* appears to be an unexceptional diatom, its return to the Atlantic will probably not modify the functioning of the pelagic ecosystem there.

Clearly, the composition of the 'ecosystem' of any region, whose boundaries have been specified for convenience or for some special interest, is the result of the interaction between the taxa that accidentally and progressively reached that region, and those that evolved there. The extreme cases demonstrate this: there is no possible ecological logic to the distribution of horseshoe crabs (*Limulus* spp.), survivors of the Triassic fauna, restricted as they are to the western Atlantic and the eastern Pacific: they somehow survived the greater part of the period of rearrangement of continental land masses, and their present distribution is purely accidental. Nor is there any ecological logic to the survival of Cretaceous coelacanths (*Latimeria* spp.) in just the Indian Ocean.

So, it would be a mistake to discuss regional ecosystems as if each represented an ideal state, and had evolved as the unique response to the characteristic exigencies of this or that region. Rather, of course, they represent the response of those organisms that happen to have been present in each ocean and coastal region; their degree of specific evolution within each region is determined by their ability to occupy a niche offered within the habitat, or the flexibility of their genetic response to new conditions. In the case of the North Atlantic, the ancient *Limulus* (like other very old organisms) probably has little potential to evolve new form and function, so it either finds a suitable habitat as ocean basins evolve or it does not survive.[28] On the other hand, the wide spectrum of gadoids that originated in the Atlantic relatively recently in geological time has evolved there and has dispersed into other oceans only to a rather limited extent.

THE UNDERSAMPLING PROBLEM AND SIMULATION OF ECOSYSTEM FUNCTIONING

Any discussion of the nature and functioning of the marine ecosystem inevitably encounters two fundamental difficulties. First, marine scientists are shackled in their thinking because the general ecological literature is dominated by studies of terrestrial habitats; much of what we know (or think we know) about the fundamental principles of ecology is based on observations of terrestrial or littoral habitats, and these have limited relevance to life in the oceans, for the reasons discussed in the previous sections. Second, because of the relative inaccessibility of the oceanic habitat, and the rapidity of changes of state within it, observations there are both quite difficult and very expensive: undersampling is endemic and long-term time series of observations are especially hard to maintain.

So, inadequacy of relevant data is a problem in all branches of marine science from stock assessment to ocean circulation, and is sufficiently serious that it merits special consideration. Since an adequate description of the physical environment of fish would seem to be an essential prerequisite to any serious progress in fishery science, it may be instructive to review very briefly the present situation.

Quite early in the history of oceanography, it was realised that comprehensive and quantitative description of oceanic circulation would be essential to progress, and would require very high-quality data to be useful. But progress was very slow, and until the 1980s, almost half of the $5^\circ \times 5^\circ$ rectangles in the ocean, representing far more than half of its surface, were not represented by even a single oceanographic profile of sufficient quality for inclusion in the global data set. The 1980s saw the development and diffusion of electronic sensors for profiling, of satellite navigation for accurate positioning, of satellite sensors for several parameters observable at the ocean surface, and of drifting buoys equipped with sensors and satellite communications. Finally, too, data could be transmitted, processed and archived by electronic computers. At this time, too, the possibilities of CO_2-induced climate change caught the attention of governments and their science agencies – not to say that of the public at large – and, because any response to the problem would require the development of global models of atmospheric and oceanic circulation, it quickly became obvious that we lacked the fundamental description of ocean circulation on which such models must be based.

This set the stage for a massive effort by several international agencies (WMO and UNEP leading the charge) to assemble scientific teams from about 25 nations in order to go to sea, both to get a comprehensive data set for the global ocean, and also to establish a continuing mechanism to service the requirements of the climate modelling community. The new data were to be provided by a one-off major physical oceanographic survey of all ocean basins to full depth, because the earlier concept of a sluggish and negligible deep-ocean circulation was finally understood to be wrong. The World Ocean Circulation Experiment (WOCE, led, I am happy to remember, by the late George Needler of our Bedford Institute of Oceanography) involved long sections, seasonally repeated to cover all ocean basins, together with about 50 recording current meter moorings at strategic locations. All this was not accomplished without dissent between those who advocated rational and uniform station sections, and those who advocated concentrating on especially dynamic regions. Yet agreement was reached on three core objectives: (i) the Global Description based on *in-situ* and satellite observations, (ii) inter-ocean exchanges between the Southern Ocean and the remainder of the global circulation, and (iii) the testing of basin-scale dynamic models in preparation for the planned global-scale models. Rapid data availability and exchange was initiated and the WOCE data now form a cornerstone of a global oceanic observing system based at IOC UNESCO, which coordinates a distributed system of data assembly and storage, and the processing of physical data in near-real time for users.[29] The major users of operational oceanography are the meteorological community, who require predictions of the sea-surface temperature field, and the offshore petroleum industry, which requires predictions of storm surges and unusual swell conditions. It has also proved possible to use coupled atmosphere–ocean models for longer-term prediction of the evolution of ENSO conditions.[30]

This briefest of synopses of the recent evolution of operational oceanography will serve to emphasise the lack of any comparable development in operational biological oceanography. Apart from the greatly more formidable technical problems, no potential clients for a comparable delivery of ecological prediction services are endowed with resources comparable to those of the military or the offshore oil industry. This has resulted in far fewer fundamental observations of biological processes having been made than high-quality deep physical profiles. A good example of the degree to which these are lacking comes from the requirements of global climate models for distributed data on plant production at sea: the rate of primary production can be computed from satellite observations of sea-surface chlorophyll, but this requires accurate information on the depth distribution of chlorophyll: in the mid 1990s, the total count of archived high-quality profiles was no more than 21 872 to represent all oceans, in all months of the year. Vast regions of the open ocean were entirely unrepresented and must now be parametrised in any global computations of plant production at sea.

This uneven areal coverage of chlorophyll profiles is one result of the absence of any activity that even approached the equivalent of an ecological WOCE. Only two basin-scale regions of the ocean have been surveyed methodically across a rational and repeated grid to obtain high-quality data integrating physics, chemistry and biology: the eastern tropical Atlantic and the Pacific, and that almost 50 years ago now. These surveys were seasonal, but each only covered a single year[31] and, even if the data might be useful today, they were obtained in the pre-electronic era and today probably exist only on paper, or as maps in atlases.

Many other smaller regions have been studied in an extensive but less rational manner: of these, one of the most important is the CalCOFI project that has been operated by university, state and federal agencies from 1950 to the present day. For the first 20 years, an almost monthly occupation of a grid of stations covered the entire coast of California and out to the edge of the California Current but, unfortunately, the extent and frequency of occupation were progressively reduced. Nevertheless, an ichthyoplankton time-series going back to the mid 1930s is available for the Los Angeles Bight and is one of the longest available for any area. The Cooperative Studies of the Kuroshio (1965–77) were intended to produce similar results, although the period of study was shorter and the station grid occupied was less uniform. Such cooperative studies are really essential before serious progress can be made in understanding the ecology of fished species in any region of the oceans.

In the same way, the JGOFS and GLOBEC studies of the Somali Current upwelling system during the 1990s resulted in a step increase in our knowledge of the region but, since they were studies of ecological processes, they were not based on a grid of observations, regular in space and time.[32] During the latter part of the last century, this became progressively the typical pattern of cooperative work at sea: consequently, the observations are very difficult to organise into a unified and accessible data archive. Perhaps the most successful of such compilations are those of the Benguela Current region, of the Gulf of Alaska and of the Kuroshio area, although all are useful.[33]

This pattern of study has favoured recourse to the use of conceptual models to try to understand important processes that we infer must be universal, but for which we have at best minimal observational data. Unfortunately, conceptual (or diagnostic) models of processes may be attractive, but wrong: the high nutrient, low chlorophyll (HNLC) myth, only recently laid to rest, was a good example. In 1988, Martin and Fitzwater suggested that oceanic phytoplankton could receive adequate supplies of the essential element Fe only through wind-blown dust, even though oceanic Fe profiles resemble those of other nutrients, indicating supply from below.[34] This idea (and it was no more than that) was enthusiastically taken up, and became part of the popular culture of oceanography, because it seemed to offer a technique for some resolution of the excess atmospheric CO_2 problem: it was rejected by very few. A recent demonstration that the pattern of productivity in the Southern Ocean in no way matches that of aerial deposition of Fe has finally shown how much ink and ship time has been spent in chasing a rainbow during the last 20 years.[35]

Another myth, only very recently refuted, has been that the massive populations of anchovies that build up off Peru are the result of a very direct link between production of phytoplankton and its consumption by filter-feeding anchovies: accounts of the feeding of clupeids (including my own) have discussed the apparent singularity that a very few species of clupeids – led by *Engraulis ringens* – feed directly and principally on diatoms, especially chain-forming species. Now, thanks to extensive studies of the stomach contents ($N = 21{,}203$) of this species the length of the Peruvian coast, we know that we were in error.[36] More than 90% of the carbon content of the food of anchovies comprises euphausiids and copepods and in only 1 of 23 individual surveys (from 1996 to 2003) did phytoplankton contribute more than 10%. These results provide, as the authors of the study suggest, 'a new vision of the Humboldt Current system'.

Diagnostic models may be translated, by those who are skilled in the art, into numerical simulations intended to represent regional ecosystems or individual processes within ecosystems: the ready availability of simple software, capable of fitting models to available time-series, such as ECOPATH with ECOSIM (EwE), has enormously facilitated this activity, and earned many a doctorate.[37] Such models do not question the original assumptions made by the conceptual modeller, because they do no more than quantify what originated in his or her imagination: but, in so doing, they bring credence to natural history.

Recourse to these modelling tools requires, of course, the drafting of a diagram of the trophic linkages between producing and consuming organisms that can be held to represent the flow of energy or material represented by some common currency. This task is at once too easy and too difficult: it is too easy to draft diagrams to represent what their authors' conceive – on the basis of available observations – to represent the principal fluxes within the system, and too difficult because the totality of the fluxes that existing at an instant in time in the sea can never be observed. So there are what I choose to term 'known unknowables' of ecosystems, and this category should be added to Donald Rumsfeld's aphorism that heads this chapter: there really *are* things that we know we cannot know.

All ecosystem flow diagrams published in the recent literature can be held to represent no more than a simplified sketch of the real ecosystem, whose detailed state we cannot observe but which is required as the initial conditions and variables for any realistic model; the very best illustration of this problem that I know is the diagram drawn by Alister Hardy in 1924 of the trophic relations of just one species – the herring – in the North Sea. This contains more than 100 predator–prey linkages and requires 4 compartments for growth stages of herring, 20 compartments for its zooplankton food organisms (and 4 for the zooplankton predators of its larval stages) and 18 compartments for the phytoplankton taxa that nourish the relevant zooplankton species. One of the most complex EwE models of a regional ecosystem that I have seen – that of the Humboldt Current by Tam and others[38] – assumes that phytoplankton can be respresented by just two compartments (diatoms and dino- plus silicoflagellates). In Hardy's diagram, all organisms except herring are specified only to the generic or higher level, and no predators of post-larval herring are specified. Even if a modelling language is capable of handling so many disparate variables, we have no realistic way of making the required observations within the space and time required, either for initiating or verifying the model.

Bravely ignoring such difficulties, the community involved in the study of the Benguela Current ecosystem has taken the unusual step of designating as their formal objective the useful forecasting of how the ecosystem will respond to changes in the natural forcing and to those induced by fisheries. A diagnostic model of this ecosystem, drafted by Cury *et al.* as a system diagram and arranged by trophic level, requires about 25 compartments each representing a group of organisms, connected by 25–40 flows, depending on whether the model represents the northern or southern part of the Benguela.[39] The authors use the model to investigate the question 'Are marine food webs controlled top-down, bottom-up or are they wasp-waisted?', a question to which I believe it would not be flippant to reply 'Yes, and probably all at the same time'.

Shannon *et al.* describe a simulation of this ecosystem for the period 1978–2002, using EwE software with serial data for catch and fishing effort for six fished species, together with population estimates for seabirds and seals.[40] The results of the simulation suggested to the authors that 'fishing pressure has little explanatory power with regard to the time-series' because, although fishing had been relatively constant during the period, large changes in biomass were nevertheless observed.

The authors suggest that relative species biomass in this system must therefore be controlled primarily by internal dynamics, and that it is necessary to look beyond fishing as the source of changes in relative species composition in upwelling regions (see Chapter 6). They suggest, too, that the changes observed were consistent with earlier suggestions that such ecosystems are subject to 'wasp-waist' control, so that the relative abundance of small pelagic fish controls the whole.

While the Benguela system might well function in this way, scepticism would appear to be in order concerning the capability of this model to verify that suggestion; if ever I saw one, this is an open system – usually defined as one in which intrinsic and extrinsic conditions are liable to change under the influence of processes or events beyond the modelled system. Oddly, the simulation excluded primary biomass input and cycling: phytoplankton production, benthic production, the detritus pool and three groups of zooplankton are all required by the parallel conceptual model of Cury *et al.*, but all were absent in the numerical simulation of Shannon *et al.* done with EwE software: the consequences of changes in the physical forcing of primary and secondary productivity were therefore not included in the model; consequently, the possibility of 'bottom-up' control of the ecosystem – one of the three potential mechanisms for the control of marine ecosystems of Cury *et al.* – was excluded from the simulation. Between-year changes in southerly wind stress were used, independently of the simulation, to investigate how time-series of forcing functions might be explored for the values that would generate modelled stock trajectories that fitted historical observations optimally.

The Benguela simulation with EwE software illustrates very well the distance that has been travelled since tools for trophic flux modelling became readily available, yet it has been suggested that this tool is perhaps better-suited to developing understanding of ecosystem performance than predicting the future state of an ecosystem with useful reliability. John Woods has suggested that this will require 'the philosophy and practice that has made weather forecasting so successful'[41] and will require the development of what he calls prognostic biodiversity and also of Ecological Turing Tests to determine to what extent predictions made in the past differ from observations made in the future. Given the extent to which it is acknowledged that ecosystem structure and function is forced by changing physical conditions, and the very great difficulty of forecasting the state of atmospheric circulation patterns more than a few days into the future, it is hard to avoid the conclusion that to simulate the consequences of different levels of fishing mortality on a population of fish in its natural habitat – in order to take rational management decisions – remains out of reach.

We also now have sufficient understanding of the complexity of marine ecosystems, and of the difficulty of generating predictive models of the behaviour of their component species, to know that it is unreasonable to isolate one species and to suppose that the real population in the sea of that species will behave like its image in a simple model. It is even more difficult to have confidence in procedures such as the multispecies maximum sustainable yield models that have been,[42] and still are used[43] to derive virtual predictions of the performance of multiple target species simultaneously; such procedures ignore the differential specific reactions to changing environmental conditions that would – under conditions without fishing – change the relative species mix in the region modelled.

ENDNOTES

1. Francis, R. C. (1980) Fisheries science now and in the future: a personal view. *N. Z. J. Mar. Res.* **14**, 95–100.
2. Dickinson, W. R. (2003) The place and power of myth in geoscience. *Am. J. Sci.* **303**, 856–64.
3. Pauly, D., *et al.* (2002) Towards sustainability in marine fisheries. *Nature* **418**, 689–95.
4. Fiedel, S. and G. Haynes. (2004) A premature burial: comments on Grayon and Meltzer's 'Requiem for Overkill'. *J. Archaeol. Sci.* **31**, 121–31.
5. Sinclair, A. R. E., *et al.* (2003) The patterns of predation in a diverse predator–prey system. *Nature* **425**, 288–90.
6. Prideaux, G. J., *et al.* (2007) An arid-adapted middle Pleistocene vertebrate fauna from SE central Australia. *Nature* **445**, 422–5.
7. Flores, D. (1991) Bison ecology and diplomacy: the southern Great Plains 1800–1851. *J. Am. Hist.* **78**, 465–85.
8. Victor Smetacek, in press.
9. Harvey, H. W., *et al.* (1935) Plankton production and its control. *J. Mar. Biol. Ass. UK* **20**, 407–41.
10. Longhurst, A. R. (2007) *Ecological Geography of the Sea*. San Diego, CA: Academic Press.
11. Bolin, B. (1983) Changing global geochemistry. In *Oceanography: The Present and the Future*. Berlin: Springer-Verlag.
12. The data sheets were those of the Smithsonian Plankton Sorting Center, Washington DC; see Longhurst, A. R. (1985) The structure and evolution of plankton communities. *Progr. Oceanogr.* **15**, 1–35.
13. Sheldon, R. W., *et al.* (1972) The size distribution of particles in the ocean. *Limn. Oceanogr.* **17**, 327–40.
14. Cohen, J. (1995) Marine and continental food webs: three paradoxes? *Phil. Trans. R. Soc. B* **343**, 57–69.
15. Carr, M. H., *et al.* (2003) Comparing marine and terrestrial ecosystems: implications for the design of marine reserves. *Ecol. Applic.* **13** (1) Suppl. S90–107.
16. Hunter, J. R. (1981) Feeding ecology and predation of marine fish larvae. In *Marine Fish Larvae*, Ed. R. Lasker. Washington: University of Washington Press.
17. Floeter, S. R., *et al.* (2005) Geographical gradients of marine herbivorous fishes: patterns and processes. *Mar. Biol.* **147**, 1435–47.

18. Cury, P., L. Shannon and Y.-J. Shin. (2001) The functioning of marine ecosystems. *FAO Reykjavik Conf. Resp, Fish. Mar. Ecosyst. Doc.* **13**, 1–22.
19. Dunne, J. A., *et al.* (2004) Network structure and robustness of marine food webs. *Mar. Ecol. Progr. Ser.* **273**, 291–302.
20. For example, Thorson, G. (1957) Bottom communities. In *Treatise of Marine Ecology and Palaeoecology. Mem. Geol. Soc. Am.* **67**, 461–534.
21. Longhurst, A. R. (1985) Relationship between diversity and the vertical structure of the upper ocean. *Deep-Sea Res.* **32**, 1535–70.
22. Woods, J. D. (2006) Forecasting a large marine ecosystem. In *Benguela: Predicting a Large Marine Ecosystem*. Amsterdam: Elsevier.
23. See, for example, Longhurst, A. R. (1985), *op. cit.*
24. Halley, J. M. (2005) Comparing aquatic and terrestrial environments: at what scales do ecologists communicate? *Mar. Ecol. Progr. Ser.* **304**, 274–80.
25. Steele, J. H. (1985) A comparison between terrestrial and marine ecological systems. *Nature* **313**, 355–8. (See also Steele, J. and P. Hoagland. (2003) Are fisheries sustainable? *Fish. Res.* **64**, 1–3.)
26. The term ecosystem is equally valid for the bacteria inhabiting a drop of water or the biota of an ocean basin: it is undefinable, but useful, provided it is not equated with some ideal dimension, as in the case of the so-called 'Large Marine Ecosystems', that are no more than compartments of convenience.
27. Dodson, J. J., *et al.* (2007) Trans-Arctic dispersals and the evolution of a circumpolar marine fish species complex, the capelin (*Mallotus villosus*). *Molecular Ecol.* **16**, 5030–43.
28. It is repugnant that these wonderful survivors should be used as bait in crab-pots in the Louisiana bayous, each adult chopped in half on the gunwale before use.
29. These are part of what has been called 'an abundant and necessary population of acronyms': World Ocean Circulation Experiment, Global Ocean Observing System, International Oceanographic Commission, and so on ...
30. El Niño Southern Oscillation, of which more in Chapter 6.
31. These were the EQUALANT and EASTROPAC surveys, both done in the 1960s.
32. The Joint Global Ocean Flux and the US Global Ecology programmes.
33. Good introductions to these compilations are found in the series of Elsevier volumes on Large Marine Ecosystems, and in the net-published PICES special publication on the North Pacific ecosystems.
34. Martin, J. H. and S. E. Fitzwater. (1988) Iron deficiency limits phytoplankton growth in the north-east Pacific subarctic. *Nature* **331**, 341–3 (HNLC refers to the supposed 'high-nutrient, low-chlorophyll' regions).
35. Wagener, T., *et al.* (2008) Revisiting atmospheric dust export to the Southern Hemisphere ocean: biogeochemical implications. *Glob. Biogeochem. Cycles* **22**, 2006.
36. Espinoza, P. and A. Bertrand. (2008) Revisiting Peruvian anchovy (*Engraulis ringens*) trophodyamics provides a new vision of the Humboldt Current system. *Progr. Oceanogr.* **79**, 215–27.
37. Ecopath was developed at the UBC Fisheries Center in 1984, at the initiative of Dan Pauly, and Ecosim by Carl Walters in 1995–96; the two components were packaged together later. They are available, free, from the Sea Around Us project at UBC.
38. Tam, J., *et al.* (2008) Trophic modeling of the northern Humboldt Current Ecosystem. Part 1: comparing trophic linkages under La Nina and El Nino conditions. *Progr. Oceanogr.* **79**, 352–65.
39. Cury, P., see above (note 18).

40. Shannon, L.J., *et al.* (2004) Modelling stock dynamics in the southern Benguela ecosystem for the period 1978–2002. *S. Afr. J. Mar. Sci.* **26**, 179–96.
41. Woods, J.D. Forecasting a large marine ecosystem. In *Benguela – Forecasting a Large Marine Ecosystem*. Amsterdam: Elsevier.
42. May, R.M., *et al.* (1979) Management of multispecies fisheries. *Science* **205**, 267–77.
43. Worm, B., *et al.* (2009) Rebuilding global fisheries. *Science* **31**, 578–85.

5

The natural variability of fish populations and fisheries

'By the fall of 1957, the coral ring of Canton Island, in the memory of man ever bleak and dry, was lush with the seedlings of countless tropical trees and vines ... great rafts of sea-borne seeds and heavy rains had visited her barren shores ... elsewhere about the Pacific ... the year had been one of extraordinary climatic events. Hawaii had its first recorded typhoon, the seabird-killing El Niño visited the Peruvian coast, the ice went out of Point Barrow at the earliest time in history and on the Pacific's Western rim, the tropical rainy season lingered six weeks beyond its appointed term.'

Preface – CalCOFI Report 7, 1960

The brief review of the structure of marine ecosystems presented in Chapter 4 treated these as static entities which, of course, they are not: this chapter is intended to review the processes by which ecosystems respond to the changing environment of the ocean. The strong contrast in the internal structure of relatively open marine and relatively closed terrestrial ecosystems is paralleled by equally different characteristic variability and uncertainty in the external forcing that is characteristic of the marine and terrestrial habitats.

Agricultural scientists are very well aware of the critical consequences of changing weather patterns and also of regional patterns of soils and sub-soils, because an understanding of these factors is essential for the management of farming and animal husbandry. Although fisheries are also dependent on the pattern of change in ocean conditions that is induced at all time scales by variable weather conditions, and also at all spatial scales by the geometry of land masses, the importance of the resulting patterns of variability – for some reason – was late in entering mainstream fisheries science texts (I have already alluded to John Gulland's 1974 *The Management of Marine Fisheries*, because it encapsulated the thinking of many fisheries scientists in the days when there was great confidence in the future of their endeavour). It is remarkable to what extent this book, intended to assist in the education of future fishery managers, ignored the fact that fish exist in a natural environment that is variable across a wide range of time and space scales. It is also remarkable that Gulland was able to ignore so completely the work of his colleague David Cushing, who was even then putting the dynamics of exploited fish populations into an ecological context.

Although Gulland's view of fishery science was typical of the mid-twentieth century, it is strange that a recent student text on fisheries ecology, emanating from the same laboratory in which Gulland worked, makes only minimal reference to the consequences of the temporal variability of the habitat of commercial fish species.[1] Even more remarkably, the author of a 2007 FAO paper felt constrained to remark: 'The argument ... that biomass and catches are ultimately driven by climate fluctuation ... runs counter to the conventional wisdom of fisheries management which considers that biomass and catches are driven mostly by fishing pressure'.[2]

So, even today, most fisheries continue to be managed as if they existed in a perfect, invariant ocean, and there are probably relatively few fisheries management scientists who have taken the time to study the chapter in Robinson and Brink's encyclopedic *Global Coastal Ocean* that discusses the complex oceanography of their own particular region: I think that many fisheries managers would be very surprised to learn something of the complexity of the physical processes involved, and the remaining levels of uncertainty concerning environmental forcing and the response of the ocean in even well-known shelf areas.

What is perhaps even more remarkable is that at least an important part of the fishery science community appears to have forgotten some of the basic understanding accumulated patiently by their predecessors. Consider these words, taken from a brochure introducing a PICES conference to be held in Japan in 2010: 'Climate change will have many impacts on marine ecosystems ... improved scientific support for policy and management decision-making in the face of these potential impacts is essential ... analyze data and develop models to explore ... future climate states.' Even if these words do not represent the views of the future attendees at this conference, this text is written as if climate was stable until perturbed by human activities, and that change in ecological conditions at sea was a novel problem for fishery managers. Nothing, of course, could be farther from the truth, yet this is not the first time I have encountered such statements.

For such reasons, I propose in this chapter to review very briefly how marine ecosystems, both pelagic and benthic, respond to changes in external forcing; this is intended to emphasise the consequences of the simple truth that fisheries depend on a time-varying resource base.

Whether or not we have studied terrestrial ecology, we all understand why forests and grasslands are distributed as they are observed to be, and why each biome reacts to natural climate changes only slowly: but not everybody understands that the oceanic environment is variable at scales entirely different from those that are characteristic of natural landscape ashore. The concept of place is almost irrelevant in the upper realms of the oceans, which vary at any time scale that you might care to imagine: it is only the physical features of underwater and coastal topography that induce any spatial permanence or repetitive pattern.

A most important characteristic of marine ecosystems is the extraordinary rapidity with which a change of state, equivalent of a change between boreal forest and tundra, can occur in the ocean. Such regime shifts do occur in terrestrial ecosystems, but not in the lifetimes of you and me; one of the most important during recent earth history was the termination of the African Humid Period around 5000 BP when, after several millennia of vegetative cover, the Sahara entered a period of desiccation occasioned by the southward retreat of the West African monsoon rainfall; this climate shift appears to have been induced by a series of orbitally induced North Atlantic cooling events.[3] Although climatologists use the word abrupt for this event, it required almost 500 years for the Sahel vegetation to be transformed to dry desert: regime shifts of equal significance occur in the ocean – and revert to the *status ante quo* – within less than 10 years. Some of the decadal-scale regime shifts in the ocean that shall be discussed below represent changes in ecological conditions that are almost as significant as the transformation of the Sahara.

THE PERHAPS VERY DISTANT FORCING OF FISH POPULATIONS

The recent suggestions by Klyashtorin and his colleagues concerning historical relationships between population biomass and the state of the global climate that recently caught the eye of FAO[4] are, in fact, nothing very new: we have known for a long time that the natural abundance of fish populations is dynamic and responds to decadal to secular variation in the state of the lower atmospheric and the upper ocean, and because these respond to predictable cyclic changes of state of the solar system, we may say that relative population abundance should also be predictable.

Atmosphere and oceans together form a complex and dynamic system within which the energy received on the surface of our planet (and at several levels in the atmosphere) is redistributed, conserved or radiated back into space. Complex spatial variability of this system – which would not occur on a water planet entirely without land – results from the irregular form of the continents, and their asymmetric locations. The ideal distribution of westerly winds in the circumpolar vortex, which carries most of the mass of the atmosphere, is strongly modified by the topography of the continents, principally by the positions presently assumed by the meridional mountain chains of North and South America and of Scandinavia.

In their eastward movement around the circumpolar vortices, anticyclones preferentially follow trajectories prescribed by this topography. Because the large heat capacity of the ocean is so readily transferred to the superjacent air masses, as part of the general poleward redistribution of accumulated heat, it is not surprising that persistent anomalies of sea-surface temperature (SST) should form a feedback system and act as a memory for the atmosphere, forcing or encouraging the repetition of anomalous climatic events, as Cushing and Dickson put it.[5] These anomalies do not occur randomly, but in preferred regions and seasons in the ocean: for instance, North Pacific SST anomalies occur preferentially during winter along the two continental margins and in mid-ocean at about 160° W.

Cyclic and quasi-periodic changes in the atmospheric circulation or in relevant characteristics of the solar system can be demonstrated at periodicities ranging from one year to tens of millennia; for practical fisheries purposes we can ignore periods exceeding more than a few decades, yet even in this restricted range we find great variety in the identified periodicities. Power spectrum analysis of sunspot numbers, annual rainfall totals, tree-ring indices and lake varve thickness shows coherence between periodicities in the different data sets: harmonics of the approximately 11-year sunspot cycle are frequently evident at 2+, 5+ and 13+ years with longer periodicities at approximately 90 and 100 years. These data also suggest the existence of two kinds of states in atmospheric forcing of the ocean: (i) short-duration, reversible changes of state and (ii) long-term trends sustained over decades or even centuries.

It is the former pattern that will most immediately concern fisheries managers, who must deal with the fact that many regions are characterised by more than one possible state of atmospheric circulation, and that a change of state between these patterns habitually occurs at decadal scale – or perhaps longer – intervals: at the widest spatial scale, such changes are induced when the circumpolar vortex switches between zonal and meridional patterns, depending on the degree of meandering in the westerly winds. When changes in the preferred location of a persistent meander occur, weather observations will indicate that a change in regional climate has occurred: dominant wind direction, relative cloudiness, air and sea temperatures and the distribution of plants and animals will all record this change. This is the stuff of climate history, so well told by the late Hubert Lamb.[6]

Decadal scale and shorter events are imposed on longer-term changes in regional climate that must be noticed, although they will not be directly relevant to this discussion: during the present post-glacial Holocene period, warmer and cooler epochs are clear in proxy records of climate conditions: tree-ring archives, elevation of vegetation on mountains, varved deposits in anoxic basins, and so on. There is also clear evidence that these, and some current changes of atmospheric circulation, are externally forced by very small changes in solar radiation, day length, and axial inclination. Bearing in mind how small the changes in solar radiation appropriate to each Milankovitch cycle are (about 9 W/m^2) and that these changes are sufficient to induce glacial epochs, we should perhaps be very sensitive to recent satellite measurements of changes in solar irradiance of $\sim 2 W/m^2$ that occur during each 11-year solar sunspot cycle. If such apparently small changes in external forcing are capable of inducing the decadal-scale changes of the atmospheric circulation that have been observed in the proxy record, then measurement and prediction of the relative importance of the upcoming Cycle 24, of the Earth's rotation rate (and hence length of day, usually expressed as detrended – LOD) should be of some interest to fisheries science.

Climatologists keep track of the state of the atmospheric circulation, and hence of regional climate, by the use of indices that represent the difference in atmospheric pressure between two distant locations, or – more primitively – simply the accumulated daily wind direction, rate of change of SST or atmospheric pressure at critical locations. Three such indices are in general use by the oceanographic community, and a fourth has been added recently to the fisheries literature: each of these has shown a characteristic periodicity associated with atmospheric circulation patterns during the twentieth century (Figure 5.1).

The SOI (Southern Oscillation Index) is quantified by the quasi-periodic changes in sea-level atmospheric pressure difference along the

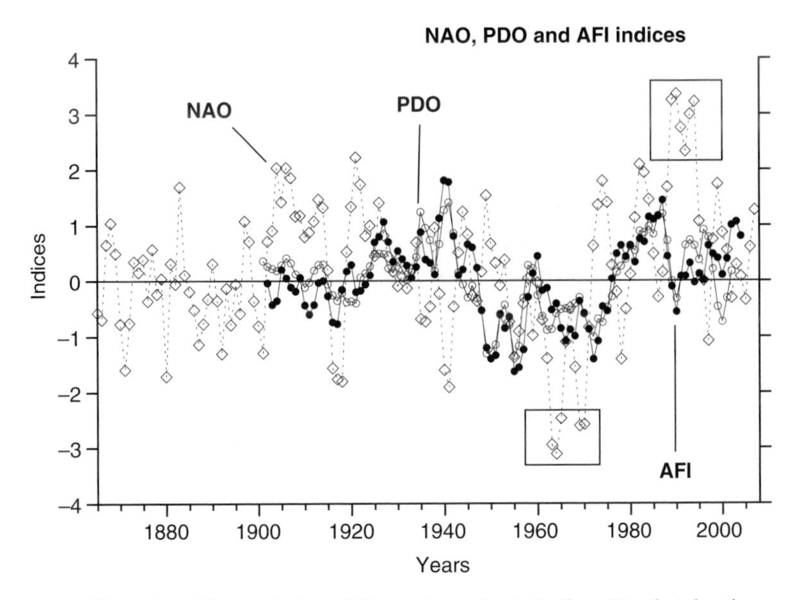

Figure 5.1 The evolution of three atmospheric indices (North Atlantic Oscillation, Pacific Decadal Oscillation and Atmospheric Forcing Index) during the course of the twentieth century to illustrate their general coherence. Note the box surrounding the 6 years at the end of the twentieth century discussed on page 86. Because digital data are not available, the meridional and zonal ASI indices published by Klyashtorin are not reproduced.

equator, between Tahiti and Darwin in northern Australia. Positive values of the SOI are associated with strong trade winds, westward down-sloping of the tropical thermocline and strong upwelling on the Peruvian coast; negative values are associated with weak trades, deepening of the thermocline in the eastern ocean and weak upwelling – an El Niño episode, often jargonised by oceanographers as an ENSO (El Niño/Southern Oscillation) event.

It is not unreasonable to trace the origins of the quasi-periodic appearance of El Niño episodes back to the biennial climatic cycle which is so evident in many data series (air temperature, rain- and snowfall) amounts, sea-ice cover and the quality of vintages over much of the northern hemisphere. The most spectacular manifestation of this biennial cycle is the reversal of zonal stratospheric winds over the equatorial regions with a periodicity of about 25 months, although anomalous cycles of 30–35 months also occur; these changes in wind stress are accompanied by alternation in intensity of the SE Pacific high-pressure region near Easter Island and the low-pressure system over the Indian Ocean.

The changes in stratification, upwelling and in the heat content of the upper ocean that are forced by the changing wind regime quantified by the SOI are now rather well-known and are global in extent, although the regime of the western coasts of the American continent has long been the focus of studies.[7]

In these eastern Pacific coastal regions, periods of weak vertical nutrient flux during an ENSO episode lead to low levels of primary production, an effect that cascades up to slow growth in fish populations, to poor recruitment and to starvation of predators such as sea birds and sea lions. These effects have become the classical frame of reference for El Niño, ignoring wider consequences: anomalously heavy and extensive rainfall in the western Pacific, unusually sustained winds in the Tasman Sea, drought conditions in India, China, Africa and Australia, anomalous nutrient conditions in Canary and Benguela Current upwelling regions, anomalous SST in the Pacific as far north as the Gulf of Alaska – such were just some of the anomalies observed during the 1982–3 ENSO, one of the first to have been studied comprehensively. It was at the 1960 CalCOFI conference, held to discuss the 1957 event (as suggested at the head of this chapter), that Warren Wooster and Jerome Namias first made the suggestion that El Niño might be part of a Pacific-wide process involving interaction between northern and southern Pacific anticylones.

Because the value of the SOI is reflected in so many phenomena, both at sea and ashore, the historical record is excellent: since 1550 to the present day, the mean return time of ENSO episodes has been about 4 (3–7) years, with a duration of 8–15 months, although there was an interstadial of almost 10 years between the El Niños of 1929 and of 1940 during the period of unusually warm climate of those decades. There is evidence, in analyses of the sequence during the nineteenth to twentieth centuries, that El Niño events occur preferentially during declining phases of the sunspot cycle, illustrating the influence of external forcing on the dynamics of atmosphere and ocean.

The NAO (North Atlantic Oscillation) represents the strength of the atmospheric winter dipole between the Icelandic low-pressure and the Azores high-pressure cells, for which barometric records of good quality extend back to 1865, enabling historical interpretation. Circulation in the ocean is sensitive to the value of the NAO; when this takes low values, the wall of the Gulf Stream comes to lie in a southerly position, the strength of the Labrador Current increases above average, and strong flow around the Norwegian Sea enhances the transport of arctic water to the eastern coast of Greenland.

The strong pressure gradient of the positive phase of the NAO between the deeper-than-normal Icelandic Low and a stronger-than-normal Azores High brings westerly weather to mid-latitudes, and stronger and more frequent winter storms across Europe on a northerly track; colder and drier conditions prevail over the northwest Atlantic and the Mediterranean region. The negative phase of the NAO brings more easterly storm tracks, wet conditions in southern Europe and dry in Scandinavia. Such changes are reflected in the ecology of marine fish: the warming period in the early twentieth century that terminated in the 1940s was accompanied by a population shift of Icelandic cod up the west coast of Greenland starting in about 1917, reaching to 73° N by the early 1920s, after which regression southwards again set in. During the period of rapid cooling after 1945, progressively early onset of drift ice north of Iceland, and a corresponding shift of the spawning pattern of Icelandic herring were seen. During the climatic regression of 1950–70, some arcto-boreal species of benthos replaced warm-water species around the British coasts.

Hurrell suggested that the end of the twentieth century was a period of positive values of the NAO, 'unprecedented in the observational record',[8] although each period in the record is, in a sense, unique within a general pattern of alternate domination by generally positive or generally negative values, and concomitant changes in the vigour of the atmospheric circulation. It is true that during a group of 6 years (1989–95, see Figure 5.1) the NAO did take anomalously high positive values, but this occurred during a longer period very like that of 1900–50, which was dominated by positive values and interrupted only by brief negative episodes; this pattern was associated with the generally warm climate of the North Atlantic region in those years (see below). The generally negative values (and cooling temperatures) of the period *c.*1945–*c.*70 resembles conditions prior to 1900, and included a group of 5 years of exceptionally low values. This sequence also demonstrated that spells of both positive and negative NAO (and of warming and cooling temperatures) have occurred during the period of rapid increase in atmospheric CO_2 as indicated by the Keeling curve.

Since the present period of generally positive values does not yet exceed the length of the generally positive period that occurred in the first half of the twentieth century, I see no logical reason to declare it exceptional; I note also that the extrapolation to earlier years by Lamb, using the westerly weather over Britain index (which follows the same pattern as the NAO during the twentieth century), suggests a 20-year period of generally positive values around 1870–90, preceded by a very long period of generally negative values – associated with easterly wind and Dickensian winters. Thus, there seems to be no good reason yet to suggest that the extended periods of generally positive and generally negative NAO of the twentieth century represent anything other than natural events.

The ACI (Atmospheric Circulation Index) is derived from several indices that together characterise periods dominated either by zonal and meridional winds in the Atlantic–Eurasian region (30–80° N and 45–75° E) during the last 110 years; it is widely used by Russian meteorologists and climatologists[9] and forms the basis of the new analysis of the co-variability of global fish populations of Klyashtorin and Lyubshin.[10] The ACI is expressed as two zonal indices (W and E) and one meridional index (C) that describe the dominant direction of air mass transfer. These indices – plotted over the period 1880–2000 – are anti-phase and have a periodicity of approximately 60 years; it is most important to observe that the ACI shows no long-term trend over the same period. The cyclical values of the ACI closely match periodic changes in Arctic temperature (dT) and also in the detrended global dT; appropriately smoothed, the changes in values of the ACI also closely match the PDO (Pacific Decadal Oscillation) and the Aleutian Low Pressure Index (see below). These Russian scientists propose that these coincident series demonstrate the existence of a 'global climatic signal, the simultaneous development of climatic processes having approximately 60-year periodicity, observed, at least, for the whole northern hemisphere'. They show that the ACI is associated with periodicity in Greenland ice cores, and in dendrochronologic records from California bristle-cone pines and Arctic pines at millennial time scales; these data suggested dominance of 52-, 76- and 54-year periodicities, respectively. Comparable analysis of periodicity in fish populations will be discussed below.

The PDO is not, like the other three discussed here, an index of atmospheric pressure, although it expresses the results of changes in the distribution of atmospheric pressure over the North Pacific, specified by a North Pacific Winter Pressure Index[11]; this is the inverse of the Aleutian Low Pressure Index, also often used, and both characterise the intensity (and dimension) of the Aleutian Low Pressure system. The PDO is therefore defined[12] as the leading principal component of North Pacific monthly SST variability, poleward of 20° N, and relative to the long-term mean. Shifts between warm (or positive) and cold (or negative) phases characteristically occur after much longer periods of stasis than the NAO and SOI, each characteristically of about 20–30 years: the positive phase of PDO is associated with warmer water along the entire American coast and colder water in central and western Pacific mid-latitude regions, so this index quantifies the consequences on the ocean of atmospheric forcing, rather than representing the nature of the forcing itself. In recent years, the PDO has not presented a persistent regime and has, since the early 1990s, fluctuated between negative and positive state.

The changes of state described by the NAO and other climate indices take place against a background of lesser and apparently random change. Most pass unnoticed, but some are so singular as to be investigated; such was the so-called Great Salinity Anomaly, a large surface pool of fresher water that passed into the North Atlantic from the Arctic Ocean through the Denmark Strait in the late 1960s, reached the Labrador Sea two years later and subsequently followed Labrador Current flow back into the Norwegian Sea in 1976. A later irruption of freshwater and ice that passed through the Denmark Strait remained within the Norwegian Sea because it occurred when the NAO index was high, and was thus under the influence of the strong westerly winds associated with this condition. These events are associated with anomalous pattern in the general surface circulation of the ocean: anomalies of similar scale in the atmospheric circulation, and their predicted evolution, are displayed on your TV screen nightly, but evolution of pattern in the ocean surface flow is predictable only in the sense that it evolves very much more slowly.

Within the surface layer of the ocean, mesoscale eddies (of scale 50–500 km) dominate the mean flow and have important consequences for pelagic biota, including both adult pelagic fish and the planktonic larvae of both pelagic and demersal fish. Vertical motion associated with eddies induces change in the local productivity of phytoplankton such that there is a correspondence between cold (warm) temperature anomalies at the sea surface and higher (lower) pigment anomalies. Vertical motion of water centrally within cold-core cyclonic eddies stimulates phytoplankton growth, while the high-velocity zone around cyclonic eddies similarly supplies photic zone cells with enhanced nutrient levels.

That these pelagic foci of biological enrichment are themselves transient, responding to the eddying motion of the upper ocean, is one of the strongest distinctions between terrestrial and oceanic plant ecology and a powerful argument against the concept of sustainability of fish population biomass even under pristine conditions prior to fishing. Although prediction as to their future movement and state remains beyond reach, the location and ecological state of these mesoscale features and the fronts associated with their formation and evolution are now readily accessible through satellite imagery: TOPEX-POSEIDON sea-surface elevation and SeaWiFS sea-surface chlorophyll may now be superimposed for any region, for almost any specified period in recent years, to indicate the state of the mesoscale pelagic production system.[13]

REGIME SHIFTS, CHANGES OF STATE OR PUNCTUATED EQUILIBRIA

The rapidity with which changes occur in the characteristic ecology of marine regions, and their characteristic frequency, is a function of the physical dynamics of each region. Change is slower and occurs less frequently in the great subtropical gyral regions of the open ocean, compared with rates and frequency of change in coastal upwelling regions, in the vicinity of major ocean frontal systems, or in very high latitudes. Evidently, responding to simple rules concerning the biotic complexity of ecosystems, the more stable regimes are the more taxonomically diverse, and those in which regime shifts are expected to occur naturally are less diverse. The extent to which major shifts in ecosystem composition and function are cyclical is discussed below, but first I propose a brief examination of the nature of regime shifts, their response to external forcing, and the role of species comparable to those that have been called keystone species (lemmings, snowshoe hares) by terrestrial ecologists.

In most cases that have been studied, the ecological consequences of variability in environmental forcing is a simple change in the status of a species, or of a group of ecologically related species, in relation to the remainder of the ecosystem, and such a change in relative abundance generally occurs only progressively so that the structure of the ecosystem and the balance of trophic flux within it remain essentially recognisable. A general model for this process was proposed by Francis and others that probably matches the case where the ecological shift is induced by changes in distant forcing very well: changes in radiation or wind forcing modify stratification and mixed-layer temperature, inducing a cascade of changes that progress from autotrophic plankton, and up through herbivores to higher trophic levels and large predators.[14] This is usually a linear and gradual change from one state to another.

Sometimes, however, the effect on the ecosystem represents no less than a catastrophic change of state that may interrupt any gradual changes already taking place. Such drastic shifts to a contrasting state have now been observed in lakes, coral reefs, ocean ecosystems, forests and arid lands, as Scheffer and Carpenter have discussed recently.[15] These do not necessarily, as one might imagine, result from a catastrophic change of external forcing, as by a volcanic eruption. On the contrary, the shift into an alternative steady state may – as in the model of the tipping boat – result from a very small increment in the external forcing that causes a critical threshold, or critical bifurcation, to be exceeded. The classic case in the ocean is that of Caribbean reefs, many of which shifted into an algal-encrusted state because of the removal of large predatory fish by sports and other fishermen; this, and other ecological consequences of fishing, will be discussed in later chapters. Hysteresis is not always easy to distinguish from a time-dependent change in regional conditions as observed, for instance, at a fixed offshore monitoring station; such changes may simply reflect a geographical shift of a boundary between two different ecological steady-state regimes.

A prominent case of naturally induced alternative steady states is the domain shift in the ecosystem of the North Pacific Subtropical Gyre, which has been discussed by Karl and others on the basis of observations made during the period 1988–97.[16] This event was associated with a prolonged period of negative values of the SOI and illustrates the difficulty of quantifying ephemeral changes in, for instance, growth rates of populations of autotrophic cells, although these form a primary source of energy for the entire oceanic ecosystem. The multi-year data suggest that there was a progressive decrease in diatoms and an increase in the relative dominance of very small cells, dominated by *Prochlorococcus* and the co-occurring nitrogen-fixing cyanobacterium *Trichodesmium*; these changes induced a shift to a more productive regime and to a major downshift in the size distribution of herbivores. This led to the establishment of a generally more complex pelagic food web.

However, there are many questions concerning what should, or should not, be termed a regime shift. Is it an effect of shifting equilibrium between two relatively stable steady states, and is it cyclic? Assuming that the shift from one state to another responds to internal dynamics, is the forcing top-down, bottom-up, or is it a consequence of wasp-waisted structure in the ecosystem? If bottom-up forcing is involved, is this generated as a response to external physical forcing and, if so, is it cyclic to sufficient extent to be predictable? To what extent is a regime shift at any location in the ocean merely the result of a geographical shift of biota induced by changes in physical transport? To what extent is the planktonic food web at the base of the trophic pathways of each state in equilibrium? Finally, to what extent are observed changes of state forced by fishery modification of the original ecosystem? I do not propose an attempt to answer all, or perhaps any, of these questions, which were raised at a workshop on regime shifts held in Villefranche in 2003.[17] What I wish to emphasise is that the range of phenomena and mechanisms that we observe in the variability of fish populations is sufficiently varied and complex that no single theory or model can be expected to fit all observations.

When we discuss recent regime shifts, we should keep in mind the scale of the natural changes of climate that have occurred during the period with which we are concerned – say, the last millennium. Recall that the consequences of the Mediaeval Warm Period, centred in the two centuries AD 800–1000, and of the Little Ice Age of AD 1550–1850,[18] were very significant in the region of the developing European fisheries, although global in scope. The coast of Iceland, for example, was invested by ice for only 2–3 weeks each year from 1550–1600, but from 1790–1810 ice was present for 20–25 weeks. Further south, the cod fishery at the Faeroes failed entirely from 1675–1704 and became very poor in the Norwegian Sea; in the same period, there were reports of the southward spread of ice and polar water north of Britain. As Lamb records, recovery was not complete until about 1840 and, even as late as the 1880s, sea ice was again reported around the Faeroes in winter. In the interim, there was a brief period of unusual warmth in the North Atlantic, some years being sufficiently ice-free that a cod fishery at West Greenland could be prosecuted from 1845 to 1851. It has been suggested that one of the difficulties of clearly associating different kinds of evidence (glacier advance, tree rings, varved sediments, etc.) from diverse regions is the evident problem of defining a normal climate on the surface of a planet subject to variable external forcing.

That the continual changes in regional and global climate over the last 2000 years will be reflected in the local abundances of fish species is evident, and many shifts in relative species abundances have been recorded during the modern era: we must expect that such shifts will be reflected in historical and archaeological data, but they may, of course, be very easily confused with the effects of human intervention by fishing.

A classical example of ecological changes wrought by a minor climatic shift, of course, is what was observed in the western English

Channel in the first half of the twentieth century. The Russell cycle,[19] as it came to be called, was one of the very earliest regime shifts to come to the notice of marine biologists. About 80 years ago, routine sampling in the western Channel by the Plymouth laboratory detected a shift in dominance between species of arrow worms: the neritic, cool-water *Sagitta setosa* was replaced by a more southerly species, the Atlantic *S. elegans*. In the same years, the nutrient content in the water column declined and cold-water herring were replaced in the Channel by more southerly pilchard. It was not until the 1950s that the situation began to revert to the previous state so that, during the 1960s, the plankton off Plymouth again resembled that of the 1920s. It now appears, thanks to a new analysis by Drinkwater, that the Russell cycle was but one item in a catalogue of change that affected the entire North Atlantic, and especially its cooler regions.[20] This regime shift is not entirely comparable with similar events in the North Pacific Subtropical Gyre, because the ecological changes there are more synchronous with the physical changes than is the case in the Atlantic.[21] The 1920s and 1930s, when the Russell cycle was initiated, was a period of generally positive NAO values (see above) and all weather archives describe a very strong warming of both atmosphere (4–6° C cumulated positive anomalies 1920–40) and surface ocean around the northern regions of the Atlantic Ocean, from Scandinavia to Labrador. These conditions relaxed only in the late 1950s and 1960s when the NAO again took negative values.

During this period, there was strongly increased inflow of Atlantic water to the northern coasts that induced major shifts in fish distribution, including the spread of Icelandic cod to western Greenland, and their occupation of 1200 km of that coast; this induced a major expansion of the previously very modest fishery to one yielding landings of 500,000 tons annually. During the warmest years of the cycle, the zooplankton biomass of northern regions (as indicated by data from routine CPR routes[22]) significantly increased and many other fish expanded into more northerly regions: *Melanogrammus*, *Reinhardtius*, *Anarchichas*, *Salmo*, *Squalus*, *Tachypterus*, *Brosmius*, *Molva* and others. The warm period also saw a major shift in the feeding migration routes, and an increase in the abundance of some herring populations; the cycle of sea temperature along the Kola meridian (one of the classical indices of climatic conditions in the NE Atlantic) was faithfully followed by the pattern of biomass of the Norwegian spring-spawning herring stock. The sudden increase in recruitment success of several gadoid species in the North Sea (the so-called gadoid outburst) that occurred in these years was a natural phenomenon probably caused by changes in the timing of peak *Calanus* abundance.[23]

These consequences of a climate shift induced the holding of an ICES meeting on Climate Change in 1949.[24] The preparatory papers remarked that 'The problem of climate changes in the Arctic ... one of the most pressing problems in the Council's area' and remarked on the urgent need for relevant data and information, especially concerning the northward displacement of marine organisms. Hans Ahlman, in his introductory address, discussed increasing air temperatures, especially in mid high latitudes of both hemispheres, receding glaciers in Europe, North America, Africa and South America, decreasing extent and thickness of ice in the Arctic Ocean, decreasing water levels in lakes through increased evaporation in Asia and Africa, and increasing sea-level elevation, citing evidence from all continents and latitudes. These effects, in the ICES area, he attributed to changes in the pattern and strength of the Icelandic Low and strongly increased flows of southerly air into northern Atlantic regions, entirely consistent with the pattern of what we now call the NAO shown in Figure 5.1. The news media in those years, as today, indulged in somewhat alarmist reports of receding glaciers and sea-ice extent. Somewhat parenthetically, it is interesting to recall that from 1960 to 1980, and the period of generally negative values of the NAO, both public and scientific concern for the now cooling climate focused rather on the likelihood of the inception of a new Little Ice Age ...[25]

Once again, during the 1990s and the early years of the new century, the pattern of the Russell cycle repeated itself and, since then, *Calanus helgolandicus* has tended again to replace *C. finmarchicus* in a North Sea that supports lower biomass than during cool periods. At yet another ICES symposium on the effects of climate change on North Atlantic fish populations held in 2004, the conveners suggested that 'While variability in the NAO up to the 1950s can be accounted for by internal atmospheric dynamics, changes in the NAO since that time are consistent with anthropogenic forcing.' It should be interesting to watch this new manifestation of the Russell cycle unfold – because, if it follows the cyclical pattern of the NAO and other climate indices, it will presumably return to the *status ante quo* in some years time. Only if it does not do so can we be sure that it is indeed a manifestation of our interference with global climate dynamics.

These recent observations bring to mind the much older records of the recurrent changes of state and distribution of the Norwegian and Swedish herring stocks, whose catches (and fisheries) have historically alternated in abundance according to the extent of ice cover in the feeding area of the Norwegian stocks to the north of Iceland. One also recalls the 'ballet' of North Sea herring stocks on the Swedish Bohuslan coast: these habitually migrate between the southern North Sea towards the shelf to the east of the Shetlands in spring.[26] On their return, in some years when easterly winds are exceptionally strong (NAO index negative), the shoals may enter the Baltic instead of regaining the southern North Sea; these pass into the Skaggerak on the anomalous subsurface inflow, which is then stronger than in other years to compensate for offshore wind drift of surface water. In such years, the population then winters along the Bohuslan coast of Sweden. The incursion may last longer than the initial conditions that induced it, due to the establishment of spawning within the Skagerrak, but such a pattern is induced only when an unusually strong year-class of young fish that yet lacks fidelity to North Sea spawning grounds has entered the Skagerrak.

In some years, herring enter the Skagerrak to overwinter but do not approach the Bohuslan coast: such episodes have been termed 'Open Skagerrak' periods. The past history of Bohuslan herring periods has been very well documented because of their regional fishery importance: 1307–62, 1419–74, 1556–87, 1660–89, 1748–1808, 1878–96, with 'Open Skagerrak' periods in 1907–20, 1943–54 and 1962–65: such events in earlier centuries probably passed unrecorded. However, the return of the herring populations to the southern North Sea after a Bohuslan escapade did not escape the notice of Campden, the sixteenth-century historian, who wrote: 'These herrings, which in the times of our grandfathers swarmed only about Norway, now in our times ... swim in great shoals around our coasts every year.'[27]

Although these events may be classed as regime shifts, it is what occurred in the final decades of the twentieth century in the North Pacific that have become the paradigm for that process: 'A regime implies characteristic behaviour of a natural phenomenon over time. A shift suggests an abrupt change, in relation to the duration of the regime, from one characteristic behaviour to another.'[28] You will note that this definition specifically excludes the replacement within a trophic level, of one species by another, if functions within the entire biological regime remain relatively unchanged. The environmental and ecological changes that occurred in the pelagic community of the North Pacific gyre in 1977 (noted above) and the return to the previous condition in 1989 have become a paradigm for such shifts. A model for a regime shift in the ocean was proposed by Francis and others[29]:

during the positive phase of the PDO, increased wind stress around an enhanced Aleutian Low spins up the Alaska Gyre and creates stronger upwelling centrally in this feature. Primary and secondary productivity is enhanced both coastally and in the central gyre, resulting in a high biomass of copepods in the spring; these are the essential prey of salmon smolts at the season when year-class strength is determined.[30]

The change of the PDO from negative to positive phase in 1976–7 came to our attention because of its effects on salmon populations: those of Alaska strengthened, while those further to the south along the American coast weakened. It was only later realised that a major increase in recruitment of groundfish populations had occurred, recruitment being unusually strong in halibut and sablefish, but also somewhat strengthened in flounder and turbot. It should be noted, however, that these were not necessarily step-increases in stock performance because the recruitment to Pacific halibut populations, for instance, climbed progressively from a low of around $3–4 \times 10^6$ recruits annually in 1965–75 to a plateau of $7–8 \times 10^6$ in the period 1979–90, after which they declined again strongly: the collapse of recruitment was significantly more rapid than its increase.

Small-mesh trawl surveys showed that the entire ecosystem was transformed from domination by capelin, shrimp and sand-lance to domination by higher trophic-level predators, gadoids and flatfish. Other changes in the ecosystem were recorded – from the level of primary production and zooplankton herbivores right up to marine mammals and sea bird populations, both anecdotally by several authors but also very formally by Hare and Mantua, who referred to 100 environmental time-series (41 physical, 59 biological of which 20 referred to recruitment or population biomass of commercial fish); all were of appropriate length and were treated by principal component analysis after normalisation. The results of this empirical analysis provided clear confirmation of the observational studies done in the region from California to Alaska.

Ecological regimes along upwelling coasts have long been the model for regime shifts in the ocean, and the frequency and causation of such shifts will be discussed below; in the Benguela Current, recent ecological events offer us a somewhat different kind of regime shift from that of the North Pacific. Unfortunately, the recent Benguela literature must be read with great care, because it is not always easy to separate the wheat from the chaff, since the output of models is often described in the same terms as observations. This is an example of what

Professor J.Z. Young warned us about so many years ago[31]: confusion between model output and the real world.

Inter-annual variability in the Benguela region differs from the eastern Pacific, exhibiting relatively small inter-annual variability and less frequent and less predictable major changes in atmospheric forcing. The principal sources of large-scale variability are intrusions of Agulhas Current water and the so-called Benguela Niños; the latter result from sustained weakening of the Atlantic trade winds, at the decadal time scale, with the consequent production of a warm water anomaly in the eastern ocean that results in the Namibian region (northern Benguela) being flooded with warm, highly saline surface water. Intrusions of Agulhas water are less predictable, major events having occurred in 1957, 1964, 1986 and 1997–8.

The ecology of the Benguela Current, both in its southern part off South Africa and in the northern, Namibian section, is dominated by two mid-trophic level clupeids, *Sardinops ocellatus* and *Engraulis capensis*, so this region is consequently thought to represent an ideal wasp-waisted ecosystem in which neither species dominates indefinitely. Perhaps the most important observation is that the alternating dominance of anchovies and sardines in this region 'appears not to have had major effects on the overall functioning of the ecosystem', so here the changing dominance between sardines and anchovies is a matter of simple species replacement rather than a regime shift.[32] More than that, the observations appear to show that turning points are difficult to define and changes in ecosystem structure appear to have preceded, rather than followed, changes in physical regime. Nevertheless, the atmospheric circulation does show that the negative EW wind anomaly from 1973 to 1993 differed strongly from positive conditions before and after that period. The period characterised by a negative anomaly corresponds approximately to a period of anchovy dominance and regression of sardines – although the effects of fishing distort the natural response very strongly and make it very difficult, here as elsewhere, to observe with clarity the environmental signal. At end of the twentieth century, after the return of positive wind-run anomalies, the biomass of pelagic fish in the southern Benguela was unusually large, while in the northern regions it was small, both anchovy and sardines having been so severely reduced in the 1970s that they were replaced by a variety of zooplanktivorous species (e.g. *Trachurus*, pelagic gobies and mesopelagic nocturnal migrants) and a great abundance of the medusae *Chrysaora* and *Aequorea*. There have been suggestions that this represented a novel, fishery-induced regime whose potential permanence was unknown.

CYCLICAL CHANGES IN THE ABUNDANCE OF FISH POPULATIONS

One of the few 'known knowns' of fisheries science is the natural cyclical abundance of the clupeids that dominate the pelagic biomass in upwelling coastal regions, and the independence of these changes from any human activity. There are two principal sources of evidence from the distant and historical past: (i) the relative abundance of scales of sardines and anchovies in the varved sediments of anoxic basin off California, for which there is an excellent record of almost 2000 years, and (ii) historical records of the activity of fishing villages around the Japanese islands that extend back for about 400 years.

Simple inspection of these series reveals cyclical changes in abundance of each species in each region and shows that these changes occurred in the absence of heavy fishing mortality. It is also clear from the California sediments that sardines are more likely than anchovies to be reduced naturally to very small population size even though, in the long term, they are the dominant clupeid fish of this coast. Off Japan, 7 or 8 periods of high fishing activity for sardines were centred around 1650, 1720, 1790, 1820, 1880, 1930 and 1990; however, it is not clear whether these represented overall increase in population biomass, or periods when population shifts made sardines more available to small coastal fishing craft.

The historical collapse of the California sardine population around 1950 and its replacement by a large population of anchovies has been much studied, initially with the general conclusion that the collapse was largely caused by excessive fishing mortality and that, in some way, the abundance of anchovy was related to the absence of sardines by competitive exclusion. However, later studies of the relative abundance of scales in the sediment cores suggested a rather different relationship: that of a background abundance of anchovies against which periods of great sardine abundance occurred periodically. To determine the period of recurrence, and the duration of sardine periods, was not simple and spectral analysis was necessary for its resolution.[33] This demonstrated a complex pattern of relative abundance of sardines and anchovies: for almost a millennium prior to AD 1000, scale deposition suggests almost continuously large anchovy populations. From AD 1000 to 1500, there were five periods of progressively increasing scarcity, although subsequently this quasi-regularity has broken down.

Since the collapse of the California sardine fishery just 15 years after the peak abundance that occurred in the season 1936–7 and of the abrupt collapse of the Peruvian anchovy population after the

1972–3 ENSO event, research into the ecological dynamics of sardines and anchovies in upwelling regions has been pursued very actively; it has became progressively clear that, as Herrick and others put it,[34] 'accumulated anomalies of physical indices are proportional to California sardine landings and that the accumulated anomaly curves change the sign of their slope showing maxima (minima) when climate is favourable (unfavourable) to successful completion of the sardine life cycle'. In recent years there has been a progressive recognition of the fact that the collapse in 1950 was not simply caused by the lack of regulation of fishing effort, but was also a consequence of changed ocean conditions: specifically, those that created cooler conditions in the California Current than during the boom years of the fishery in the 1930s and 1940s.

The search for connection and causality between fluctuations of fish populations – both in their size and location – and changes in environmental conditions has been a central theme of one section of fishery science for many years. The sequence of Bohuslan herring events was investigated as early as the nineteenth century because of the regularity of its return: the full cycle occupied 110–120 years in the historical record and the average duration of events was about 55 years. Many attempts have been made to associate this periodicity either with the cycle of solar activity or with a cyclical pattern in the weather systems over NW Europe. Studies of other repetitive changes in the NW Atlantic are also a major part of the patrimony of classical fishery science of the twentieth century; the fluctuations of cod and herring stocks of the Norwegian Sea were related to a range of regional environmental indices (e.g. water temperature 0–200 m on the Kola Meridian, regional air temperatures at Iceland, Jan Mayen and Spitzbergen, and others) were invoked as candidate proxies for population biomass, especially of Arcto-Norwegian cod by both Russian and western scientists. Over a longer period than permitted by instrumental data, the fluctuations of the same population were investigated in relation to air temperatures that were inferred from the dendrochronology of trees growing on the Lofoten Islands.

Not surprisingly, changes in coastal fish populations, particularly of clupeids on upwelling coasts, became a central theme of fisheries oceanography in the closing decades of the twentieth century.[35] Emphasis was placed on concurrent population changes in three populations of *Sardinops* spp. (Japan, California and Chile) and the shifts of their distribution centres that occurred in response to the SOI cycle, together with the striking collapse of the Peruvian anchovy population that started in the early 1960s and culminated with the El Niño events of the 1970s; it was noted, however, that these did not change the relative distribution of the anchovy population along the coast.

Although it was recognised early that it was problematic to assign all observed changes in population size to the consequences of climatic regime shifts, even 50 years of research in the California Current failed to demonstrate unequivocally the mechanism by which physical changes were translated into species shifts. This failure will serve as an excellent example to illustrate the depth of our remaining ignorance of the mechanism of even such apparently simple and natural regime shifts that are nevertheless critical to industrial fisheries. In fact, the mechanism was eventually revealed only during the year in which I have been occupied in writing this book.[36]

The key to the mechanism is a recognition that upwelling in eastern boundary currents, including those off California, is not simple because two different physical processes are involved: (i) equatorward, alongshore winds induce strong divergence and vertical motion at the coast, and (ii) wind-stress curl induces either weak upwelling or downwelling, patterned by the dominant regional wind pattern, over much larger areas offshore. As is well known to biological oceanographers, strong upwelling will induce a plankton dominated by large particles (diatoms, large calanoids), while weak upwelling will induce the growth of small organisms (flagellates, protists, small calanoids). Our new understanding – contrary to received wisdom – that sardines have small mouths and fine gill rakers, while anchovies have relatively large mouths and coarse gill rakers[37] completes the explanation of the observation that sardines are the dominant species in a wind-curl upwelling regime, anchovies in a coastal divergent upwelling regime.

The 50-year record of observations off California demonstrates shifts between these two regimes, and a surplus production model for the well-monitored sardine population shows that the biomass of this species tracks the strength of wind-stress curl upwelling offshore sufficiently closely to convince at least this reader of the relationship. But the authors of this study have a last shot in their locker: they point out that there is a strong relationship between latitude and the vertical motion induced by similar wind-stress curl due to changes in the Coriolis parameter; off Peru (at 3° S), this will be more than three times the motion induced by the same wind-stress in the California Current at 10° N. It is therefore entirely predictable that the Peru Current should be an ecosystem dominated by anchovies, while sardines dominate off California.

More recently, the whole question of natural cyclical abundance of fish populations, and the possibility of prediction of their future states through prediction of the future states of climatic regimes, has become more generalised with the publication in English of the work of Klyashtorin and Lyubshin of the Federal Institute for Fisheries and Oceanography in Moscow.[38] This work broke the mould of the largely regional analyses that had previously occupied fishery oceanographers, and its consequences are yet to be fully assimilated, because they are global in scope. As you would expect, there are those who prefer the accustomed way of doing things, and recently I stumbled across some web-published lecture notes that warned students at a major California university that Klyashtorin's research 'was regarded with caution in some quarters' and that it 'was typical of Russian work'. We are fortunate that FAO wisely decided to make an earlier and abbreviated version of it freely available as a Fishery Technical Paper.[39] It will be useful to examine, at least briefly, the external forcing of cyclical abundance because – as the same authors suggest – prediction of future abundance at some level useful to industry might emerge if clear understanding could be reached. It is not clear if the prediction that their sardines would return in 1975–80 would have been of much use to Cannery Row in 1950, but it might have saved the California Fish and Game Department a lot of anxious research!

As I noted earlier, the ACI of Russian meteorologists specifies atmospheric mass transfers zonally (the W and E indices) and meridionally (the C index); each index is defined by the number of days each year that it was dominant over Europe and Asia, and is expressed as an anomaly so that $(C) + (E) + (W) = 0$. The long-term indices show that the dominant atmospheric circulation pattern falls into roughly 30-year periods or circulation epochs, dominated either by meridional or zonal flow. As Hubert Lamb remarked long ago,[40] zonal circulation (W or E) epochs are associated with increases in global dT, while meridional (C) circulation epochs are associated with cooling. Very early English weather records, prior to organised meteorology, often included a record of wind direction, and these have been used to demonstrate the periodicity of westerly weather over the British Isles – an early forerunner of formal atmospheric indices.

The similarity of the patterns of detrended ACI and the global dT and –LOD records is remarkable during the second half of the twentieth century, when we have excellent records of each. The 60–70-year periodicities match very closely, as do the brief and minor changes of slope that occur on the principal cycles; for this reason, it is suggested that these three geophysical indices are adequate tools for examining the possible occurrence of cyclical changes in the biomass of fish populations. The only significant deviation from direct correspondence is that the ACI cycle runs about 6 years ahead of the dT cycle, and this fact may offer an additional possibility of prediction of conditions likely to affect population dynamics.

Cyclical change in this group of classical regional climate indices also corresponds rather closely to the changes in LOD (as already noted above for the NAO index), taking principal maxima in the 1930s and 1980s and an intermediate maximum in the 1960s; global dT and the ACI lack the 1960s peak. Reaching farther back in time, spectral analysis has been performed on millennial-scale data series representing (i) air temperature, obtained from Greenland ice cores, (ii) summer air temperature from the dendrochronolgy of spruce and cedar trunks at high latitudes, and (iii) the fish scales in varved sediments off California discussed above. Comparison with the much shorter series available for ACI, dT and −LOD suggests matching between these and ice-core temperature and the periodicity of sardine abundance off California (average maximal period was 56 years) and between anchovy abundance in the same area and tree rings (51 and 70 years); recently, the standard index of wind-driven upwelling strength on the California coast was found to be correlated with LOD over the period 1950–2000. It also appears very probable (although the evidence is not watertight) that the historical Japanese sardine periods over the last 400 years occurred preferentially during periods when global surface temperature was warming rather than cooling.

Such results have encouraged a wider examination of the relationship between the three geophysical indices and the population dynamics of fish species that dominate the global landings today. Klyashtorin pointed out (somewhat to my personal surprise) that today only 12 species comprise between 40 and 50% of all landings.[41] His analysis of the population fluctuations of these since about 1950 shows that there is a remarkable correlation between changes in their population sizes and synchronous changes in the ACI, dT and LOD indices. These 12 species fall into two groups whose population fluctuations are in opposite phase: Group 1 (Atlantic cod, Atlantic and Pacific herring, South African sardine and Peruvian anchovy) exhibited maximal catches in the 1960s, with minima in the 1930s and 1990s, while Group 2 (Japanese, Californian, Peruvian and European sardines, Pacific salmon, Alaskan pollock, Chilean jack mackerel and Peruvian anchovy) had the inverse pattern, with minimal population size in the 1960s. It is also to be noted that during the 1960s when, for instance, the North Atlantic cod stocks were expanding rapidly, the NAO was persistently and strongly negative and cooling conditions prevailed; decline of cod stocks began in the same years after 1970 when the NAO entered a long and persistent period of positive values. The effects of changing climate conditions on these fish were perhaps a logical result of the disaster on the Grand Banks of the 1980s, as will be discussed in the next chapter.

The match between the fluctuation of biomass of these groups and the global ACI is intriguing, catches from Group 1 species closely matching the pattern of the zonal (W or E) ACI, peaking during warm circulation epochs (Figure 5.2a), while almost all of Group 2 species correlate very well with the pattern of the meridional ACI, peaking during cool epochs (Figure 5.2b). Only the Peruvian anchovy is anomalous, apparently due to the occurrence of strong El Niño events, which may abruptly terminate a period of high abundance and so distort the symmetrical periodicity that would otherwise be expected. The 1972–3 event[42] caused a complete spawning failure in September 1971, just as the first physical precursors of an El Niño event could be detected in water temperatures; this failure led to a sudden 80% fall in anchovy biomass off Peru that did not recover until growth began again in 1985–86 – in conformity with expectations derived from the general relationship of this species with the ACI index. Suggestions that the attribution of the collapse to climatic conditions was made so as to enable the effects of very large landings in previous years to be ignored, and to allow business as usual in the fisheries, downplays the historical evidence of repeatedly changing population abundance of this species prior to industrial fishing.[43] In fact, a non-linear additive model using parent biomass and offshore transport of surface water at Trujillo explains 75% of the variance in recruitment.[44]

Unfortunately, such analyses cannot be carried far back in time with fishery data because, prior to the twentieth century, fishing effort was so low everywhere that catches bore little direct relationship over the long term to changing biomass. This is evident in the analysis of cod landings at Newfoundland during the nineteenth century to be discussed in the next chapter. Nevertheless, it is relatively easy to explore informally such apparent cycles in population abundance in other populations than those used by Klyashtorin, in the regionally stratified version of the FAO global landings statistics which are readily accessible online.

In selecting competent data, it obviously cannot be assumed that the relationship between total catches and population biomass is linear, and care must be taken to eliminate those data sets in which trends are principally due to changing fishing intensity: this is

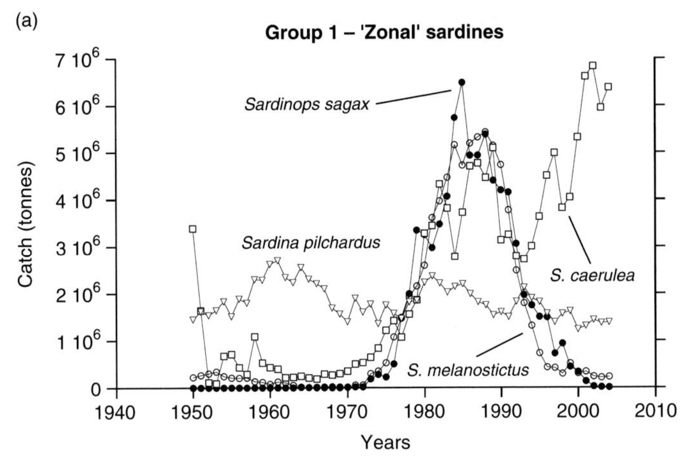

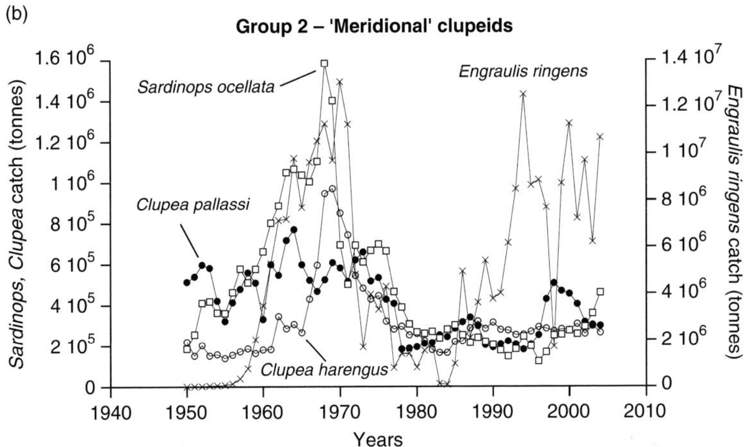

Figure 5.2 Progression of reported landings of (a) four Group 1 (zonal ACI, *sens.* Klyashtorin) sardines and (b) four Group 2 (meridional) clupeid species during the second half of the twentieth century to illustrate the coherence of their cyclical abundance and the disturbance of these cycles by fishing pressure.

sometimes easy, as in the case of recently exploited deep-water species, and sometimes not, as in those for which the data appear to represent a boom-and-bust fishery. A brief exploration of FAO data shows that the pattern for Group 2 species can be extended with some confidence individually to Atlantic and Pacific halibut populations (Figure 5.3a), to haddock in the NW Atlantic, to the NE Atlantic stocks of anchovy (but not to those of the Mediterranean, or of FAO region Atlantic East–Central), to Atlantic herring, and to Japanese and European jack

(a)

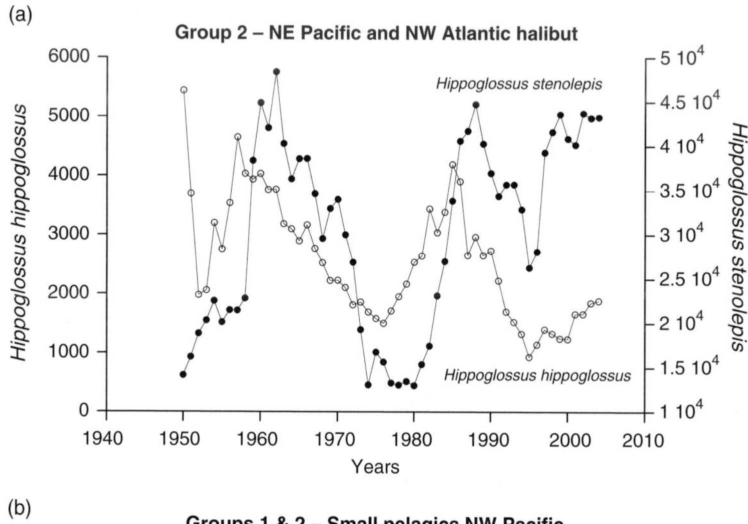

(b)

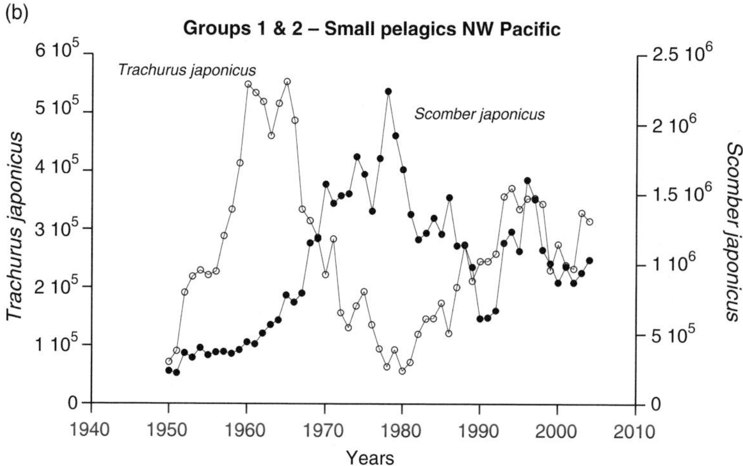

Figure 5.3 (a) Progression of reported landings of Pacific and Atlantic halibut in the second half of the twentieth century, to illustrate their coherence; (b) landings of two small Northwest Pacific pelagic species having opposite trends of abundance.

mackerels and perhaps to Atlantic pollock and South Pacific hake. On the other hand, Atlantic ling, pouting, both NE and NW Atlantic cod, and also NE Atlantic haddock all conform to the Group 1 model, with a single peak around 1970–75. In the NW Pacific (Figure 5.3b), the two dominant small pelagic scombroids appear to partition their habitat in the temporal sense – *Trachurus japonicus* responding to meridional and *Scomber japonicus* to the zonal ACI conditions.

The main problem with such apparent correlations is that the data on which they are based are reported catches, not population biomass, and it is rather difficult to be convinced that at least some changes reflect anything other than fishery effects on population size. Some aspects of the biomass fluctuations are clearly fishery-induced: the small trough and subsequent peak in the biomass trend of Atlantic cod is the well-known consequence of the establishment of Canadian management rights over the 200-mile zone and the introduction of some control over foreign trawlers. The customary interpretation of the general downward trend of cod biomass after the mid 1960s is that it reflects no more than the disastrous effects of distant-water fishing by Europeans and it will certainly take a regime shift in fishery science before the progression of cod landings is interpreted in any other way. Then, it has to be admitted that simple curve-matching is not a rigorous procedure, although very seductive; to deduce causation with a little more confidence requires the application of formal trend analysis and this has not yet, to my knowledge, been done with these data.

It is also probable that the simple relationship between climate indices and population size proposed by Klyashtorin will prove to conceal many complexities, especially as the population dynamics of a fished species can no longer be expected to react to environmental change in the same way as a pristine population: population biomass may become more variable. This has been termed the age-truncation effect, because the integrated population growth rate of a population dominated by young fish tracks the consequences of changing environmental conditions more directly than the pristine population[45]: older fish, generally lacking in heavily fished populations, should integrate short-term environmental fluctuations better than young individuals because of the greater flexibility of their reproductive strategies.[46]

Nevertheless, the synchrony of changes in the three global geophysical indices and in the biomass of some important fish stocks does hold a little hope for some level of practical prediction of long-term population biomass in both zonal-dependent and meridional-dependent groups of species. The average climate period indicated by ACI, dT and LOD is 55–56 years at the present time, although this periodicity has evolved during the last millennium from 120 to 140 years around 900 AD, then 70–75 years until the 1700s. For the first time, predictions of future cycles are being made: in 2006, NSF/NCAR predicted the start of Solar Cycle 24 in late 2007 or early 2008 and, right on schedule, the first reversed-polarity spot appeared on 4 January 2008! Unfortunately for fishery predictions, the probable strength of Cycle 24 has been variously predicted to be both stronger and weaker than preceding cycles; I have seen predictions from serious sources at NASA and NCAR ranging from 'the weakest for several centuries' to '30–50% stronger than Cycle 23'!

During the period 1985–2000, Klyashtorin suggests that a part of the declining trend in landings of the major species was the result of a natural and gradual decrease in population size in the meridional-dependent group of species during that period. As to the future, I will quote his words written a decade or more ago: 'After 2000, the present zonal ACI epoch will come into its final stage and ... the new meridional epoch will start. Extrapolating from past experience (1950–1980s) ... Pacific and Atlantic herring, Atlantic cod, South African sardine and Japanese and Peruvian anchovy are likely to increase in the incoming meridional epoch (2015–2030). This will result in a gradual increase ... up to 23–24 million tons by 2015–2020, followed by a decrease ... by 2030.' Similarly, he suggested that the catches of zonally dependent species will decrease to a minimal catch in 2015–20 before again increasing; the relationship observed between catches and geophysical indices should permit modelling of the probable biomass of each population in each group in future years.

I offer a final caution, however, because not all major populations have behaved in the last half-century as you would expect them to have done from Klyashtorin's thesis. Consider, for example, the Indian oil sardine (*Sardinella longiceps*) that exploits the seasonal diatoms blooms in the upwelling region off western India, and yields catches of about half a million tons annually; in this fishery, almost entirely directed at the 0-group fish, recruitment (and hence catch) is closely associated with relative sea level, indicating a direct response to changes in upwelling strength on this coast. In the late nineteenth century, an industry to extract sardine oil was established in Cochin, despite the catch being described as strongly variable; for almost 75 years, landings remained below 10,000 tons, until 1953 when a major increase in landings was initiated, peaking at around 600,000 tons by the late 1980s: they have subsequently been sustained at about the same level (Figure 5.4). The sudden increase of catches in the mid 1950s was confidently described (by myself and a co-author[47]) as a regime shift involving oil sardines and *Rastrelliger* mackerel, but it is now clear that this was not what occurred; more likely, it was no more than the result of an expansion of industrial investment that enabled the population to be fished throughout its range, rather than just coastally in regions from which it was periodically excluded by anoxic upwelled water. The fisheries on this coast continue to exploit two abundant

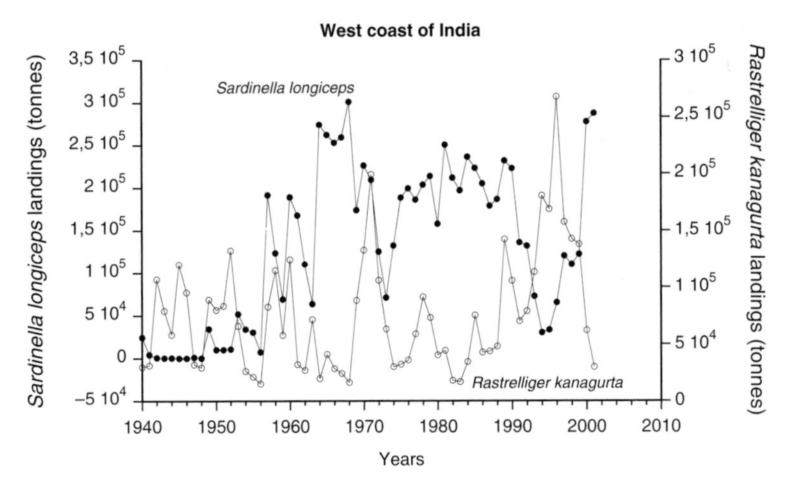

Figure 5.4 Reported landings of oil sardine and mackerel on the west coast of India to illustrate the lack of major abundance cycles, but rather a differential response to episodic changes in upwelling conditions.

populations of pelagic fish, both of which appear to lack consistent response to changing values of the atmospheric circulation indices but only to sporadic changes in upwelling intensity.

As always, the devil is in the details ...

A GENERAL CONSEQUENCE OF OCEAN INSTABILITY: RECRUITMENT VARIABILITY

That each population of marine fish is not always equally abundant, and that this is at least partly due to variable recruitment, appears to have been understood even in antiquity. Greek tuna fishermen believed that the relative abundance of young fish was reflected in the later abundance of adults, and early Hindu fishermen observed that the abundance of fish on the shelf was related to the strength of flow from the major rivers. These two ancient observations interestingly reflect the present-day divergence of opinion concerning the relative importance of a stock–recruitment relationship or the effect of environmental variability.

Changes in adult abundance were long assumed to represent the effects of migration of centres of maximum population density, but early in the twentieth century it came to be understood that strong annual variability in recruitment was habitual in many species and was reflected in adult abundance. Johann Hjort famously suggested in 1914 that relative year-class strength in the adult population was determined by the relative survival of eggs and larval forms, prior to recruitment; he also proposed two mechanisms for this process – larval starvation, or larval drift beyond habitat suitable for recruitment. These two concepts survive today in Lasker's *critical period* observations[48] and (at least partially) in Cushing's *match–mismatch* hypothesis.

Lasker observed that successful larval feeding, at least in sardines, depends on one of the main physical attributes of the present-day ocean: strong density stratification. The distribution of planktonic organisms – the food of fish larvae – is too often misunderstood because it is commonly assumed (especially by modellers) that the mixed layer, separated by a pycnocline from the denser water below, is a homogeneous habitat in which plankton is more or less uniformly distributed. This assumption is quite false, because organisms suitable for larval fish food usually exist in much higher concentrations in two layers, one just below the surface (the neustonic species) and another – and much denser – layer at the density gradient where the larger algal cells are concentrated, together with herbivorous copepods and their predators (see above).

Thus, it was really no surprise to find that sufficiently dense food concentrations for anchovy larvae in the California Current occur only in the relatively dense aggregation of plankton that is associated with the pycnocline. The larvae must locate this layer, and find sufficient food for growth within it, during a very brief critical period – or else they will starve. This they do more or less successfully each year depending on the seasonal evolution of stratification in the upper water column and the initiation of a seasonal bloom.

Cushing's hypothesis, on the other hand, was the natural outcome of studies of the recruitment consequences of variable transport of larvae within the eddying mesoscale circulation off NW Europe, in which the names of Ellertsen, Sundby and Dickson are prominent. Emphasis was placed by these authors on the coincidence between the occlusion of fish larvae and the spring bloom, with its associated population of small copepods, so that implicit in their studies is a simple trophic relationship between algae, copepods and fish larvae; unfortunately, reality is somewhat less logical than that. Consider the life cycle of a dominant copepod of the North Sea, *Calanus finmarchicus*, which includes an overwintering generation that spends the cold months in diapause in deep water, such as the Landsort Deep, well below continental shelf depths. Not only does the simple trophic match–mismatch model require perfect temporal coordination of the return of this generation to the surface, to match the onset of stratification and the start of the spring algal outburst, but spatial coordination is also required between the patchy onset of the bloom in response to local meteorological conditions, and the return to the surface of copepods after a long period in which horizontal transport at depth will have been imposed on them. It also requires that the survival rate during the winter diapause of *Calanus* should be approximately constant between years, but there is no reason to suppose that this is the case: many processes, both physical and biological, determine the density of individuals that seed the new copepod cycle in the mixed layer each spring.[49]

So simple models of the match–mismatch hypothesis, that must be based on trophic relationships and forced by environmental physics, can therefore be expected neither to validate the match–mismatch hypothesis nor to predict seasonal events in the real world. This may be a difficult truth for some to accept. Nevertheless, the apparent elegance of the match–mismatch hypothesis has led to its general acceptance in fishery science, although this has meant that other potential mechanisms that determine annual recruitment have been neglected, and we tend to forget that other mechanisms are perfectly possible. The demonstration by Francis and Bailey that recruitment in Bering Sea pollock depends on a totally different mechanism is one of these: larval pollock are able to satisfy their energy requirements from environmentally realistic low concentrations of food particles within the mixed layer (unlike Lasker's anchovy), so that a critical period concept is inappropriate for pollock, as is match–mismatch theory: in reality, it is much more likely that a differential annual rate of cannibalism by adults on juveniles is the cause of differential recruitment. Year-class strength in these pollock is determined by the extent to which, each year, the water column structure permits the adults and younger fish to mingle: in years of heavy stratification, juveniles remain largely in the warm surface layer that is avoided by adults, but when stratification breaks down seasonally – or is weak during the summer – the age groups mingle and heavy cannibalism occurs.[50]

Although the problem of stochastic recruitment has not been examined so extensively in benthic invertebrate populations having planktonic larvae, these organisms appear to react very similarly to inter-annual variability in the ocean as do teleosts. Although most authors have been concerned with settlement in the littoral and intertidal, at least one study of shelf benthos (at the 100 km and annual scale) in the California Current showed that inter-annual recruitment variability of Dungeness crab (*Cancer magister*) and red sea urchin (*Strongylocentrotus franciscanus*) populations is forced by ENSO events that modify both regional productivity levels and transport vectors. It was also found that transport by mesoscale eddy flow from larval concentration areas behind prominent capes was the determinant of the subsequent spatial pattern of settlement on the sea floor.[51]

This is obviously not the place for a major discourse on the recruitment problem in commercial fisheries, nor am I equipped to provide one, but a case may be made for a close relationship between the strong heterogeneity in space and time of today's ocean and the strong (and unpredictable) variability of recruitment. There is, of course, some relationship between spawning stock size and subsequent recruitment, and this relationship is sought for use in population management models, but it is nowhere near so simple as is suggested by a recent textbook for fishery science students (which I shall not cite): four graphs, carefully selected from among the hundreds available that I shall discuss below, are not competent to demonstrate a simple and universal stock–recruitment relationship.

That recruitment was a critical factor in the prediction of commercial fish stock size was early confirmed by the observation that occasional very large year-classes could subsequently dominate the adult population over many years. The classical case is, of course, the Atlanto-Scandian herring stock from 1907 to 1963: during this period, 8 major year-classes can be followed through the annual year-class abundance diagrams for this stock, each for between 10 and 15 years. In one case, a single year-class comprised >50% of total population abundance across a period of 9 years. A similar, although not so extreme, example occurred in cod at the Lofoten Islands: the 1939 year-class formed >40% of total numbers of cod in this stock from 1946 to 1949. In fact, where a suitably long time-series of year-class structure is available for any fish population, irregularities in the contribution of individual year-classes reveal the extent of recruitment variability.

The phenomenon of dominant year-classes is not restricted to high latitudes, although there are significant effects of latitude (or, rather, of the attributes of the ocean at different latitudes) on the reproductive patterns of regional fish populations. Everywhere, the problem to be solved by each species is to ensure correspondence in space and time between the emergence of first-feeding larvae and of their food. At high latitudes, the brevity of the spring outburst of phytoplankton and small herbivorous crustacea combines with the extremely slow development of fish ova and larvae to provide a very uncertain conjunction; reproduction is therefore concentrated into a very brief period. In low latitudes, the availability of suitable food is spread over periods of several to many months each year and spawning occurs over an extended period. In the pelagic habitat of the subtropical gyres, tunas spawn almost continuously, whenever and wherever water temperature is above a critical minimum, while continental shelf demersal fish spawn in almost every month, but preferentially when local meteorological conditions are relatively calm: thus, on Indian and West African shelves, peak spawning occurs during inter-monsoon periods. The Sciaenidae and Polynemidae that I studied on the West African shelf produced cohorts at about two-monthly intervals, but these were strongest during the dry season (June–September) and could then be followed through length–frequency distribution tables for the first two years of their five- to six-year lives.

Although one of the earliest preoccupations of fishery science was to obtain an understanding of the consequences of variable recruitment, the concept of a formal and universal spawner-recruit relationship has haunted fishery science since the early intervention of Ricker, who suggested that if one could be satisfactorily established it would greatly simplify the use of stock management models. Subsequently, it was confidently assumed that such a relationship existed and could therefore be formalised; the literature describing investigations of individual populations and their recruitment is legion. There was some hope that the matter might be satisfactorily resolved during the 1990s when the late Ransom Myers assembled a larger series of spawning stock and recruitment data than had previously been put together: it is publicly available. What can only be described as a flood of analyses of these data by Myers and his collaborators was published during the next decade or so, and these exposed some very important generalities. And yet, as I shall suggest, the principal problem is still with us: that is to say, the unpredictable nature of environmental forcing and the consequent variability of the survival rate of very young fish.

In 2001, Myers usefully summarised the generalisations that he had obtained by meta-analytic techniques from his database of almost 700 populations of fish and invertebrates from fisheries in all seas, although predominantly in mid- to high-latitude seas. His generalisations fell under the eight following headings.

- *Depensation* – of 129 populations examined, only 9 showed some evidence of a decline in reproductive success at low population sizes; the effect was weak and Myers suggested that it could not be separated from effects of environmental variability; nevertheless, depensation is still with us and will be discussed in Chapter 9.

- *Inter-cohort density-dependent mortality/survival* – data for some stocks suggest that recruitment is influenced by the relative abundance of the previous cohort, but that this is unlikely to result in strong biennial oscillations of abundance, as had been suggested previously.
- *Effect of spawner abundance on recruitment success* – because he found that 'almost without exception' the largest recruitment in each series occurs when spawner abundance is high and the smallest recruitment occurs when spawner abundance is low, Myers suggests that spawner abundance cannot be ignored in stock management.
- *Effect on recruitment of fecundity of spawners* – no significant effect on subsequent recruitment levels of variable fecundity of spawners could be detected, although this possibility had been suggested by others. This effect has, however, been observed in benthic invertebrates by Hughes and others: temporal and spatial variability in fecundity accounted for 72% of the variability of recruitment of *Acropora* spp. on the Barrier Reef.[52]
- *Effect of oceanic stability on recruitment* – because recruitment is more variable in populations of banks in the open ocean than of the same species in coastal regions, Myers suggested that environmental stability enhances stability in recruitment.
- *Edge-of-range effect on recruitment* – recruitment is more variable at or near the edges of a species range than in the centre. This is especially clear at the northern edge of the range of high-latitude species.
- *Effect of environment on recruitment* – generally, the consequences of environmental variability between-year and between-decade variability appear to be weak. Dome-shaped relationships between recruitment success and upwelling intensity in eastern boundary currents are recognised, but were judged not very useful for stock assessment modelling.
- *Range of maximal reproductive rate of stocks* – based on *c.* 700 stock–recruitment relationships, the ratio between maximum and minimum rates of reproduction appears to be have a very limited range, falling always between values of 1 and 7.

This is very useful information, but there is a sense in which it conceals as much as it reveals, and I recommend the reader to scan carefully the 272 stock–recruitment relationships that illustrate Myers' 1995 technical report obtainable from DFO, Canada; in this publication, they are presented both as time-series plots and as stock–recruitment regressions with fitted recruitment functions. Among these, if your eyes see them the same way as mine, you will detect several repetitive patterns that do not accommodate themselves readily to the above generalisations, even though many plots do show the expected positive relationship between recruitment and spawner population. However, imposed on this, I identify four patterns that do not all conform to Myers' generalisations and which appear to characterise the response of recruitment to other factors.

- *Very large annual recruitments may occur at low, or relatively low, population sizes* – despite Myers' first generalisation, a scan through his time-series plots reveals many cases where one or several consecutive years of high recruitment levels occur when spawner stock is at a relatively low level; this occurred in 1984 in the NAFO region 1 stock of *Gadus morhua*, when this was at an all-time low level after a 30-year steady decline. The same species, in the ICES VIa area, produced a very large 1986 year-class from a relatively very low parent population. Icelandic haddock *Melanogrammus aeglefinnus* produced consistently strong recruitment in the years 1970–78 from a stock smaller than in previous or subsequent years. A scan through the 272 time-series plots will readily reveal many other similar cases to you.
- *Unusually large annual recruitment may initiate population growth in subsequent years* – it is easy to locate this pattern in the Myers data series; however, it has to be said that this is a less convincing pattern than others discussed here, because many significant positive deviations in recruitment are not followed by growth in parent stock. However, this pattern is already well understood in respect of some classical events, such as the South African sardine resurgence in the mid 1950s. The same pattern was discussed by Hennemuth and others on the basis of data for 18 populations of 11 species of demersal and pelagic fish, of which 6 showed a serial correlation with a 1- or 2-year lag that was significant at the $\alpha = 0.05$ level. As these authors put it: 'significant deviations in catch closely followed significant deviations in recruitment'. It is also relevant to note that the resurgent Peruvian anchovy stock grew from a negligible biomass in 1984 to about 5 million tons in only two years and continued from that population size in another two-year period after 1995 to more than 12 million

tons: growth that is otherwise only dreamt about on the trading floors of our big banks!

- *Except at very low stock size, no relationship exists between recruitment and spawner stock size* – this pattern is almost characteristic of herring and other clupeid populations, but also occurs in some other groups. Here, the plots reveal no relationship except below population sizes that are less than 5–10% of the maximum population observed: above that minimal population size the relationship is flat, so that when data are available for very low-population densities, the whole takes an inverted hockey-stick trajectory. Among 45 herring and sardine stocks in Myers' 1995 data I find that 32 match this pattern (although many lack any data at very low stock size), while only 13 show a positive, quasi-linear relationship across the entire range of stock sizes observed. Similar patterns are, of course, also observed in other groups: for example, cod in NAFO areas 4TVn, 4VsS, 4x and at Faeroe; haddock in the North Sea, at Iceland and in ICES area Vic; whiting in the North Sea; South African hake and chub mackerel off California.
- *Recruitment levels and population size closely follow the same trajectory over a long period of years* – this pattern occurs infrequently but is very striking in a small number of cases, for example in the trajectories of cod in NAFO area 2J3KL and herring in the southern Baltic Sea. In these two cases there is an almost exact correspondence between progressive downward trends of stock and recruitment. The cases of the Peruvian anchovy and the California sardine are somewhat different, because stock size and recruitment are closely matched, both during stock size increase and decrease.

It would be vain to attempt any generalisation of these observations and conjectures except to remark once again – as so many others have done – that the environmental forcing of recruitment variability through the consequences of instability and heterogeneity of the larval environment must be taken seriously. During the evolution of reproductive strategies of benthic invertebrates and teleosts, one might say that it had to be taken very seriously indeed by them in order to achieve their present levels of occupation of the marine environment.

To support that statement, I return briefly to an analysis of the ability of benthic invertebrates to evolve reproductive strategies in response to extreme conditions, strategies that are apparently denied to teleosts, but have enabled invertebrates to maintain relatively higher taxonomic diversity in high latitudes than teleosts – this despite the obviously greater difficulty of successful recruitment in polar environments with very brief productive seasons.[53]

The basic observations to be explained are that while demersal fish faunas become progressively attenuated in species richness towards polar latitudes, this does not occur in the invertebrate benthic communities of the sandy and muddy substrates of continental shelves. That fish faunas are attenuated polewards in comparable habitats will surprise nobody, but that communities of benthic invertebrates have similar composition and diversity (again in comparable habitats) at equatorial and polar latitudes is not so well understood. Nevertheless, this pattern was described many years ago by Thorson, who found that while the total biomass of benthos increases poleward, the communities of invertebrates off eastern Greenland were as diverse as those in the English Channel. Others then confirmed his observations, showing that comparable communities in the Gulf of Thailand, in Costa Rican estuaries and off Antarctica were rather similar in diversity to those off NW Europe. My own surveys of benthic communities on the Sierra Leone shelf turned up equivalent numbers of species of (for instance) molluscs or annelids in each community as are found in parallel communities in the North Sea.

The key to this surprising situation was provided by Thorson, who observed that although many genera have global distribution (e.g. *Macoma*, *Venus*, *Mactra* and so on), they modify their reproductive method in high latitudes, substituting the release of planktonic larvae by brooding the larvae until metamorphosed and ready to enter the soft substrates directly from the parent brood pouch. Attributing this as a response to the unreliability of plankton production – and hence of larval food – at high latitudes, Thorson further noted that at mid-latitudes, annual recruitment variability is lower in molluscs which brood their larvae than in those which release eggs into the plankton: this result should not surprise you if you remember that a cod egg in the Labrador Current requires almost a month to hatch (27 days at 2° C), while in tropical seas hatching occurs overnight. I think it goes almost without saying that, had teleosts been able to follow the same evolutionary track as the far more ancient molluscan genera, the latitudinal gradient of teleost diversity that we observe in today's ocean would have been very different.

* * * * * * * * *

The range of variability inherent in fish populations in a pristine, unfished sea that has been discussed in this chapter carries a chilling message for the proponents of the sustainability of fisheries and for those who might still advocate quota management by single-species population models, with no input concerning the consequences of environmental variability. It is very difficult to avoid the simple inference that populations of fish are not, by and large, a reliable resource base for a modern industry, as will be discussed in later chapters. It is quite clear that natural fluctuations in population abundance and location of fish populations are sufficient to ensure that sustainability – in the sense of continuity of catch levels – cannot be achieved.[54]

ENDNOTES

1. Jennings, S., M.J. Kaiser and J.D. Reynolds. (2001) *Marine Fisheries Ecology*. Oxford: Blackwell Science.
2. In 'Highlights of special FAO studies' concerning *FAO Special Publication* **410** of 2001 (see below).
3. Brooks, N. (2004) Beyond collapse: the role of climate desiccation in the emergence of complex societies in the middle Holocene. Report of First Joint Meeting of 'ICSU Dark Nature and IGCP 490' panels, pp. 26–30.
4. See Sharp, G.D. (2005) Future climate change and fisheries: a comparative analysis. *FAO Fish. Techn. Paper* **452**, 1–74.
5. Cushing, D.H. and R.R. Dickson. (1976) The biological response in the sea to climatic changes. *Adv. Mar. Biol.* **14**, 1–122.
6. See, for example, Lamb, H.H. (1995) *Climate History and the Modern World*, 2nd edn. London and New York: Routledge; I note (as one obituarist pointed out in 1997) that 'now the world is acutely aware of global climate change, Lamb ... maintained a guarded attitude to the importance of greenhouse gas warming ... he felt that there was too much reluctance to consider the full range of other, natural, causes of change. Right to the end of his life, he was promoting his different view'.
7. To which a good entry point would be Sharp, G.D. and D.R. McLain (1993) Fisheries, El Nino-Southern Oscllation and upper ocean temperature records: an eastern Pacific example. *Oceanography* **6**, 13–22.
8. Visbeck, M.H. (2001) The North Atlantic Oscillation: past, present and future. *Proc. Natl Acad. Sci.* **98**, 12876–7.
9. Girs, A.A. (1971) Multiyear oscillations of atmospheric circulation and long-term meteorological forecasts. *Gidrometeroizdat*, 480.
10. Klyashtorin, L.B. and A.A. Lyubshin. (2008) *Cyclic Climate Changes and Fish Poductivity*. Moscow: VNIRO Publications (in English).
11. Trenburth, K.F. and J.W. Hurrell. (1994) Decadal atmosphere–ocean variations in the Pacific. *Clim. Dynam.* **9**, 303–19.
12. Mantua, N.J., *et al.* (1997) A Pacific interdecadal climate oscillation with impacts on salmon production. *Bull. Am. Met. Soc.* **78**, 1069–79.
13. Ocean colour data and images may be obtained from http://oceancolor.gsfc.nasa.gov/ and matched TOPEX-POSEIDON/SeaWiFS images from http://argo.colorado.edu/~realtime/welcome/.
14. Francis, R.C.S., *et al.* (1998) Variability of the marine ecosystems of the North Pacific. *Fish Oceanogr.* **7**, 28–45.
15. Scheffer, M. and S.R. Carpenter. (2003) Catastrophic regime shifts in ecosystems: linking theory with observation. *Trends Ecol. Evol.* **18**, 248–56.

16. Karl, D.M., *et al.* (2001) Long-term changes in plankton community structure and productivity in the north Pacific Subtropical Gyre; the Domain shift hypothesis. *Deep-Sea Res. II* **48**, 1449–70.
17. Steele, J.H. (2004) Regime shifts in the ocean: reconciling theory with observation. *Progr. Oceanogr.* **60**, 135–41.
18. See, for example, J.M. Grove. (1988) *The Little Ice Age*. London: Methuen.
19. Named for Fred Russell, later director of the Marine Biological Association laboratory in Plymouth: a fine scientist.
20. Drinkwater, K. (2006) The regime shift of the 1920s and 1930s in the North Atlantic. *Progr. Oceanog.* **28**, 134–51.
21. Ken Drinkwater, pers. comm.
22. CPR – Continuous Plankton Recorder, now operated from Plymouth Marine Laboratory.
23. Cushing, D.H. (1984) The gadoid outburst in the North Sea. *J. Cons.* **41**, 159–66.
24. ICES. (1949) Contributions to Special Scientific Meetings 1948. Climatic changes in the Arctic in relation to plants and animals. *Rapports et Procès-Verbaux des Réunions du Conseil Permanent International pour l'Exploration de la Mer* **125**, 5–51.
25. See Science section in *Newsweek*, dated 28 April 1975: 'The Cooling World'.
26. Corten, A. (1999) A proposed mechanism for the Bohuslan herring periods. *ICES J. Sci.* **56**, 207–20.
27. From H. Lamb (see note 6 above).
28. Hare, S.R. and N.J. Mantua. (2000) Empirical evidence for the North Pacific regime shifts in 1977 and 1989. *Progr. Oceanogr.* **47**, 103–45.
29. Francis, R.C., *et al.* (1998) Effects of interdecadal climate variability on the oceanic ecosystems of the North Pacific. *Fish. Oceanogr.* **7**, 1–21.
30. Hare, S.R. and N.J. Mantua. (2001) *Report of twentieth NE Pink Salmon and Chum Workshop*, Seattle.
31. see Chapter 1.
32. Cury, P. and L. Shannon. (2004) Regime shifts in upwelling systems: observed changes and possible mechanisms in the northern and southern Benguela. *Progr. Oceanogr.* **60**, 223–43.
33. Baumgartner, T.R., *et al.* (1992) Reconstruction of the history of Pacific sardine and northern anchovy populations over the past two millennia from sediments of the Santa Barbara Basin. *Cal. Calif. Coop. Oceanic Fish. Invest. Rep.* **33**, 24–40.
34. Herrick, S.F., *et al.* (2007) Management application of an empirical model of sardine–climate shifts. *Mar. Policy* **31**, 71–80.
35. See, for instance, Sharp, G.D. and J. Csirke. (1983) Proceedings of the Expert Consultation on changes in the abundance of and species composition of neritic fish resources. *FAO Fish. Rep.* **291**, 1294 pp.
36. Rykaczewski, R.R. and D.M. Checkley. (2008) Influence of ocean winds on the pelagic ecosystem in upwelling regions. *Proc. Natl Acad. Sci.* **105**, 1965–70.
37. van der Lingen, C.D., *et al.* (2008) Trophic dynamics of small pelagic fish. In *Climate Change and Small Pelagic Fish*, eds. E. Checkley *et al.* Cambridge: Cambridge University Press.
38. Klyashtorin, L.B. and A.A. Lyubshin. (2005) *Cyclic Climate Changes and Fish Productivity*. Moscow: VNIRO Publishing.
39. Klyashtorin, L.B. (2001) Climate change and long-term fluctuation of commercial catches: the possibility of forecasting. *FAO Fisheries Technical Paper* **410**.
40. Lamb, H.H. (1972) *Climate: Past, Present and Future*. London: Methuen.

41. Parenthetically, this demonstrates the error in treating the collapses of all fished species as equivalent events, as was done by the authors of the recent *Science* paper that predicted loss of all fisheries by the year 2048.
42. Prior to this event, the stock was under exceedingly heavy fishing pressure and many still ascribe the population crash at this time entirely to this cause.
43. Salvatecci, R., *et al.* (2005) Decadal- to centennial-scale variations in anchovy biomass in the last 250 years inferred from scales preserved in laminated sediments off the coast of Pisco, Peru. AGU Fall Meeting, 2005, abstract #PP51D-0628.
44. Mendelssohn, R. (1989) Re-analysis of recruitment estimates of the Peruvian anchoveta in relationship to other population parameters and the surrounding environment. In *The Peruvian Upwelling Ecosystem: Three Decades of Change* (ICLARM Studies and Reviews No. 3), eds. D. Pauly, *et al.*, pp. 294–306.
45. Hsieh, C.-H., *et al.* (2006) Fishing elevates variability in the abundance of exploited species. *Nature* **443**, 859–62.
46. Anderson, C. N. K., *et al.* (2008) Why fishing magnifies fluctuations in fish abundance. *Nature* **452**, 835–9.
47. Longhurst, A. R. and the late W. S. Wooster. (1990) Abundance of oil sardine and upwelling on the southwest coast of India. *Can. J. Fish Aquat. Sci.* **47**, 2407–19.
48. See Chapter 2.
49. These processes were the subject of a masterly review by the late David Cushing, in his *Population Production and Regulation in the Sea* (Cambridge University Press, 1995, pp. 354).
50. Francis, R. C. and K. M. Bailey. (1983) Factors affecting recruitment of selected gadoids in the NE Pacific and East Bering Sea. In *From Year to Year*, Ed. W. S. Wooster. Seattle, WA: Washington Sea Grant.
51. Wing, S. R., L. W. Botsford and J. F. Quinn. (1998) The impact of coastal circulation on the spatial distribution of invertebrate recruitment, with implications for management. In *Proceedings of the North Pacific Symposium on Invertebrate Stock Assessment and Management: Canadian Special Publication of Fisheries and Aquatic Sciences*, Eds. G. S. Jamieson and A. Campbell, **125**, 285–94.
52. Hughes, T. P., *et al.* (2000) Supply-side ecology works both ways: the links between benthic adults, fecundity and larval recruits – statistical data included. *Ecology* **83**, 2241–9.
53. Longhurst, A. (1999) Does the benthic paradox tell us something about surplus production models. *Fish. Res.* **41**, 111–7.
54. See, for example, Steele, J. and P. Hoagland. (2003) Are fisheries sustainable? *Fish. Res.* **64**, 1–3.

6

Has sustainability in fishing ever been achieved?

'Among the ancients, the accipenser was esteemed the most noble fish of all. At the present day, however, it is held in no esteem, which I am the more surprised at, it being so very rarely found.'

Pliny, Natural History 9, 27

Before discussing what we know, or think we know, about the present state of fish populations globally, it may be useful to enquire whether fishing was sustainable at the subsistence level of early coastal societies. It is generally reported that fish populations are everywhere failing to maintain themselves, but is this a novel situation? To what extent did earlier societies, having much narrower horizons than ours, perceive that they had exhausted or depleted the fish populations to which their techniques gave them access? What is the response to subsistence fishing of a previously unfished population in a pristine ocean?

Answers to such questions may give us a better understanding of our present situation. This enquiry will lead us back in time to some consideration of the earliest fisheries and the consequences of the progressive evolution of fishing methods in local fisheries, and the spread of these methods to other seas. What follows is heavily biased towards the evolution of fisheries in the North Atlantic, but I believe that this region will serve very well as a general model for what happened in other seas. Because the response of fish populations to periodic environmental change was discussed in the previous chapter, this aspect of the problems of sustainable fishing will not be discussed again here, but it is the background against which all the other processes take place.

ANCIENT FISHING AND THE DESTRUCTION OF PRISTINE POPULATIONS

From the beginning, we have taken part of our diet from the sea. Neanderthals, living in caverns on the limestone cliffs of Gibraltar about 40 000 years ago, used open hearths, enjoyed clam bakes, fish and seals, the bones of which bear the marks of butchery with simple stone flakes.[1] The bones of small porpoises were also recovered from these caverns, in midden deposits that were formed in periods when the shore was several kilometres distant.

We do not know if Neanderthal food-gathering had any significant effect on the coastal habitat, but we do know that pristine littoral ecosystems are very sensitive to human occupation. The arrival of the first people on some Caribbean islands almost 2000 years ago had consequences that strangely recall some of the present currents of environmental concern – fishing down the food web, growth overfishing and resource switching.[2] The first immigrants from South America to the unpopulated Leeward and Virgin Islands brought with them the tradition and techniques of horticulture, animal husbandry and fishing, and the osteo-archaelogical record suggests a heavy exploitation of reef fishes adjacent to settlements between 1850 and 500 BP. The estimated biomass of reef fish populations declined from early to late deposits, as did the mean size of obligate reef species, although lutjanids and serranids were less affected than most. The trophic level of reef fishes was progressively reduced, and small coastal clupeids replaced large predatory fish in the diet as these latter became scarce. The remains of coastal pelagic species in the middens (e.g. *Caranx*, *Euthynnus*) suggest that these were little modified by fishing. This example demonstrates that even early subsistence fishing was capable of so significantly modifying its resource base that a switch to alternative resources became essential.

However, the Caribbean reef fisheries should not be taken as a general model because not all coastal habitats are as sensitive to disturbance as are coral reefs: extensive data from middens around Passamaquoddy Bay do not demonstrate any significant changes in the diet of the Native American hunter-gatherers from 2200 to 500 BP. There was no significant change in the species of fish (cod, haddock, pollack, herring, sculpin) or invertebrates (clams, barnacles, mussels, whelks) taken by them, nor in the average individual size of the individuals of these species, prior to the settlement of the area by Europeans. The sedentary fish that inhabit the strongly structured habitat of coral reefs, as in the Caribbean, are more sensitive to fishing pressure than the more mobile species of temperate coasts. In the Quoddy region, significant changes in the coastal fish populations cannot be positively identified until the nineteenth century.[3]

One of the problems in understanding early fishing impacts is that we cannot be sure how the pristine coastal ecosystems were structured, although it appears to be agreed that there may have been as much as an order-of-magnitude difference between pristine fish biomass and what was present in the mid-twentieth century in the North Atlantic and, most likely, also in other regions. The question of how pristine ecosystems were structured and able to support such a large biomass at higher trophic levels has not been answered definitively, but the negative relationship between metabolic rate and body size in fish may be part of the answer. Unit biomass at low trophic levels can theoretically support a larger biomass of old individuals at high trophic levels than would be the case if the latter were all younger, faster-growing fish.

Steele and Schumacher were among the first to suggest that to obtain balance, we should assume that energy flow across trophic levels in modern ecosystems must differ from their pristine state because they exhibit relatively greater flow to medusae, small invertebrates or small planktivorous fish.[4] It is generally supposed that in pristine ecosystems there must have been a lower level of flux to populations of small pelagic fish than we observe in contemporary ecosystems.

For the NW Atlantic, we do have some documentary evidence for the composition of the pristine marine fauna – pristine, that is, except for the consequences of the fishing activities of the native Americans prior to the arrival of Europeans. We are all familiar with the accounts of the early European fishermen, amazed at the abundance of cod of great size, and with John Cabot's 1497 report that these were so abundant that 'they could be caught with baskets' let down and weighted with a stone. But what is not so well known is that similar reports were made for a wide variety of marine organisms. We have enthusiastic accounts from the sixteenth century onwards of the extraordinary abundance of oysters, squid, salmon, sturgeon, striped bass, mackerel, herring, bluefin tuna, herring, shad, smelt and capelin that could be found on the other side of the ocean. Sturgeon of several species 'in great plenty' and up to 3 m in length were reported in the seventeenth century and used extensively for food. Of turtles, now essentially extinct in the NW Atlantic, Cartier wrote in 1553 'In this bay, and around this island, there are inestimable numbers.'[5] Archaeological data from middens also confirm the great size of individual fish prior to industrial exploitation: around the North Atlantic from Greenland to the Barents Sea, the mean length of cod reconstructed from measurements of pre-maxillary bones was consistently in the range 90–100 cm during the period 1450–1800.[6] The largest fish were consistently up to 200 cm in the earliest period.

The consequences of interventions by humans in kelp forests and seagrass meadows have been reviewed recently in some detail by Jackson and others,[7] while Roberts has provided a chilling account of the wholesale destruction of shallow-water populations of large fish for sport, and of invertebrates for the table.[8] It is, of course, in this habitat that many marine species of vertebrates are already extinct: the great auk, Steller's sea cow, the Atlantic sea mink and grey whale, and the rest of them. The abundance and variety of large mammals and birds that inhabited the Atlantic littoral in the early Holocene would have amazed even mediæval seamen: their populations were almost all destroyed by hunting before the historical period.

In those coastal environments where the habitat is highly structured in the spatial sense, the removal – or a sufficient reduction in numbers – of a single species can dislocate the balance of the entire habitat so as to change it totally. There are several good examples of this having happened very early in the human occupation of a coastal region and also during the early period of artisanal exploitation of the habitat by fishing or other activities.

The best-known example is perhaps the destruction of the pristine kelp forests of the Gulf of Alaska and neighbouring regions of the North Pacific, by simple removal of one top predator. In the pristine state, three key kinds of organisms maintained the structure of this habitat: several species of kelp, several sea urchins and the sea otter. Subsidiary players were killer whales and, perhaps, Steller's sea cow. A dynamic balance was maintained between the growth of kelp and the consumption of algal fronds by herbivorous sea urchins, but only because otters restricted the growth of urchin populations. Sea otters, in turn, were prevented from increasing unduly by killer whale predation.

The problem faced by this coalition was that sea otters have very fine, warm pelts that are desirable in cold climates, like that of coastal and insular Alaska. Already by 2500 BC, archaeological evidence suggests that the aboriginal peoples of the region had significantly reduced the sea otter population; this is indicated by a progressive increase in the size of individual urchins at this time. However, it was the arrival of fur traders, looking for supplies for the eastern markets, in the late nineteenth century that destroyed the dynamic balance of the kelp forest habitat. So-called urchin barrens were created in the littoral zone as soon as the removal of sea otters allowed the urchin population to increase sufficiently to graze the individual plants of the kelp forest down to their stipes. This process was repeated during the twentieth century after some recovery of sea otters and regrowth of kelp, when the absence of large whales – decimated by whaling – caused killer whales to turn to sea otters in the littoral zone of the Aleutian Islands for food.[9]

Seagrass meadows formed the dominant habitat in the pristine state of many bays, lagoons and sheltered inshore regions of tropical continental shelves and were maintained by the grazing of a diverse community of large herbivorous mammals and reptiles: dugongs, manatees and turtles. These are among the organisms that were most vulnerable to the arrival of Europeans or the development of local societies: they are good to eat and easy to capture. Sirenians disappeared very rapidly, while turtles held on a little longer, but are now in danger almost everywhere except where very rigorously protected. Deprived of the effects of grazing, seagrass meadows are profoundly changed. The individual plants become elongated, causing the near-bottom water movement to be sufficiently slowed that suspended particulates sink to the sediments; the resulting accumulation of organic debris then causes the progressive die-off of the seagrass plants themselves.

There also exist accounts by the earliest European navigators further south on the Pacific coast of their astonishment at the abundance of marine life that they encountered, and particularly at the size of the individual organisms. I am struck by a report from one James Colnett, who commented in 1798 after a visit to the Gulf of California that he 'was ready to confess that I was deceived respecting the species of ... humpback whale (that) was so much larger ... that the most experienced seaman that I had on board believed them to be black whale' – by which he referred to the larger species of sperm whale.

Andrea Saenz-Arroyo and others[10] have described and, moreover, quantified this phenomenon to perfection for three generations of fishermen in the Gulf of California. This gulf is richly diverse and still retains significant abundance of large marine organisms (hammerhead sharks, manta rays, billfish and cetaceans) because intensive fishing is much more recent there than in other comparable regions, such as the Caribbean. Three generations of fishermen were interviewed and asked to name species that they considered to have been depleted by fishing; although almost all (84%) recognised that fishing had led to the loss or depletion of some species, the older group (55 and above) thought that 11 of 34 species listed had been depleted, the middle-aged group (31–54 years of age) listed 7 species, while the youngest (15–30 years of age) thought that only 2 had been depleted. Subjectively, the old men described the past abundance of large species of both fish and invertebrates and recognised that depletion had occurred during their working lives, but the middle-aged men showed less recognition of past abundance and the young men 'seemed unaware that such species had ever been common'. The evidence of these three generations of fishermen, the oldest of whom probably began to fish around 1955–1960, suggested that the accounts by American sports fishermen of the unimaginable numbers of large predatory fish in the Gulf of California during the 1930s were no fishermen's tales: only the very oldest men of those interviewed remembered those fish. A regression of the year in which they remembered catching their very largest fish and the size of that fish showed a decline from 85 kg in the 1950s to 65 kg at the end of the century; the largest day's catch of groupers that these fishermen remembered was 25 groupers in 1940 but only 1–4 fish in the 1990s. Dynamite fishing on the offshore banks during the 1950s assured that outcome: a 72-year-old remembered how the San Bruno seamount was crowded with Gulf grouper until one year *La Avecita* from Guaymas used dynamite, and killed thousands of fish in a single trip: his boat alone carried away 70 tons from that day's work. But, of course, all of this only concerns what happened in the twentieth century: the authors of this important study also pointed out that even the old men seemed unaware of the existence of earlier abundant fisheries, such as that for pearl oysters, that had flourished from the seventeenth to the early twentieth century and are now commercially extinct.

These remarks serve as an excellent example of the very great difficulty of knowing with any certainty anything about the fisheries of the past, and the populations on which they depend: Pauly famously described this as the 'shifting baseline syndrome', because each generation of fishermen will tend to regard the fish populations that he grew up with as a young man as being the norm. In parenthesis, I suggest that this syndrome applies to very much more than fishing: as we age, the social structure within which we grew up seems intangibly more real than the one in which we now find ourselves ...

THE EVOLUTION OF FISHERIES INTO THE LATE MEDIÆVAL PERIOD

Fishing techniques evolved over long periods and independently in several regions, of which the most important were the Far East and the Mediterranean. Although the problems associated with the capture of the various kinds of fish are universal, generic solutions appropriate to each fishery appeared independently in each region: fishing boat design, for instance, evolved quite differently in the Mediterranean and the Sea of Japan.

Fortunately, we are still able today to study examples of fishing methods that resemble those used millennia ago, the better to understand their effects. It is a very remarkable fact that during my lifetime I have been able to observe examples of all stages of this evolution, from papyrus canoes on Lake Chad to mega-seiners on the high seas, representing perhaps 4000 years of fishing history. So we can still study at first hand at least some examples of artisanal fishing methods, and in this way appreciate the level of craftsmanship and technique inherent to each that must closely resemble the skills deployed in the distant past elsewhere: we quickly learn, for example, that a canoe is not just a canoe.

Consider *ali lele*, the Fante canoe used extensively in the Ghana fisheries to handle the 500-m drift nets that are deployed in fishing for the shoals of *Sardinella aurita* that abound in this upwelling region; this canoe is, according to Brown,[11] 'a fine, buoyant sea-boat, with an easily-driven hull and a high speed ratio for her length ... rides surf well ... easily righted and baled out ... this eulogy makes her a paragon among boats, and so I believe she is'. She was the product of a long evolution dating back at least to the mid-nineteenth century, when the Fantes migrated to the coast of Ghana, and before the use of sails on canoes in tropical western Africa. However, *ali lele* would be useless for trolling for large pelagic predators, such as *Sphyraena* or *Caranx*, or for cast-netting for *Ethmalosa*. Trolling is done by a single fisherman in a *kru* dugout, which is as light as a kayak, is very handy in big seas and is easily driven with a single-bladed paddle having a sharp tip to give long reach. To use *odunka*, the 6-m radius (yes, radius) cast net for *Ethmalosa*, requires a narrow, more manœuverable canoe than *ali*, carrying one cast-netter standing in the bow and a crew of 4–5 paddlers astern.

This is but a sampler of the diversity of artisanal fishing techniques used on the African coast during the mid-twentieth century, but is probably typical of fisheries everywhere at this stage of their evolution, and from the very earliest days. One did not catch Mediterranean bluefin tuna in the fourth or fifth centuries BC, for sale in organised fish markets, without gear evolved for that particular fishery. Egyptian and classical wall paintings and other images make very clear that cast nets, drifts nets, boat-seines and beach-seines were among the diversified gears in use in the Middle East 2000 years ago. They would be quite familiar to fishermen of the same region today, at least in the essentials of their design and functioning.

The texts of Oppian (his *Halieutica*), and of Aristotle and Pliny make it clear that, already in antiquity, there was wide knowledge of the natural history of fishes, how to catch them and how to prepare them for the table. *Garum*, a condiment made from the salted offal of tuna and other fish, was widely known in the Roman world and differed according to provenance; that from Pompeii was highly appreciated, and was sold in special one-handled jars. Ridiculous prices were paid by the rich for large fish of high quality to embellish their banquets. It is recorded that Apicius 'an exceedingly rich voluptuary', once sailed all the way to Libya in search of particularly large prawns; not finding any to his satisfaction among those that were brought out to his ship, he returned to Campania without going ashore.[12] So, even in antiquity, fishing had become part of a complex demand-driven industry, appropriately capitalised, whose resources around the Mediterranean were by no means pristine: the quotation from Pliny at the head of this chapter seems to suggest that some species were already so depleted as to have disappeared from the markets.

Nearer our own times, the progressive evolution of fishing methods in the North Atlantic from small planked boats, through to the elegant and seaworthy schooners used with hand-lining dories on the Banks in the early twentieth century, and finally to the steam- and diesel-powered successors of the sailing vessels, is so well known as to require no repetition here.[13] What is difficult for us to grasp today is the confidence with which sixteenth-century fisherfolk from coastal ports in Devon or the Basque country set out westwards each year in what would appear (to us) to be very small, square-rigged ships with the intention of establishing temporary fishing camps, progressively on the Irish, Icelandic and North American coastlines; nor does John Cabot's little *Matthew*, of only 50 tons burden and a crew of only 18, seem an adequate ship in which to set off westwards from Ireland in search of Asia and the Spice Islands ...

Although we are accustomed to reading accounts of the amazement of the first European sailors at the abundance of fish and marine mammals that they found when they reached the American shores for the first time, we seldom ask why they should have been so surprised by what they encountered, compared with the fish that they knew in their home waters? Was the difference really so striking? After all, this was only the fifteenth century, and long before the fishery developments in European seas that cause us so much concern 500 years later.[14]

Cabot's astonishment at what he found on the other side of the ocean (if not feigned, as a come-on for future sponsors) is readily explained: by his time, the European coastal habitat was far from pristine and its fishery resources significantly depleted. Already by the 3000 BP, the Bronze Age farmers of what is now the Dutch coast had begun to modify the coastal wetlands by clearing the elm–ash tidal forests. Thus, by the Roman period, the coastal landscape of Europe was not pristine, and the destruction of forests continued apace during the subsequent mediæval period. This had the effect here, as elsewhere, of destabilising regional hydraulic regimes while siltation, caused by episodic expansions of arable cultures, progressively blocked the lower river channels and modified the coastal habitat. During the thirteenth century, for instance, the Vistula(Baltic) built out a delta to occlude the bay that had previously received its discharge. As early as the fifteenth century, eutrophication of the drainage from developing urban centres can be detected in archaeological data and, in 1415, a royal ordinance sought to remedy the 'infectious and corrupted' effluents from Paris into the Seine.[15] Chesapeake Bay became highly eutrophic during the eighteenth century because of forest clearance to create arable land in the drainage basin of the bay.[16] The importance of this observation is that it demonstrates that the collapse of the oyster population, and the subsequent loss of biological filtering capacity, cannot be attributed simply to mechanical harvesting during the nineteenth century, as has been done.

The evidence from historical records suggests that the early navigators had good reason to be surprised at what they found on the other side of the Atlantic; before the end of the late mediæval period, fishing in European seas and estuaries had passed through all three characteristic stages of development – from subsistence to artisanal, and thence to the industrial fishing for export to distant markets. The populations available to European fishermen in their own waters had already been much modified.

The development of fishing to support a food-supply industry was part of what has been termed the Commercial Revolution of the mediæval period.[17] Prior to this, the western Christian practice of substituting fish for flesh on the table for about 30% of the year induced the development of artisanal fishing, to replace the coastal and riverine subsistence fishing that went back to the earliest human occupation of the European coasts. Until the end of the first millenium, freshwater fish supplied the bulk of the needs of Europeans, but as these could no longer supply the growing population, fishing turned progressively towards the coastal and offshore species. In England, the clearest changes occurred after AD 600, with the most rapid development of sea fishing around AD 1000. Quite early on, near-coastal spawning aggregations of herring (*Clupea harengus*) were fished successfully, and other species taken included cod (*Gadus morhua*) offshore, while oysters, clams and crustaceans were all heavily fished in shallow water. The intensification in sea fishing around the turn of the millennium occurred during the Mediæval Warm Period and so at a time when survival at sea in small craft may have been relatively easy but when, paradoxically, we would expect that populations of cod and herring would have been relatively small in the North Sea.[18]

Both salmon (*Salmo salar*) and sturgeon (*Accipenser sturio*) were readily accessible in the estuarine regions, and very desirable for the tables of important men; unfortunately, these fish are highly vulnerable to river modification, both by diking to regain arable land, and by damming to run watermills for grinding cereal crops, so there are many accounts from the twelfth to fourteenth centuries of concern over falling abundances associated with changed river regimes and pollution: reduction in recorded sizes of fish taken during this period suggests over-exploitation. Because these fish were so valuable, traders began to look farther afield for supplies, and international trade in salted and barrelled salmon developed to supply German towns, while Paris began to receive salmon from Scotland and Ireland during the fifteenth century.

Already in England in the thirteenth century, fishing regulations were promulgated for the protection of resources, sometimes specifically with conservation and sustainability in mind: so minimum size limits, seasonal closures and gear restrictions are nothing new. During the mediæval period, resource privatisation by the English Crown progressively constrained subsistence fishing in favour of artisanal, and placed taxes on the latter. Complex property rights developed in shallow water so that, under English law, coastal fisheries were either exclusive or common. In the former case, the right of fishing belonged to the owner of the soil while, in the latter case, the common fishery was expressed as the right of the public to fish anywhere in tidal waters *not appropriated as exclusive fisheries*, a right that extended back to times immemorial. Royal prerogative was exercised over sturgeon which, when caught, became the property of the Crown. Similarly complex legal arrangements were undoubtedly in place in other nations; indeed, the ancient *droit de pêche* in France has become today a source of friction between fishermen and the European fishery regulators.

The always-increasing demand for fish in Europe was responsible for major investments in both coastal and offshore herring fisheries. As early as the eleventh century, herring were being taken by coastal artisanal fishermen to supply local markets while, during the thirteenth century, major investments were made in the offshore herring fishery off Norway, East Anglia, in the Baltic and off Denmark. By the fourteenth century, it is recorded that some hundreds of millions of herring yearly entered the inland trade, barrelled. These were partly supplied by the Dutch fishery, using three-masted, flat-bottomed *buss* drifters with fleets of 50 m hemp drift nets, which were able to process herrings at sea in brine; these drifters were thus the first factory ships in any fishery. It is recorded that fleets of 400–500 sailed to the Dogger and to the Shetland herring spawning grounds annually, much to the chagrin of the English government. The Dutch became dominant in the herring business after the demise of the Hanseatic League and by 1525 were supplying herring (salted, in barrels) to the fish markets of Europe, as far afield as Rome. Later, this fishery became subject to serious interference during European wars, starting with the 1702 War of Spanish Succession: despite naval escorts, hundreds of drifters were sunk in several naval interventions.

All this is very interesting, but what concerns us here is the fact that the evolution of these fisheries did not occur without serious damage to the herring stocks. The Baltic herring fishery had to shift to sprat (*Sparttus sprattus*) and to a dwarf, coastal herring stock after the main population collapsed in the late thirteenth century. Once again, this herring stock has never fully recovered, especially with regard to mean adult size, although some of this damage may have been caused by a very heavy period of land clearance and consequent run-off from watersheds.

However, such could not have been the cause of the collapses of the southern North Sea herring populations after 1360 (at the end of a Bohuslan period, as discussed in Chapter 5) and of those off western Sweden in the early fifteenth century. Once again, we have some difficulty in attributing this solely to the effects of fishing, because this was the time of the climatic transition from the Mediæval Warm Period to the Little Ice Age, a period of great climatic instability and rapid cooling during the summer season when, it has been pointed out, year-class strength in herring is determined.

The consequent shifts of populations of marine organisms that occurred at this climatic transition confuse our interpretation of the evidence; it is clear that major natural meridional shifts in salmon abundance occurred, while a trans-Atlantic shift of the North American sturgeon to Europe in response to differential water temperatures appears to have been associated with a revival of catches of sturgeon on the North Sea coast during the fifteenth century. This period also saw the extinction of the European grey whale (*Escherichtius gibbosus*) and the near-extinction of grey seals (*Halichoerus grypus*); the latter species re-colonised the Dutch coast only in the twentieth century, from a remnant population on the northeast English coast.

There is thus abundant evidence to suggest why John Cabot, and subsequent navigators, should have been astonished at what they found on the American coasts. It is also clear that industrial fisheries did not have to wait for steel hulls, for motor propulsion by steam or diesel, or for hydraulic winches in order to inflict unsustainable mortality on target fish populations. Fishing mortality caused by simple sail-powered fishing smacks was capable of causing some populations to crash: that this occurred almost 500 years before the introduction of powered fishing vessels should cause us to look at what is presently happening at sea in a new light.

EARLY FISHERIES OF THE NORTHWEST ATLANTIC

Here, fishing developed very differently from that for the North Sea herring, although we take the same message from its history: pre-mechanised fisheries may cause population collapses. The indigenous population in the Late Mediæval Period was small along the coasts of eastern Canada and New England, and their subsistence fishing effort was negligible; it was in this vacuum that an industrial fishery was developed to supply the markets of Europe with cod, either salted in very strong brine, or air-dried as stockfish – a product whose peculiar odour can still be recognised today in rural towns in southwest France.

The cod fisheries of Newfoundland appear to have been the well-preserved secret of Basque fishermen for some centuries prior to Cabot's voyage: it is likely that ships from the Basque country were routinely sailing to the Northwest Atlantic each summer by the fifteenth century: this is a remarkable indication of how hermetic

European society was during the mediæval period. Nevertheless, some people in the fish trade knew where the source of Basque stockfish lay, so it is very likely that Columbus and perhaps Cabot, before setting out, were well aware that land existed on the other side of an ocean that could be crossed by the ships of their day.

Within just a few years of Cabot's triumphant return to Bristol in 1497, his advertisement of the rich cod resources that he had found encouraged the first landings in their home ports of transatlantic cod by Portuguese (1501) and then French (1504) boats. Within 50 years, almost 500 ships were operating each summer on the coasts of Newfoundland; the French preferred the offshore banks fishery, while the others concentrated on the inshore populations, where stockfish could be wind-dried on flakes ashore.

The English did not enter the fishery off Newfoundland in significant numbers until later than other nations and, in fact, the English fishery off western Iceland intensified in the decades following Cabot's landfall, so that by 1528 almost 150 ships from Grimsby and Lowestoft were working there.[19] This fishery began to decline 50 years later, and collapsed entirely in the seventeenth century, largely because of longstanding jurisdictional problems with the Hanseatic League, and the consequences of strong ice years: both had much to do with the progressive removal of English fleets from Iceland to the Grand Banks. The English migratory fishery off Newfoundland was subsequently prosecuted largely from English West Country ports, such as Brixham, Falmouth and Plymouth, some of which sent out as many as 80 ships each summer. The development of the fishery was much influenced by foreign affairs that had nothing to do with fishing.[20] The European wars in which England was involved after 1620 meant that seamen were pressed into the Navy, and new wartime taxes on salt upset the economics of the fishery. In the same century, there was much harassment of the fishing fleet by pirates, renegade seamen of all nations, sailing out of North African ports: these were the so-called 'Sallee rovers'.

For the shore fishery, the barren coast of Newfoundland provided excellent conditions for wind-drying lightly salted cod on wooden platforms (or flakes); in many places, split cod could be dried directly on rocky slopes. The success of the Newfoundland shore fisheries was initially due to the great abundance of summer-spawning cod along the shoreline, so that hand-lining from small boats, or from the fishing vessel itself, was an efficient method of taking fish. Permanent fishing settlements slowly developed, according to need, on that coast even as the Gulf of St. Lawrence was progressively opened to settlement and to fishing in the mid-sixteenth century. Technically, the fishery was very conservative and netted enclosures, or fish-traps, were developed only much later, in the 1880s.

The pristine resources that so surprised the early navigators were quickly reduced in abundance. As early as 1683, concern was expressed in London not only for shore-space for fish processing on the Newfoundland coast, but because of an apparent over-capacity in the fishery itself, that risked (it was thought) making 'so great a destruction one year as to prejudice the next year's fishery'. By 1715, the inshore fishery had become so bad that a Devon skipper, fishing along the South Shore, sent a shallop to investigate the offshore banks, with very satisfactory results; previously, the Grand Banks had been largely neglected by the English, but this changed rapidly when those resources were needed.

This test voyage led to the rapid development of the offshore fishery in which salt-cod was prepared at sea aboard bankers, small ships of 80–100 tons that arrived crammed with crew, provisions for the summer and that critical ingredient, salt. Larger vessels, the sack-ships of around 200 tons, made several voyages a year, transporting barrels of salted cod home to Europe and also engaging in other trans-Atlantic trade; this practice makes it very difficult now to extrapolate estimates of the production of the fishery from the number and size of the bankers themselves.

Off Cape Breton in Nova Scotia, the French shore fishery (that had moved there from Placentia in 1713) experienced very significantly reduced catches in the early 1740s, with catch rates falling by a factor of about six. The Newfoundland inshore fishery also continued to collapse, especially on the southern coast and in Conception Bay, catch rates being reduced by about 60% by 1750, although >2000 boats were still employed; it was only after a period of reduced fishing effort during the war years (1775–83) that fishing briefly improved; it is recorded that up to 50,000 tons were taken annually in the 1780s. But by 1790, inshore catch rates were again falling off Newfoundland, this time by about almost half, so that trips to the Labrador coast were made, and a regular schooner fishery was soon established there. Much later, in the 1860s, during the very cold episode at the end of the Little Ice Age that is recorded in tree ring data,[21] the entire Newfoundland and Labrador fishery for both cod and herring collapsed, and caused great social distress in St. John's and the outports. This was a natural phenomenon, because Lear describes reports of unusually clear offshore water ('you could see objects on the bottom in 20 fathoms of water'), while 'slime' reported on nets and cod-traps perhaps indicates that a massive population explosion of salps or medusae had developed. Nevertheless, the calculated total production of cod from the entire NW Atlantic region increased continuously throughout the nineteenth century, from around 200,000 tons in 1800 to around 350,000 tons in the 1930s[22]; in part, this must have been the result of progressive shifting from exhausted inshore stocks to the offshore population.

The coast of New England was settled after the discovery of Cape Cod in 1602 by a ship from Falmouth, England: here, as is well known, they 'took great store of codfish' and so changed the name of 'Shoal Hope' given by them when it was first sighted. Already in the following year, Bristol merchants financed the first fishing vessel to try the grounds off New England, a voyage that proved a great success: commerce moved fast, even in the seventeenth century! The developing colonies took enthusiastically to fishing and within 20 years, as many as 50 vessels from Gloucester alone were hand-lining cod, haddock and pollock for the export trade to England and the West Indies. New England fishermen spread widely and fast, so that by the 1780s they had more than 50 ships fishing annually on the Grand Banks.

Georges Bank, off New England, was not fished seriously until 1821, and the halibut fishery there was not established until a decade later. By 1850, Lear comments that the halibut population had been so seriously depleted by simple hand-lining from schooners that the fishery collapsed not only on Georges but also on the New England and Nova Scotia shelves. After this, the acquired taste of New Englanders for halibut had to be satisfied by importing fish from Iceland.

Although a fishery for cod developed strongly on Georges Bank, only to decline at the end of the nineteenth century, this decline may have been at least partly due to reduced demand by New Englanders for salt-cod. In the adjacent Bay of Fundy, there were early signs of reduced individual size of fish, and reduced population abundances, although these became serious only during the nineteenth century: typical was the progressive shift in dominance from large to medium and then to small herring in the coastal fishery between 1860 and 1960.

It is evident that here, as off the European coasts, and prior to the arrival of mechanised vessels, fishing was not a sustainable activity, even though the population crashes were not as extreme as those that had afflicted European herrings. Nevertheless, the large anadromous and estuarine fishes, characterised by salmon and sturgeon, went the same way as their European cousins, although more gently because the transformation of coastal lands and waterways was (and, in many parts of the Atlantic regions of North America, still is) far less fundamental than had occurred on the European coasts.

The progressive degradation of the Quoddy region (on the New Brunswick and Maine coasts) was probably typical of what occurred elsewhere: populations of diadromous and coastal species were reduced by as much as an order of magnitude from pristine biomass levels by both habitat modification and fishing.[23] Already in the late eighteenth century, river courses were strongly modified by dams for saw- and pulp-mills (there were 30 such dams on the St. Croix River alone by the mid-nineteenth century!) that not only modified run-off dynamics but also polluted estuarine and coastal habitats with organic and chemical effluents. The strong diadromous populations of Atlantic salmon (*Salmo salar*) and gasperau (*Alosa* spp.) found by the earliest settlers have been reduced to very small fractions of their initial abundance.

GLOBALISATION OF INDUSTRIAL FISHING: FROM PUSHKINS TO RUSTBUCKETS

Even if reductions in stock biomass and, in extreme cases, population collapses can be associated with earlier fishing activities, it was only after the globalisation of the industry, supported by the technological advances of the twentieth century, that we risked facing a serious crisis of fisheries resources of our own making. So it may be useful at this point to recall briefly the reasons for the surge of additional pressure on fish stocks that originated in the early 1950s by the fishing industries of half a dozen nations, and which rapidly extended, over a period of just a few decades, to a global exploitation of all important fishery regions and fish stocks. It was largely this phenomenon, rather than the coastal fisheries off the shores of fishing nations, that has left the resources – not without some important exceptions – in the complex, but generally depressed, state that shall be discussed in the next chapter.

The reader may recall my earlier evocation (see Chapter 1) of the expansive mood of fisheries planners in the middle years of the twentieth century and their great confidence that fishing could be globalised without endangering populations, provided that simple management based on MSY was established in each case. Policies that favoured the expansion of fisheries to global status were driven by a variety of motives: in the West, reaction to the extension of national jurisdiction beyond the territorial sea and straightforward commercial motivation and, in the East, State planning to ensure sufficient food supplies for their population during the post-war reconstruction of the Soviet Union and its satellite nations. The same held true for Japan, where the occupying US military administration looked to fishing and whaling to assist them in ensuring that the civilian population was fed. If these were the prime movers of the globalisation of fisheries, a host of smaller initiatives, some by individual fishermen, were induced by the coming of peace in 1945, and the renewal of commercial activity together with the increasing ease of movement and communication in the following decades.

Although it is usually the appearance of *Fairtry I* on the Grand Banks in 1954 that is credited with starting the revolution in the fishing industry that led to globalisation, the process was, in fact, much more complex than that. It included not only the expansion of industrial-scale trawling to continental shelves and shallow banks in all oceans and at all latitudes, but also the expansion of pelagic fishing by purse-seine and long-lining for tuna and related species over the entire extent of the tropical and subtropical oceans – the largest living space on the planet – and the expansion of specialised trawling for shrimp from tropical to polar seas. This globalisation, initiated in the 1950s, was largely completed prior to the end of the century; its most recent manifestation is the exploration of deep benthic habitats beyond the shelf edge and the development of trawling techniques for the fish that live there.

The so-called factory trawlers, of which the *Fairtry* class and similar ships built in Japan were the forerunners, brought many radical innovations to fishing technology – in fact, more than they are usually given credit for. It is easy to forget now that until the introduction of the stern ramp – perhaps suggested by the stern ramp of whalers – side trawling was a complex and dangerous operation that required great skill on the bridge. While the otter boards, the warps and the trawl itself were streamed outboard from the fore and aft gallows, the ship had to turn continuously in that direction: to arrange all this so that the final heading was in the desired fishing direction was not simple. At the end of each tow, all deckhands were required to claw the trawl in by hand over the weather gunwales. But on stern trawlers, like *Fairtry* and all subsequent factory ships, not only is the shooting and hauling process much simpler and more mechanised, but once the trawl is shot, the fishing skipper – using sonar – can more easily target individual shoals of fish on the sea bed because of the relative ease of controlling a stern trawler during the tow. Without the development of novel materials for nets and warps, and of electronic instrumentation, none of this would have been possible.

However, it was the processing and freezing facilities of the early factory ships that took the attention of industry everywhere, and nowhere more so than in the Soviet Union: these trawlers have been described as a 'latter-day industrial revolution'. As Moiseev noted in 1965, the Soviet sea fisheries were almost non-existent after the war, yet state planners called for a supply of $12–13 \times 10^6$ tons of fish products to supply all nutritional, agricultural and industrial needs of the nation: this would require at least an order of magnitude increase in catches. To achieve this, by the 1950s planning put was in place 'for the construction of a large deep-sea fishing fleet consisting of various types of fishing, fish-processing and transportation vessels with long operational ranges'.

Things moved quickly. In the two years following the appearance of *Fairtry I* on the Grand Banks, the Soviets had ordered 24 *Pushkin*-class stern trawlers, very similar to the *Fairtry* design, from German shipyards. Rapidly, several hundred of the larger *Mayakovski*-class were built, followed by many others so that by 1975 there were 900 factory trawlers fishing worldwide: 400 Russian, 125 Japanese, 75 Spanish, 50 East German and so on. Many ships of the Soviet fishing fleet in the Northwest Atlantic were factory trawlers, each exceeding 1000 tons, accompanied by even more numerous, smaller side trawlers and the factory ships used to process the catches of these. At the same time, similar ships from almost 20 other nations were also fishing the Grand Banks, even as the Japanese factory trawlers were pioneering the gadoid fisheries in the Gulf of Alaska, and the Spanish were using fleets of side trawlers with mother ships in the South Atlantic. It was only in 1982, after the establishment of the UN Law of the Sea Convention, that the access of these fleets to the major shelf regions was in any way limited.

Moiseev described graphically how the new fleet of factory trawlers progressively exploited new regions during the 1960s: the north and northwest Pacific, the Norwegian Sea, the Grand Banks, the shelf off western Africa, the northwest Indian Ocean, the Caribbean and Gulf of Mexico and other places.[24] He commented that 'most of these regions had either been utterly ignored by other countries or else had been exploited on a relatively small scale. These resources thus entered the commercial balance of the oceans as a result of the activities of the Soviet fishing fleet.' This was the period of most rapid expansion of global catches: in the 30 years following 1955, the catch of gadoid fish from all continental shelves increased by a factor of 3, from 4.7 to 12.5 million tons. It is important to remember that these catches were almost entirely taken in the absence of any regulation, apart from whatever discipline was established within each national fleet: neither total maximum allowable catches nor minimum mesh sizes was imposed by any authority.

Globalisation is now general and international relations in the industry have become, quite literally, chaotic. As an example, consider some recent events in the North Pacific fishery for Alaskan pollock (*Pollachius pollachius*), which is now a widely used whitefish. The annual quota in US waters is taken in less than three months' fishing, so US companies respond as best they can to this resource restriction: for instance, American Seafoods has entered into partnership with a Russian company based in Vladivostock and is consequently able to fish in Russian waters, using nine Spanish-built factory trawlers. Other Spanish flag trawlers, having received EU subsidies to cease fishing off Europe in the early 1990s, then turned to exploit the outer parts of the shelf off Argentina. These arrangements often involved re-registering under flags of convenience, resulting in problems that are to be discussed in Chapter 8.

And so it goes: very unfortunately, small factory trawlers are rugged ships and, provided that the motor, the trawl winch and the freezer continue to function, they can go on fishing even when reduced otherwise to rustbucket status, many decades after their launching. These, not representing very large capital investments, are widely used in illegal fishing: unfortunately, the shelves of many nations are inadequately policed and, in many regions, fisheries protection ships are entirely lacking, and it is in such places that many delapidated factory trawlers continue, even today, to fish for the international market. The ultimate use to which these ships have been put is almost unthinkable: they are used as motherships by the Somali pirates, who recently attacked a modern French tuna-seiner that was incautiously fishing the northern Indian Ocean, not far from its base in the Seychelles.

There can be few who would disagree that the hopes expressed by fisheries scientists and planners in the 1950s have been fulfilled in only a single respect: the global catch has reached, and perhaps temporarily stabilised, at a level greater than was hoped for in the mid-twentieth century. Unfortunately, the cost has been very great in terms of the health of the stocks, although perhaps not as great as suggested by some studies: however that may be, it is generally agreed that the current level of fishing cannot be sustained indefinitely.

To review the extent of the problems caused by rampant globalisation and technical innovation during the latter decades of the last century it will be sufficient to discuss three examples with which I have had some personal acquaintance – the continental shelf stocks in the western Gulf of Guinea in the tropical eastern Atlantic, the great and apparently eternal fishery for cod on the banks of the NW Atlantic, and the global hunt for high-priced tuna. These will illustrate the range of resource problems arising from globalisation.

What is clear is that the impetus for the globalisation of fisheries was not, as has been suggested by some, simply that the coastal fisheries off Europe and North America were generally exhausted, and that we had to turn elsewhere. It was a much more complex development than that, and it rapidly developed into nothing less than the globalisation of the capture, distribution and sale of the products of the fishing industry. This evolution from artisanal level fishing appears to have represented the normal development of an industry that progressively explored the resources and markets that were open to it.

I observed this process in the Gulf of Guinea where, towards the end of the colonial period, the fresh fish markets in the large coastal cities began to be supplied by landings from enterprising European trawlermen who explored personally, in their own boats, the possibilities offered by resources and markets distant from their home ports. Previously, supply had been entirely in the hands of individuals and specialised guilds of local fishermen and market women, while the importation of stockfish was in the hands of traders, often from Lebanon. The first of the foreign boats arrived in Freetown in 1956. She was a small Italian wooden trawler of perhaps 15 m (the *Ombrina Verde*, as I recall) and, in the years that followed, similar ventures occurred the entire length of the Guinea coast: in Nigeria, things progressed rather quickly so that by 1961, 10 small trawlers were landing about 3500 tons of fresh fish in Lagos annually, all from the inner-shelf community of demersal fish that is dominated by sciaenids, drepanids and polynemids. European owner-skippers and engineers operated these small diesel trawlers, while deckhands were recruited from among the local canoe fishermen. This invasion did not quickly kill the canoe fishery, at least in Sierra Leone, where in the mid 1970s less than one-third of the total landings of around 60,000 tons was supplied by foreign trawlers fishing locally; the rest (largely the coastal clupeids *Ethmalosa* and *Sardinella*) was still supplied by the indigenous canoe fishery. In the mid 1980s this comprised about 18,000 fishermen using 8000 canoes and, for instance, about 250 beach-seines; by then, however, the percentage of total landings produced by the indigenous fishery had fallen to well below 50% of total landings.

Already at this time, a few distant-water factory ships were based in West African ports, and fishing far to the south, off Angola, but then things changed rapidly. By 1988, as many as 25 Soviet trawlers were operating with motherships on the wide shelf regions off Sierra Leone, and as many as 51 ships had been authorised by a long-term agreement under which Sierra Leone deckhands were to be 'trained' aboard Soviet ships: in fact, they were no more than a cheap labour pool. Official statistical data in the mid 1990s suggested that commercial fisheries production was around 130,000 metric tons, with clupeids constituting over 75%. At the same time, up to 130 largely illegal shrimp trawlers were active in a for-export pink shrimp (*Penaeus duorarum*) fishery on the Sherbro Bank, yielding 2000–4000 tons annually.[25]

A similarly rapid expansion of this export fishery occurred in the eastern Gulf of Guinea where a survey by the Nigerian federal fisheries services in 1961–1963 had found that shrimp biomass (as recorded in standard trawl catches) significantly exceeded fish biomass on soft grounds of the shallow continental shelf above the thermocline.[26] Only the fish were then of serious commercial interest but, 50 years later, about 250 specialised shrimp trawlers are registered in Nigeria together with 65 others fishing on the adjacent Cameroon shelf. One of the frozen products from the Gulf of Guinea currently on sale in European supermarkets is 'tropical sole' (*Cynoglossus canariensis*) which was noted in the 1962–1963 surveys as being abundant inshore off Nigeria on the same grounds as those that yielded good catches of shrimps, both *Penaeus duorarum* and *Parapenaeopsis atlantica*. Unfortunately, the distribution of commercial concentration of penaeid shrimp generally coincide with that of the inshore community of warmwater demersal fish (dominated by Sciaenidae, Polynemidae, Ariidae and Drepanidae), of lesser commercial and export value. Because this inshore community of fish has always supported the canoe and small-boat fisheries for local consumption, it is probably now impossible to implement any measures – such as cod-end mesh regulation – designed to conserve this supply to local markets. So much for the research on how these populations might be managed that we undertook at Lagos in the early 1960s!

As the numbers of boats fishing off Sierra Leone built up, illegal fishing by unlicensed boats rapidly became a serious problem, but was eventually controlled rather effectively by privatising the licensing process, in such a manner that the company had incentives to ensure that licences were obtained prior to fishing, and by instituting patrols at sea operated by the same company. Unfortunately, during the civil war after 1992, these arrangements began to fall apart: the fisheries offices and research institute (including my 1950s reference collection of fish and benthic invertebrates) were torched in 1999 by the rebels.

In recent years, Sierra Leone has again offered fishing licences for sale to vessels fishing offshore, to a value of US$225 million. Under this system, against a licence costing US$71,000 and the obligation to hire 14 local deckhands, the Norwegian owners of a 300-ton capacity freezer-trawler obtained the rights to take unlimited amounts of penaeid shrimps from the Sierra Leone shelf in 2003. Taking advantage of such licensing arrangements obtainable not only in Freetown, but also in Guinea and Guinea-Bissau, and of a total lack of policing at sea (except by organisations such as Greenpeace), a significant fleet of decrepit factory trawlers is currently fishing for deep-water sparids off the coasts of Sierra Leone, Guinea and Guinea-Bissau: more than 50 trawlers belonging to the China National Fisheries Corporation and other foreign corporations have been located in this region. Such operations are a textbook case for the illegal, unreported and uncontrolled (IUU) fishing that shall be discussed later, and this case may be generalised to many other regions.

The price that the resources of the Guinea shelf have had to pay for this and other injuries is a radical transformation of ecosystem structure both of the inshore sciaenid fauna and the offshore, cool-water sparid fauna[27]; loss of large predatory fish has changed the balance between invertebrates and fish, and between demersal and pelagic fish. Clupeids and penaeid shrimps are both relatively more dominant in the shelf ecosystem than previously. Perhaps because recruitment is relatively invariant in the tropical shelf fish fauna, there appear to have been no complete collapses of species, but the relatively unrestricted fishery of the last half-century, and the progressive domination of the capture sector by foreign ships, has clearly represented an unsustainable level of catch. Parenthetically, I note that the history of this region parallels that of the more talked about Gulf of Thailand, where trawling also began about 1960 with less than 100 Thai-registered boats, rising to 3300 trawlers within only 11 years! Catches increased over the same period from 63,000 to more than 1.1 million tons and, predictably, catch per unit effort dropped from 400 to only 20 kg/h at the present time. This region has now become the paradigm for over-utilised tropical demersal resources; however, as shall be discussed in the next chapter, official global landing statistics appear to be blind to these realities.

A rather similar pattern of evolution occurred further north along the western coast of Africa off Morocco, where foreign boats had fished since the sixteenth century and where sardine and cephalopods are among the most valuable resources. The coastal fishery, based on wooden 8–10 m dories that operated not off the beaches, but from small ports, had reached almost 3000 units by the 1950s and three times that number by the end of the century, based largely on an *Octopus vulgaris* fishery.[28] A coastal fishing fleet (<150 units) of purse-seiners, trawlers and long-liners had developed during the early twentieth century which, after 1970, grew very rapidly and extended from a few traditional ports so that, by 1990, there were around 2500 long-liners, trawlers and purse-seiners: sardine, anchovy, mackerel and *Caranx* and some demersal species being the main targets. More remarkable was the rapid increase from the early 1970s of an indigenous industrial fleet, growing from four ships based in Las Palmas in 1973 to around 450 ships at the end of the century, all landing in Moroccan ports. The growth of the fleet was then frozen by the government because of over-capitalisation of the fishery. But already, the rustbucket syndrome had set in and many ships were laid up for lack of maintenance.

During the period of the expansion of the Moroccan fleet, foreign ships were flooding in from Japan, South Korea, USSR, Poland, Romania and the EU: a 1988 accord permitted 800 EU ships (90% Spanish) totalling almost 100,000 registered ton to fish in the Moroccan zone. Subsequent agreements progressively reduced this number and allocated zones to each type of vessel: sardine seiners from Cap Ghir to Cap Blanc, shrimp trawlers from Tangier to Cap Ghir, cephalopod trawlers from Cap Bojador to Cap Blanc, and so on. The declared catches of the foreign fleets greatly exceeded those of all domestic fleets combined: the Spanish catch alone almost matched that from Moroccan ships, and small pelagic fish greatly outweighed all other catches. Unfortunately, discards and unreported catches were heavy, as will be discussed in Chapter 8.

The evolution of fishing in the remainder of the non-industrialised world was probably very similar to what occurred in the mid-twentieth century in the eastern Atlantic, and occurred everywhere with no thought for the social consequences for artisanal fishing societies. To some extent, and in some countries, the total number of people employed in the activity increased significantly, which was helpful for the national economy, but this occurred only at the expense of great social changes. I have seen suggestions that this should have been prevented and the traditional fishing societies should have been preserved: all resources within the reach of artisanal fisheries should have been reserved to them and industrialised fishing permitted only on populations that inshore fishermen could not reach.[29] However, such suggestions ignored the political realities of the countries concerned, and the clock cannot be turned back now: seafood has become an international commodity of great importance in human nutrition everywhere, extracted and traded internationally, just as Wilbert Chapman foresaw.

Turning now to the much better-known recent history of the Northwest Atlantic cod stocks of regions largely beyond the reach of coastal fishermen, it is now clear that the Grand Banks fish were the inevitable victims of the globalisation of the fishing industry and the consequent international tensions that this engendered. Their history is extraordinarily complex, involving not only problems with management science and implementation, but also the international tensions between Canada and other fishing nations, and between different levels of government in Canada.[30]

This is a case history about which much has been written and about which much more is surely still to be written. Several accounts have underestimated the complexity of what happened in the NW Atlantic by laying the blame entirely on the shoulders of a single agent, usually the Canadian fishery bureaucracy.[31] But it was a disaster that had many causes: DFO scientists who got things wrong, DFO administrators who wanted to please the minister, ministers who wanted to preserve their seats in the house and jobs in their constituencies, Newfoundlanders who pressured everybody to keep the fishery open, Canadian fishermen who cheated right and left in the Exclusive Economic Zone (EEZ), foreigners who poached inside the EEZ, foreigners who raked the banks outside the EEZ, Canadian External Affairs and Defence departments who didn't want to rock any boats, and a changing flux in the Labrador Current that would have modified the distribution and abundance of even unfished stocks in ways difficult to predict.

There were three quite separate general problems, any one of which would alone have been quite sufficient to cause the fishery to fail or, at the very least, be unable to sustain catches at the accustomed levels: (i) a management regime for the offshore regions that can only be described as ridiculous; (ii) a scientific structure that produced (and worked with) misleading or inadequate data, with inappropriate management models, and was subject to political interference; and (iii) the unanticipated consequences of natural environmental changes that occurred at the regional level, initiated in 1984 and 1985 in response to a strong shift of the NAO into positive phase. In the discussion that follows, it will be also useful to remember that, as discussed earlier in this chapter, the terminal collapse in the early 1990s was not the first time that such a problem had hit the Newfoundlanders: during the 1870s, catch rates in the cod fishery declined so severely that a cod hatchery was established in the hope of rebuilding the populations.

When it first became clear in the middle of the twentieth century that major international fishing pressure was imminent in the NW Atlantic, the five nations most concerned (Canada, USA, UK, Denmark and Iceland) created an apparently competent body, ICNAF,[32] to 'investigate, protect and conserve the resources for maximum sustainable yield'; finally, 18 nations adhered to this body, but *it was not until 20 years later that ICNAF introduced its first catch quotas* – voluntary quotas, be it noted – on a species already commercially extinct (haddock on Georges and Browns Banks) and its first regulation for cod-end mesh size. Well before then, the effects of the hundreds of factory and pair-trawlers working the offshore banks had caused serious depletion of the inshore fishery. Moreover, it was only in 1973–6 that the first estimates of F_{max} for northern cod were obtained from formal stock assessments by ICNAF nations; a permissible catch of 650,000 tons was then established by ICNAF, of which Canada received a quota of around 20%.

Because of the failure of the ICNAF arrangements to control offshore fishing, in 1976 Canada unilaterally declared a 200-nautical mile fishery zone[33]; ICNAF agreed to Canada's right to manage this zone, and was disbanded, to be replaced by NAFO in 1978.[34] This organisation was to be responsible for the management of the major regions of the Grand Banks that lay outside the Canadian zone, with the extraordinary result that some unit stocks (like that of the northern cod) were to be managed by two agencies at different times of the year as they migrated! Further, those nations that chose not to join NAFO were legally free to fish as they chose on the Nose and Tail of the Banks, beyond Canada's fishery zone!

Despite the fact that there was very little hard evidence of their real status, Canada proceeded to sign 13 bilateral agreements for the fishing of under-utilised stocks in the zone; that same year, Canada initiated its first stock assessment surveys based on randomly selected sites in each area using commercial trawl gear, and allocated 44% of the computed allowable catches to its own ships. Confidence that the populations would rapidly rebuild within the protection of the fishery zone led to optimistic predictions and planning for the future of the Newfoundland fishery economy – already in crisis because of overinvestment – but it is now clear that the assessment surveys performed by Canadian DFO scientists were giving over-optimistic results because (i) the aggregations surveyed did not represent the real state of the dwindling populations of northern cod, and (ii) the data obtained from commercial catches failed to account for progressively improving technical efficiency of the offshore foreign fleet! NAFO assessments, moreover, ignored data from the inshore fishery within the territorial sea of Canada …

By 1981, economic and employment problems in Newfoundland were so acute that a Task Force on Atlantic Fisheries was established by Ottawa, headed by Michael Kirby, having the impossible mission of proposing viable solutions for increasing employment in the fishery, ensuring year-round fish supplies for the processing sector, enhancing product quality control and Canadianisation of the industry. Little attention was given to the amount of raw material that was assumed to be available at sea for the brave new industry. But suspicion was growing in some quarters that the stock assessments done by DFO scientists were altogether too optimistic and that the annual catches set by $F_{0.1}$ logic had perhaps been too generous; one DFO stock assessment biologist wrote, in a now notorious internal report, that the DFO northern cod stock assessments were 'not worth a rats rear-end',[35] and that stocks had been consistently over-estimated since 1977 because of too great a reliance on reported commercial catch rates: the increase in population size had been totally imaginary.

During this period of introspection and planning in Ottawa, quotas were routinely exceeded. In 1986, for instance, NAFO scientists recommended a total catch of 13,000 tons outside the Canadian fishery zone, but at the formal annual meetings it was set at 26,000 tons. Ships of NAFO nations nevertheless took 110,000 tons, while non-NAFO ships took a further 25,000 tons – a total catch one order of magnitude higher than had been recommended by the science teams! In the same year, EU ships received an allocation of 9500 tons, but took 68,000 tons. Very unfortunately, and for reasons apparently never clarified, the 1986 assessment surveys were rather positive and subsequently Canada rejected NAFO assessments and produced its own through an internal DFO committee (CAFSAC),[36] and began to conceive of greatly increased quotas within the fishery zone from around 50,000 tons to as much as 330,000 tons for 1988. It was only later understood to what extent progressive aggregation of the declining population had given misleading information on stock abundance; in reality, the area occupied by the population was positively correlated with stock abundance.[37] This possibility appears to have been neglected in stock assessments during the late 1980s, although such a range contraction had been observed in a wide variety of fish species as their overall stock size declined. In the case of northern cod, hyper-aggregation of individuals occurred preferentially in the southern parts of the range.

A succession of independent scientific studies of the stocks had been commissioned during the 1980s – one by the Newfoundland Inshore Fishermen's Association, and two by the federal fisheries minister – and each concluded that the DFO assessments were over-optimistic, stocks were much smaller than had been computed by government scientists and, consequently, fishing mortality rates had been far greater than anticipated – and would have been even if legal catch limits had been respected. In addition to these problems, there was uncertainty concerning changing circulation patterns in the Labrador Current and the consequences of the variable influence of cold intermediate water on cod distribution – and hence on its apparent stock abundance. Rather naturally, DFO administrators hopefully ascribed stock collapse to environmental factors rather than to the effect of fishing: reports on stock status off Newfoundland hardly mentioned overfishing, but discussed environmental change extensively.

Nevertheless, the 1988 assessment surveys suggested that the northern cod stock was rapidly dwindling, and this collapse continued over the next few years against a background of agonised decision-making in Ottawa over necessarily falling quotas. The independent report commissioned by DFO on the status of the fishery by Leslie Harris, president of Memorial University of Newfoundland, who suggested that, as an agency for conservation, ICNAF had been 'a total failure' and NAFO was toothless and that its 'authority counts for less than nothing in the world of realpolitik' of the fishing nations. This confirmed the findings of the earlier report by Lee Alverson and others, also commissioned by DFO, who found serious errors in the DFO assessment procedures and concluded that populations had been declining fast for several years. Perhaps these studies were indeed commissioned by DFO to deflect critical evaluation of past decisions of the minister, as has been suggested to me, but their conclusions cannot be ignored on that account.

Harris recommended an immediate and severe reduction in fishing levels, but this was rejected by then DFO minister Valcourt, who proceeded to establish multi-annual quotas of around 200,000 tons for 1990–3 – prior to receiving any formal advice from departmental scientists. The 1991 assessment surveys were again optimistic (perhaps because of the hyper-aggregation of individuals, so that there was talk of a coming bonanza) even though Canadian trawlers could not catch their quotas; but, in 1992, new assessments revealed that all stocks really were very small indeed and, in July of that year, the cod fishery in the Canadian territorial sea and fishery zone was finally closed, and CAFSAC was disbanded by the then minister, Newfoundlander John Crosbie: he had the toughest job of all, but had the courage to meet the angry fishermen at home, face to face, to explain why the closure was necessary. Not surprisingly, Spanish and Portuguese ships ignored the closure and continued to fish in NAFO area 3L.

Throughout this tragedy, for that is what it was, the rational setting of catch quotas was either non-existent (in the early years), or compliant to political motives (throughout). After the establishment of the fishery zone, there was continual pressure on DFO from the Canadian Department of Foreign Affairs to allocate *quid pro quo* quotas for cod, turbot and other species to nations with which they were negotiating other matters. International tension culminated in the *Estai* affair in 1995, when this Spanish trawler, fishing turbot, refused to be boarded for inspection when challenged by RCMP officers. This affair reveals the real and extraordinary complexity of what should be the simple process of fisheries inspection at sea, and explains why it is so seldom effective. Canadian fisheries minister Brian Tobin immediately took the matter to Cabinet and requested that the *Estai* should be seized, no matter what, and inspected: against his request were ranged the departments of Foreign Affairs (not worth upsetting friends for a mess of fish), of Justice (jurisdiction of national offshore fishery zones is murky), of Defence (Spain is a fellow NATO member) and of the RCMP (civil suits against officers might ensue, which would be an embarrassment).

Prime Minister Chretien, trying to head off a row in Cabinet, thought that he had reached agreement with EU fisheries authorities for a 60-day turbot moratorium; but when he was told that the Spanish fleet was ignoring the moratorium, he gave his approval for the seizure. *Estai* was immediately boarded by DFO fishery inspectors, after a .50 calibre machine gun burst across her bows had finally induced her to heave-to, although her crew could not be prevented from dumping her trawl; when *Estai* was brought in to St. John's she was greeted by jeering crowds and by EU diplomats (France, Germany, Spain) who came to protest, but who were jostled by the crowds. Almost the entire catch was of small fish. The bridge had been keeping two logs – one for the inspectors, one for the owners – and part of the hold was concealed from the inspectors. Newfoundlander Brian Tobin was an instant national hero who later faced off to EU anger (personified by Emma Bonino) at the UN by hoisting the recovered trawl net on a barge on the East River to demonstrate its illegal cod-end meshes to UN delegates.[38] Within days of the *Estai* incident, the Spanish patrol boat P-74 *Atalaya* appeared among the remaining Spanish fleet that were fishing illegally and contravening agreements signed by Spain. And that, as you may suppose, was the end of further intervention by Canada.

In fact, undiplomatic activity may be the only way that a coastal state can keep any control over what occurs in its EEZ, as was demonstrated by Iceland's confrontation with the Royal Navy during the Cod Wars with Britain. Unfortunately for the stocks on the Grand Banks, the *Estai* boarding was not repeated by Canada and such tactics had not been used in earlier years and before the offshore cod populations were fished out, largely by the same fleet of European vessels. Such activism at sea will only happen if a well-placed administrator or politician has sufficient energy and initiative to make it happen; Brian Tobin did just that by energetic and skilful lobbying in Cabinet at Ottawa but, even so, he was unable to maintain the requisite activism by Canadian patrol vessels at sea: Canada had other important business with Spain that was not to be jeopardised for a few fish. I will return later to this general problem.

In retrospect, it would be easy to conclude that the scientific evaluation of allowable catches of cod and other species on the Grand Banks was essentially irrelevant in the management process, so consistently were catch quotas set higher than the levels computed by the DFO and NAFO scientists; this occurred in response to political pressures from the Canadian fishing industry and from Newfoundland provincial politicians, who were desperate to avoid the massive unemployment that the final closure of the fishery eventually involved. This tension was translated down within the DFO structure as a very strict control on scientific studies and conclusions, with the result that a peer-reviewed paper entitled 'Is scientific enquiry incompatible with government information control?' was published by three academic scientists[39]; they suggested that 'there is an urgent need for public scrutiny of the influence of senior-level bureaucrats in the management of Canada's natural resources'.

The resulting scramble in Ottawa is best left to the imagination, but was inevitable in a situation in which scientists had been pressured to withdraw papers in press because they reached conclusions contrary to the official interpretation of the situation: DFO scientists were enjoined by at least one senior official 'to support the Minister's position', since it is the responsibility of the Minister, not of public servants, to establish policy decisions. Custom dictated that senior bureaucrats should ensure that uncertainty and choice not be expressed in advice to the Minister, and so it would be very little exaggeration to suggest that the fishery, from start to finish, lacked any real scientific basis for its management.

The final result might have been avoided if Canadian federal–provincial political relations had been less confrontational, and if the DFO had been a more open organisation. On the other side of the Atlantic, fortunately for Norwegian fishermen, politicians are apparently able to respond more pragmatically and directly to difficult scientific advice. Around 1990, as Harris relates,[40] a major collapse of the capelin population occurred in the Barents Sea because of a conjunction between predation by a series of large year-classes of cod and very heavy industrial fishing of capelin themselves; the 2–3-year-old cod then turned to cannibalism and were, in turn, attacked by seals that now lacked capelin as food. The entire Barents Sea ecosystem became unbalanced and fishermen noted the symptoms of starvation in their cod catches: large heads, thin bodies. The message from scientists and fishermen was understood by the fisheries minister, who acted immediately with a programme of fleet reduction through vessel buy-out, and with a fishery closure that was imposed as soon as the problem had been identified. Within only three years, results could be seen as the Barents Sea stock rebuilt and rapidly exceeded biomass levels observed for many years.

Finally, one has to ask whether a collapse – or at least a very significant reduction – of the northern cod stock off Canada would have occurred at the end of the twentieth century even in the absence of the heavy fishing pressure that was laid upon it. Although at the time there was little or no consideration of the consequences of changing climatic conditions on the stock, in retrospect there are indications that support such a suggestion. The entire period between 1970 and the end of the century, as we noted earlier, was characterised by positive values of the NAO, reaching highest values (+3.1) in 1990; the only comparable period since 1850 was in the first three decades of the twentieth century. Positive NAO values (as discussed in the previous chapter) are associated with a strong Icelandic low pressure, more northerly storms track across the Atlantic and cold, dry conditions across northern Canada and Greenland. Regional environmental information, although slight, supports a pattern of cold air temperatures and unusually heavy ice conditions at St. John's during the 1990s, with below long-term mean SST at the routine oceanographic station 27 off Cape Spear. Increased southward flow of cold Labrador Current water appears to have been involved, and fish populations moved southwards.

The population of American plaice (*Hippoglossoides platessoides*) off southern Labrador and eastern Newfoundland declined severely after 1980 and by the 1990s had essentially collapsed, although it had been fished at very low levels indeed: the annual exploitation rate was less than 20% of population biomass and there was negligible by-catch in other fisheries.[41] In the 1990s, the remaining plaice had taken refuge in deeper, warmer water than was normally the habitat of the species. At the same time, capelin stocks were reduced to about 3% of their former abundance and, in fact, were almost eliminated; being an essential item in their diet, the cod followed them south. There had been some doubt as to the reality of this southward movement of cod in the decade 1983–94, until the penetrating review by Rose and others[42] who showed that: (i) survey data indicated southward shifts where there was no fishery and (ii) that cod taken in the south in these years had characteristics of northern populations in their antifreeze production capacity, their vertebral counts and their low length-at-age. These observations are consistent with the model of Rose,[43] which quantifies interactive dynamics of cod populations, fishing and climate changes since 1505, when the population was assumed to be pristine. A surplus production model incorporating depensation and climate (quantified by regional tree-ring growth data) mimicked documentary history of the stocks, including population declines in the cold years of the nineteenth century (1850–80) associated with lower productivity, the 1960s collapse due to fishing pressure and the final 1988–9 collapse due to fishing, depensation and – once more – cold water.

Taken together with the steadily falling recruitment to the population from almost a million and a half fish in the 1970s to about half a million in 1980 and almost none in 1990, these observations and simulation suggest that that even if the population had been fished according to good scientific advice all along, and if rapid political action had been taken, the population would still have seriously dwindled in response to the environmental changes that occurred in the 1980s. Rose recalls that in 1895 Prowse wrote of Newfoundland that 'her geographical position requires that she should always stand prepared for suffering ... (but) if her resources are liable to collapse, they are not less liable to sudden and brilliant revival'.[44] What is surprising in this context is that nobody seems to have remembered that amongst many other trenchant observations, Johann Hjort's 1914 'Fluctuations in the great fisheries of northern Europe viewed in the light of biological research'[45] describes his studies of the correlation in the period 1880–1911 of cod liver weight (an excellent indicator of condition) and the solar cycle of sunspot numbers (an excellent indicator of global climatic conditions, as discussed in Chapter 5). Hjort recorded decadal-scale variability in the fecundity, liver oil yield and total catches of cod at Lofoten and Finmarken from 1874 to 1907 and noted that these indices of stock well-being responded to the sunspot count. More recently, the code of the periodic fluctuations in biomass of cod, capelin and herring in the Barents Sea has been revealed: population growth is maximal when the 6.2 and 18.6 years Kola temperature cycles are positive at the same time.[46] Therefore, the natural variability of cod biomass at Newfoundland should have been no surprise to anybody.

The third and final example of the effects of globalisation that I want to discuss, very briefly, concerns the fishery for tuna, especially for tropical species. Like the trawling industry, tuna fishing companies have always protected their commercial interests behind a wall of secrecy, and it is by no means easy to understand exactly how they operate in the global market. Prior to WWII, the US tuna fishery was based in San Diego and San Pedro, and on the purse-seine technique, which was also used in West Coast salmon and some clupeid fisheries. In those days, a 200-ton capacity tuna seiner was considered a large ship but, with the development of freezing capabilities and of the Puretic power-block for handling the seine, vessel size progressively increased – just as in the trawling fleet. A San Diego seiner of 1950 would fit nicely on the foredeck of the ships being built today: the Ching Fu shipyard in Taiwan recently completed two super-seiners for the western Pacific, each of 2000 m^3 capacity and with a large helicopter pad ('for a bigger helicopter for longer search operations', as the brochure says, although these are illegal in many regions) and a top speed of 18 knots: on a 28-day trip, these ships can take almost 2000 tons of tuna for a consumption of 300 tons of fuel. Previously, purse-seining for tuna had been restricted very largely to the US fleet in the eastern Pacific, but now it is used everywhere and even by some European nations. During the post-war recovery period, the US tuna industry was very active in seeking resources beyond those previously exploited from southern California; the American Tuna Boat Owners Association was one of the principal proponents of the EQUALANT and EASTROPAC oceanographic surveys of those years.

Although purse-seining has largely replaced the tuna long-lining that was the favoured Japanese method, this too has developed technically, and its use began to spread to other nations once these vessels were once again given the freedom of the open seas by the US occupying authorities in the early 1950s. The long-lining technique evolved in the post-war years with the use of baiting machines that were capable of baiting as many as 20,000 hooks a day with sauries, thus enabling the use of very long lines, individually as much as 100 km in length. Hooks are set on snoods at depths that correspond to the swimming depth of the intended catch, e.g. deep for bigeye, shallower for yellowfin. The long-liners use extremely low temperatures in their freezers, enabling them to stay at sea for many months but to land their catch in perfect condition. These techniques are now almost worldwide, and are used by vessels registered in more than 30 nations.

Tuna fisheries are managed, at least in principle, by five regional fisheries management organisations[47]: IATTC (eastern tropical Pacific), ICCAT (tropical Atlantic), CCSBT (southern bluefin), IOTC (Indian Ocean) and WCFPC (Western Pacific). Each has a rather similar mission statement, being (for instance, in the case of IATTC[48]) 'responsible for the conservation and management of fisheries for tunas and other species taken by tuna-fishing vessels in the eastern Pacific Ocean. Each member country of the IATTC is represented by up to four Commissioners, appointed by the respective government.' Together, they offer integrated information to the public on, for example, lists of tuna boats from each nation authorised to fish in their region. Some, such as IATTC, have their own scientific staff while others, like ICCAT, depend entirely on meetings between scientists from member nations to generate advice on quotas. At the IATTC, scientific staff have engaged in an unusually wide range of activities that include monitoring the physical oceanography of the eastern tropical Pacific, resolution of the complex problems arising from the mortality of porpoises in purse-seines[49] and, of course, the routine assessment of tuna populations and their recruitment levels. An observer system is maintained with the support of member nations to collect data at sea on catches and by-catch levels. As shall be noted later, these organisations are subject to outrageous tensions by the behaviour of national delegations – Such as vote-swapping and voting with a larger nation to obtain favours elsewhere.

You might conclude that this was a recipe for yet another disaster, and so it would be except for the fact that IATTC (and other tuna management organisations) deal with a group of fishes whose ecology and reproductive biology is totally different from that of the coldwater cod of the North Atlantic: I have suggested elsewhere[50] that at least the large tropical tuna – bigeye, skipjack and yellowfin – may be peculiarly accommodating to fishing pressure at levels that would destroy, for instance, coolwater gadoids. These fish shall serve in my concluding chapter to illustrate one model for a potential sustainable fishery: unfortunately, cod are closer to representing the general response that must be given to the question posed in the title of this chapter.

ENDNOTES

1. Stringer, C. *et al.* (2008) Neanderthal exploitation of marine mammals in Gibraltar. *Proc. Natl Acad. Sci.* **105**, 14319–24.
2. Wing, E.S. (2001) The sustainability of resources used by native Americans on four Caribbean islands. *Int. J. Osteoarchaeology* **11**, 112–26; see also Wing S.R. and E.S. Wing. (2001) Prehistoric fisheries in the Caribbean. *Coral Reefs* **20**, 1–8.
3. Lotze, H.K. and I. Milewski. (2004) Two centuries of multiple human impacts and successive changes in a North Atlantic food web. *Ecol. Applic.* **14**, 1428–47.
4. Steel, J.H. and M. Schumacher (1999) On the history of marine fisheries. *Oceanography* **12**, 28–30.
5. See, for example, Mowat, F. (2004) *Sea of Slaughter.* Mechanicsburg, PA: Stackpole Books.
6. Amorosi, T., *et al.* (1998) Bioarchaeology and cod fisheries: a new source of evidence. *ICES Mar. Sci. Symp.* **198**, 31–48.
7. Jackson, J.B.C., *et al.* (2001) Historical overfishing and the recent collapse of coastal ecosystems. *Science* **293**, 629–38.
8. Roberts, C. (2007) *The Unnatural History of the Sea.* Washington: Island Press.
9. Springer, A.M., *et al.* (2003) Sequential megafaunal collapse in the North Pacific: an ongoing legacy of whaling. *Proc. Natl Acad. Sci.* **2003**, 12223–8.
10. Saenz-Arroyo, A., *et al.* (2005) Rapidly shifting environmental baselines among fishers of the Gulf of California. *Proc. R. Soc. B.* **269**, 1957–62.
11. Brown, A.P. (1947) *The Fish and Fisheries of the Gold Coast.* London: Crown Agents for the Colonies.
12. Athenaeus (*Deipnosophists*, I.7).
13. See, for example, Sahrhage, D. and J. Lundbeck. (1992) *A History of Fishing.* Berlin: Springer.
14. A component of the Census of Marine Life project is currently reviewing the recent history of marine animal populations and will eventually throw more light on the consequences of increasing human demand than I am able to do in this brief introduction.
15. In 2009, as I write, it is reported that wild Atlantic salmon, errants from several stocks, have again entered the Seine and have been seen at Paris!
16. Boesch, D.F. and R.B. Brinsfield (undated) *Coastal Eutrophication and Agriculture.* Paris, OECD.
17. Hoffmann, R.C. (2005) A brief history of aquatic resource use in mediæval Europe. *Helgol. Mar. Res.* **59**, 22–30.
18. Barrett, J.H., A.M. Locker and C.M. Roberts. (2004) The origins of intensive marine fishing in mediæval Europe; the English evidence. *Proc. R. Soc. B* **271**, 2417–21.

19. Jones, E.T. (2000) England's Icelandic fishery in the Early Modern period. In *England's Sea Fisheries since 1300*, Eds. D. Starkey *et al.* London: Chatham.
20. Lear, W.H. (1998) History of fisheries in the Northwest Atlantic: the 500-year perspective. *J. Northwest Atl. Fish. Sci.* **23**, 41–73.
21. Rose, G.A. (2004) Reconciling overfishing and climate change with stock dynamics of Atlantic cod (*Gadus morhua*) since 1505. *Can. J. Fish. Aquat. Sci.* **61**, 1553–7.
22. See pp. 70–6 in Cushing, D.H. (1988) *The Provident Sea*. Cambridge: Cambridge University Press.
23. Lotze, H.K. and I. Milewski. (2004) see note 3.
24. Moiseev, P.A. (1965) The present state and perspective for the development of the world fisheries. *FAO Report FAO/EPTA 1937–11*, 69–83.
25. Vakily, J.M. (1992) Assessing and managing the marine resources of Sierra Leone. *NAGA (ICLARM quarterly)* **15**, 31–5.
26. Longhurst, A.R. (1965) Shrimp potential of the eastern Gulf of Guinea. *Comm. Fish. Rev.* **27**, 9–12.
27. Fager, E.W. and A.R. Longhurst. (1968) Recurrent group analysis of species assemblages of demersal fish in the Gulf of Guinea. *J. Fish. Res. Bd.* **25**, 1405–21.
28. Baddyr, M. and S. Guénette. (2001) The fisheries off the Atlantic coast of Morocco 1950–1997. In Fisheries' Impacts on North Atlantic Ecosystems, Eds. D. Zeller, *et al. Fish. Cent. Res. Rep.*, 191–205.
29. For this suggestion, I am indebted to Menachem Ben-Yami, *Upon Reading Conversations or How to Support the Fisherfolk* – an electronic version of an essay in Samudra #36.
30. It has attracted several authors, of course, and my account owes much to the books of retired DFO scientist George Rose (2007) *Cod, the Ecological History of the North Atlantic Fisheries*, St John's, Newfoundland: Breakwater Press, and of Michael Harris (1998) *Lament for an Ocean*. Toronto: McLelland and Stewart. Both are written in anger by Newfoundlanders, and both with hands-on knowledge of the affairs. A more sober account is that of Elizabeth Brubacker (2000) pp. 161–210 in *Political Evironmentalism*, Ed. T. Anderson. Stanford, CA: Hoover Institution Press.
31. The Department of Fisheries and Oceans, based in Ottawa, was at that time responsible for fisheries research and management, oceanographic research and hydrographic surveying and chart production in three separate divisions; to these responsibilities has now been added the Canadian Coastguard.
32. International Commission on North Atlantic Fisheries.
33. It was only in 1995 that this was declared by Canada to be an EEZ, within the meaning of the International Law of the Sea.
34. North Atlantic Fishery Organisation, with Secretariat in Dartmouth, Nova Scotia. NAFO activities take place under the direction of its three constituent bodies, the General Council, the Scientific Council and the Fisheries Commission, supported and coordinated by the Secretariat. NAFO plans fishing regulations, ensures compliance, ensures adequate scientific advice and publishes results of investigations.
35. Winters, G. (1986), internal CAFSAC report.
36. Canadian Atlantic Fisheries Scientific Advisory Committee of DFO, based in the DFO Regional Office in Halifax.
37. Rose, G.A. and D.W. Kulka. (1999) Hyperaggregation of fish and fisheries: how catch-per-unit-effort increased as the northern cod (*Gadus morhua*) declined. *Can. J. Fish. Aquat. Sci.* **56** (Suppl. 1), 118–27.
38. The 'cod-end' is the terminal bag of a trawl.

39. Hutchings, J., *et al.* (1989) Is scientific enquiry incompatible with government information control? *Can. J .Fish. Aquat. Sci.* **54**, 1198–210.
40. Harris, M. (see above), note 30.
41. Bowering, W.R., *et al.* (1997) Changes in the population of American plaice (*Hippoglossoides platessoides*) off southern Labrador and eastern Newfoundland: a collapsing stok with low exploitation. *Fish. Res.* **30**, 199–216.
42. Rose, G.A., *et al.* (2000) Distribution shifts and overfishing the northern cod: a view from the ocean. *Can. J. Fish. Aquat. Sci.* **57**, 644–63.
43. Rose, G.A. (2004) Reconciling overfishing and climate change with stock dynamics of Atlantic cod over 500 years. *Can. J. Fish. Aquat. Sci.* **61**, 1553–7.
44. Ibid.
45. Hojort, J. (1914) Fluctuations in the great fisheries of northern Europe viewed in the light of biological research. *Rapp. Proc-verb. Réun., Cons. Int. explor. Mer*, **20**, 1–228.
46. Yndestad, H. (2003) The code of the long-term biomass cycles in the Barents Sea. *ICES J. Conseil* **60**, 1251–64.
47. See http://www.tuna-org.org/ for detailed information on each.
48. IATTC headquarters and offices are located on the campus of the Scripps Institute of Oceanography in La Jolla, California. I have always regarded IATTC as a model for how international fisheries commissions should operate; it is rigorously scientific, yet transparent, offering its data files for open inspection to the public.
49. Because porpoises and tuna form common aggregations there was initially heavy mortality of porpoises in purse-seines until Bill Perrin went public in a research paper which I am glad to say I didn't have sufficient bureaucratic good sense to ban!
50. Longhurst, A. (1998) Cod: perhaps if we all stood back a bit? *Fish. Res.* **38**, 101–08.

7

What is the real state of global fish populations?

'Man, attracted by the treasure that victory over the whales might afford him, has troubled the peace of their immense solitary abodes, violated their refuges ... in vain do they flee before him; his art will transport him to the extremity of the earth; they will find no sanctuary except in nothingness.'

de Lacépède, Histoire Naturelle des Cétacés, 1804

In the mid-twentieth century, before the globalisation of industrialised fishing, our main anxiety was not for the state of the stocks, but rather our uncertainty about how much we might harvest from them over the long term. There was very little information on which to base a prediction of the potential of sea fisheries, and it is not surprising that the early estimates ranged over at least an order of magnitude, from a suggestion made in 1951 of a potential of 22 million tons to the 260–350 million tons suggested in 1970. Some thought was also given to the amount that might be taken from invertebrate populations, chiefly of krill and cephalopods, which it was hoped might yield even more than the fish populations themselves. The natural variability of biomass was rarely mentioned in these discussions of potential catches, and the early investigations by Johann Hjort of the comings and goings of fish populations appear to have been forgotten.

Although most of the early predictions were no more than educated guesses, some were computed more rationally, as was Moiseev's suggestion[1] that 120–150 million tons of fish should be harvestable: he obtained this figure by the use of the simple relationship '*observed fish biomass* $\times$ *P/B* $\times$ *factor*'. A more sophisticated, multi-trophic level model was used by Ryther in 1969 to compute a potential maximum yield of the sea fisheries from estimates of global phytoplankton production: he predicted a global total for sea fish of around 100 million tons.[2]

A critique of Ryther's work suggested that the inherent levels of uncertainty in his model were so great that its output could not be taken seriously: at that time, we had very little confidence in global estimates of phytoplankton biomass, let alone of primary production and, besides, it seemed a great leap of faith to assign all fish production to a single trophic level, and 'to reduce the variable values for efficiency of transfer of material from predator to prey to a single set of values representing ecological efficiency ... involves a possible error of an order of magnitude or more, depending on the ecological efficiency factor chosen'.[3] Nevertheless, his prediction was not that far off the mark!

Later estimates, largely by authors associated with FAO, converged on global catches of around 80–100 million tons but, as early as 1971, an FAO study[4] noted that although global catches were reported to be increasing at a rate of 7% a year, and thus doubling every 15 years, for 'some stocks, e.g. the larger demersal fish in the North Atlantic, it has long been clear that catches have approached the level of the total potential of the area'. Nevertheless, concerns for the continued expansion of sea fisheries did not yet run very deep. At a CalCOFI discussion in 1969, a representative of the fishery industry expressed great confidence in the future expansion of global catches. He thought that these might reach 123 million tons by 1986, basing his projection on the evolution of market demand and on resource availability, believing that there was sufficient potential for the expansion of industrial fishing to new species to make this possible. Since these remarks were from Wilbert Chapman, who probably knew more about the fish business than anyone else before or since, I am happy to see from the record that I publicly disagreed with his projection, pointing out that of the 12 species that then produced 60% of landings worldwide, 5 were already in trouble.

In the closing decades of the twentieth century, it became apparent that global landings of sea fish had reached their limit – at least from conventional fisheries as recorded by FAO – because the reported rate of increase slowed in the late 1980s and reached what seemed to be a fluctuating plateau of 85–95 million tons in the 1990s. Unfortunately, although the FAO statistics for total world catches are widely quoted, they have a very much greater level of uncertainty than is generally understood. In fact, the most believable interpretation of these data, corrected for misreported landings (and also for the ENSO effect on Peruvian anchoveta populations), is that the reported trend of global landings is incorrect: in reality, global landings of all other fish

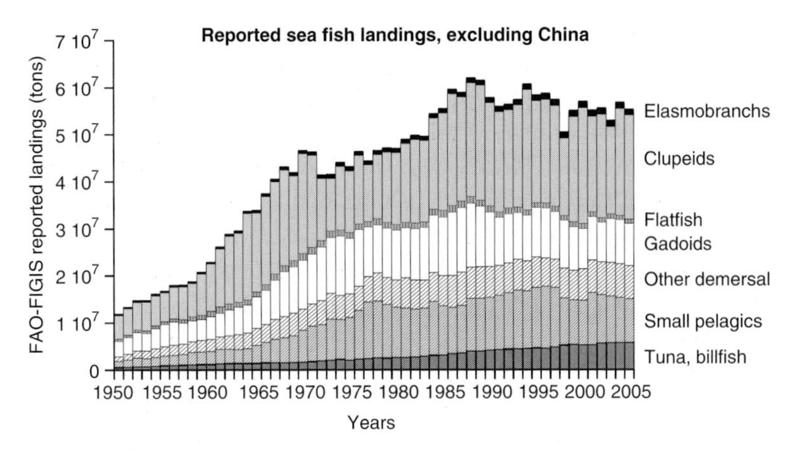

Figure 7.1 Global landings of sea fish from 1950 to 2005 reported to FAO, with those of China excluded.

peaked in 1989 and have declined slowly but progressively until the present time (Figure 7.1).[5]

HILBORN'S GREAT DIVIDE IN FISHERIES SCIENCES

Once it appeared that global catches had peaked, and before the extent of the subsequent decline had sunk in, concern for what might be the long-term potential of the sea fisheries was quickly replaced by concern for the ability of the populations to sustain the catches already being taken. This rapidly became part and parcel of the general popular concern for the global environment, with the unfortunate consequence that not all subsequent evaluations of the situation have been as objective as one would wish. Alarmist reports concerning the status of global fish stocks began to appear in the daily press, quoting papers by 'international teams of scientists' published in 'influential scientific journals'. Ray Hilborn, evoking a divide in the science, noted that these papers, published largely by ecologists (a group that he later distinguished from fisheries scientists), purported to offer very precise quantifications of the current state of the stocks and of their rate of depletion.[6] Pauly also lamented the loss of what he described as a developing consensus in fishery science concerning global stock status represented by heavy criticism of these papers.[7]

Polachek commented on one of these contributions, which proposed that a catastrophic collapse of open-ocean tuna populations had occurred; he suggested that the scientific method had been abandoned by the authors. Their chosen method was, he wrote, 'more like a set of tactics that we have come to expect from a political campaign than those underlying a scientific debate ...'.[8] He also discussed the refusal of *Nature* to publish a reasoned rebuttal of another such paper that was evidently in error. An obituary of the late Sir John Maddox, redoubtable past editor of *Nature*, reported that he fustigated during the 1970s against inadequate peer-review of such alarmist papers, written by ecologists and based on 'incomplete facts or irresponsible exaggerations likely to ... divert attention from the real problems'.[9]

However, Maddox long ago left the helm of *Nature*, unfortunately, and these papers have often been published either in *Science* or *Nature*. They represent the alarmist or 'ecological' side of the divide that appeared in fishery science at the end of the twentieth century. Views of the future of fisheries were very different on the two sides of the divide between ecologists (and I don't use the term in its classical meaning) and fishery management scientists. One such study that was published in *Science* a few years ago now[10] continues to be widely reported, and newspaper journalists continue to tell us that unless we change our ways, all fish stocks will have collapsed by the year 2048: that well-worn prediction has landed on my breakfast table twice in just the last few weeks!

My critique[11] of it was rejected by *Science* on the grounds that 'We are aware that fisheries science and ocean ecology are contentious subjects, and that areas of legitimate disagreement exist.'[12] Yet Hilborn (a far more experienced critic than I am) characterised the conclusions of this paper as 'absolutely and totally wrong', and such papers in general as 'either outright wrong or serious distortions of reality'.[13] I do not think that he exaggerated, although most critics seem not to have understood that the principal objective of this paper was to demonstrate that the relative biodiversity of ecosystems influences their resistance to disturbance so that criticism of the unfortunate 2048 projection has tended to conceal the fact the study was flawed with regard to its main objective.[14] The principal error lay in using the series of 64 large marine ecosystems (LMEs) as if they were standard experimental plots, rather than the highly diverse entities that they are, to make an inter-comparison of the kind attempted – which was, moreover, performed with data inappropriate to the task.

Rapid rates of depletion or collapse have been suggested for large pelagic fish and some of these projections, too, have been criticised for an apparent lack of rigour. Myers and Worm compiled data from which they suggested that pristine biomass, prior to industrialisation, could be estimated and in which the subsequent trajectories of the catches of fished species could be traced, as could the evolution of fishing effort[15]: data from continental shelf trawl surveys in the NW Atlantic, in the Gulf of Thailand and off South Georgia, and from global long-line fisheries for oceanic tuna in nine oceanic, mostly warm-water regions. From the evolution of apparent biomass (as catch per unit effort, CPUE) in each fishery, a rate of biomass decline was obtained for each species. The indicated rates for oceanic tuna were surprisingly high, especially in the early years of long-lining and in temperate regions where equilibrium catch rates were lower than in warmer seas.

The general conclusion of the authors – that 'the global ocean has lost more than 90% of large predatory fishes' – was challenged formally by several people who pointed out, for example, that species-aggregated CPUE is an unrealistic index of biomass, because it must evolve with the evolution of fishing technique over such an extended period.[16] It was also shown that the apparent decrease in CPUE in the first few years of a long-line fishery was not a novel finding, but had been debated for decades within the tuna science community. Subsequently, a detailed study of long-line data from Indian Ocean sub-regions found no difference in CPUE when comparing catches in the same year in previously fished and in unfished regions.

Unfortunately, the extrapolation from Japanese long-line data was also misleading for other reasons, especially because the progressive global switch from long-lining to purse-seining resulted in major increases in landings of species (such as yellowfin) that were supposed by the study to have been depleted by long-lines; these remain today some of the best-managed populations anywhere. Further, the changes in catch rates of southern bluefin in the subtropical region had little to do with the near-total population depletion proposed by Myers and Worm, but was the result of a change in the 1960s from canning tuna at sea to obtaining fish to satisfy market demand for sashimi, made possible by new ultra-deep-freezing techniques. This induced the long-liners to move into high southern latitudes where fish quality was higher in cooler seas. Albacore, a by-catch in the Japanese fishery in the subtropical zone, naturally appeared to decline rapidly in abundance in the overall data, but in the Taiwanese subtropical long-line fishery it was the target species, and showed only a very gradual decline in CPUE over a 35-year period.

A later and more indirect response to this study by Hampton and others suggested that some species of pelagic predators in the Pacific had indeed lost 36–91% of the biomass that would be predicted in the absence of fishing, and that the proportion of very large fish in each population had declined slightly.[17] These authors thought that 'substantial, though not catastrophic, impacts of fisheries on top-level predators' have occurred and that the authors of the original study had done 'a disservice to the fisheries community by applying a simplistic analysis which exaggerates declines in abundance ...'. Paul Kleiber[18] noted that it took two years to induce *Nature* to print a comment, to which he contributed, on the Myers and Worm paper. Despite all this criticism, the original suggestion – that 90% of top oceanic predator biomass is now missing – continues to be reported in the scientific literature as established fact.[19]

The US pelagic long-line data have been used to report an alarming decline in the abundance of several species of sharks in the NW Atlantic: these data, assuming that catch rate reveals abundance, suggest that hammerheads declined by about 90% and white sharks by about 80% between 1986 and 2000, while tiger sharks and coastal species of *Carcharinhus* declined by 60–65% in the same period.[20] Later, the same team concluded that two species of *Carcharhinus* in the Gulf of Mexico had declined by 99% and 90%, respectively, since 1950.[21] In these studies, a relationship between catch and effort was obtained from logbook data concerning the incidental catches of sharks during tuna fishing operations; because of the nature of these data, the authors suggested that problems should be anticipated in, for instance, distinguishing missing values from true zeroes.

Once again, these claims were discussed in the open literature in a formal sequence of challenge, authors' response and final reply that, once again, leaves the careful reader with no clear opinion of the reliability of the study[22]; in this case, not only were the validity of the logbook entries by untrained crew members challenged, but – as in the study based on Japanese long-line catches discussed above – the effects of changing capture techniques is raised. The change from steel to monofilament leaders (that sharks can bite through and so escape) could explain the Gulf of Mexico decrease, and the change from standard 9/0 J-hooks in the 1950s to a range of different hooks later would affect the catch rates according to the challengers. The reader may find it as difficult as I do to judge the validity of the authors' response, and of the final reply to that response.

We can hope that contributions such as those described in the previous paragraphs are behind us now, and that a more balanced view of fisheries management and the status of global populations can be expected in the future. I am comforted in this hope by a very recent publication in which two of the principal protagonists of opposing viewpoints have collaborated.[23] This paper discusses the evolution of the globalisation of fishing that was discussed in Chapter 6 and suggests, from the results of ecosystem modelling in 10 well-studied regions, that the average exploitation rate is in decline and that this is consistent with requirements for the rebuilding of stocks.

Some have suggested that the contributions of conservationists discussed above have been drafted principally to satisfy what has been described to me as 'the societal responsibility of scientists', conforming to the published suggestion that 'scientists have a right and, arguably, a responsibility to take on advocacy roles'.[24] Although the role of scientists in society is complex, I would suggest that our primary responsibility is not advocacy, but rather to get as close to absolute certainty as possible, and to specify the level of the uncertainty that is inherent in any prediction of future states of the natural world, including that of marine fish populations. At the very least, we have to reveal quite openly the difficulties likely to be encountered when assessing the current state of sea fish populations, and when predicting their ability to maintain present landings. These levels of uncertainty are very great, but are largely obscured in published fishery statistics.

Huxley's well-worn aphorism that 'scepticism is the highest of duties, blind faith the one unpardonable sin' is very appropriate today, because the aggregation of data from a multitude of disparate and untidy sources into a tidy and comprehensive database is now so easy, and because the product is ideal for simple predictive modelling. These new abilities have the potential of confirming the selective conjectures of those who know what answer they want to get.

WHAT DO WE REALLY KNOW ABOUT THE PRESENT STATE OF GLOBAL FISH STOCKS?

We could quantify the state of global fish populations very accurately if we had access to formal stock assessments, in the form prescribed by fisheries management science, in all seas; but it goes without saying that the number of such assessments performed annually is trivial compared with the very large number of species that are fished. Some of the major regional fisheries management organisations (RFMOs), such as ICES, have access to the results obtained by state-of-the-art resource assessment surveys performed by their member nations, but these cover only a small fraction of all fisheries. Elsewhere, formal stock assessments and resource surveys are very largely lacking, as is any formal control of fishing mortality within limits that are thought to be biologically safe.

Global assessment of the status of fish populations, therefore, has come to depend very largely on analysis of the reports of landings that are recorded nationally and aggregated for analysis at regional or global scale: this is a process fraught with uncertainty. Although changes in reported catches are habitually assumed to indicate changes in stock biomass – even in peer-reviewed articles – this assumption is valid only if we also have access to data on changes in fishing effort over the period of interest. Unfortunately, while it is relatively easy to assemble global or regional histories of the capture rates of marine fish, available data archives of fishing effort are much less comprehensive, and so it is logically impossible to compute an instantaneous biomass of global fish populations.

For the use of catch data as a proxy for the evolution of stock biomass, we have access to a large number of regional and national data archives: of these, the most useful are those provided by RFMOs, including not only ICES, but also NAFO, IATTC, ICCAT[25] and many others, together with some tens of national fisheries agencies. None of these, however, is as comprehensive as the global compilation by the fishery staff of the UN Food and Agriculture Organisation (FAO) of the data submitted to them concerning the annual catches of national fleets since 1950. FAO also offers some statistics on the size and number of fishing vessels both nationally and regionally, but these are less complete than landings data and do not extend so far back in time. The FAO files are easily and freely accessible to all interested parties and have been the foundation for many assessments of the general condition of global fishery resources. Nevertheless, as FAO officials have always recognised, the data sets are replete with uncertainties.

This should be no surprise, considering the process by which each nation gathers the statistics that are subsequently reported to FAO; each fishing unit from canoes to factory trawlers must either submit written landing reports, or its landings must be recorded by a fisheries observer on the beach or quayside; in each case, species identifications must be made and recorded, and in many cases weights will have to be estimated by eye. Large vessels have many incentives to report catches incorrectly, as shall be discussed later, and at the other end of the scale the landing points of artisanal fisheries may be too numerous to be monitored: landings are made at more than 150 fishing villages along the coast of Sierra Leone, for example.

Finally, the data must pass through the bureaucratic mill of each nation for consolidation prior to being reported to FAO, where the FishStat database is maintained. We now know (what we should have guessed) that this process may introduce very serious errors, both through negligence and intentionally. The reports from China, a major fishing nation that catches in the region of 17 million tons of sea fish annually, began to concern the FAO in the mid 1990s, and similar problems were recognised in statistical data concerning agriculture submitted by China to other UN agencies. The political liberalisation of China in the 1980s, during which fisheries were privatised and which permitted catches to be sold anywhere, even at sea, appears to have swamped the management and enforcement machinery, which is no longer capable of tracking the number of fishing boats (perhaps as many as 300,000) currently operating on the Chinese continental shelf. It has been suggested that because the state entities that monitor fisheries also have the responsibility of enhancing that sector, there is a temptation to exaggerate success (human nature being what it is).[26] The establishment of a zero-growth policy in 1998 was faithfully recorded in the China statistics, which thereafter reported similar landings each year.

FAO has been quietly addressing these problems with the Chinese authorities as, indeed, it does with the authorities of any nation concerning problems with statistics that have been submitted to it. This was a special example of a general problem, for accurate reports to FAO will only be presented by nations having a fully effective central fishery agency: many nations, unfortunately, cannot be so characterised. It is not clear if public exposure of the problem has helped to resolve it,[27] but most critical tabulations of global sea fish catches now report China separately from the rest of the world. However, it is difficult to be convinced that such problems are China's alone, and careful analysis of the reporting situation in some developing but fish-rich nations reveals problems that are difficult to resolve. Mauritania is such a nation which, unlike its neighbour Senegal, has very little tradition in fishing so that its resources have been exploited mainly by foreign fishing fleets which have not always reported their catches to FAO as originating off Mauritania. While the reorganisation of the FAO statistics by the SAUP (see below) has re-allocated most of this to the Mauritanian zone, several hundred thousand tons of pelagic fish recorded in the database maintained by the Mauritanian fishery authority 'have simply disappeared from the statistics reported to FAO'.[28] In the same nation, landings data from small-scale fishing (which despite this have produced landings of 60,000 tons in recent years) have been relatively neglected, and by-catch data ignored: from the shrimp fisheries, only shrimp landings were recorded.

The tidy tabulations of catches by FAO that are available for downloading conceal the extraordinary complexity and uncertainty of the original data: the global tables for the year 2000 include entries that represent 850 species, or species-categories, of teleosts and elasmobranchs from 150 reporting nations; this results in a total of 5113 data points that are distributed among 18 sea fishing regions representing the entire surface of the oceans. About 17% of total landings are not reported to species group, so informal taxonomic categories have been established in which to place such data: the most highly aggregated category is simply 'marine osteicthyes'. In 1996, a typical year, FAO had to make informed although arbitrary allocations in this way to the data that had been reported by 20 of the 50-odd major fishing nations.[29] In some regions, especially in warm seas, the extent of such arbitrary allocation is extensive: 25% and 36% of all landings in the east and west central Atlantic, respectively, are reported in an aggregated category in the final tables. Solutions can be, and have been, devised that will result in all landings being allocated to a single species or, at most, a genus; nevertheless, interpolations and extrapolations of the kind required must further sap our confidence in the reconstructed data table. It should be noted that the FAO recently performed a quality control on all post-1970 data, to give greater confidence to analyses performed with it. Consequently, it has become customary to restrict retrospective analysis to the last 30 years or so.

The impression of precision when the data are aggregated into FishStat is seen to be illusory when single national data sets are examined. This may be illustrated by data representing clupeid catches from Ghana, a West African nation with a strong fishery tradition and a stable administration. According to these, Ghana caught only *Sardinella* sp. from 1950 to 1968, after which four items (*Sardinella aurita*, *S. maderensis*, *Ethmalosa fimbriata* and *Engraulis encrassicolus*) were progressively added during the next 10 years; it was not until 1989 that the full range of Ghanaian clupeoid catches was reported at species level although, and even then, *Sardinella* sp. also continued to be reported right up to 2001. This makes it very difficult to evaluate the reported near-collapse of the *S. aurita* population during the 1970s, and the resulting explosion of the pelagic triggerfish *Balistes capriscus*. Unfortunately, this situation is typical of catch statistics offered by many nations.

This same region will serve to demonstrate the level of uncertainty that must be placed on FAO data that represent landings in smaller natural regions than the principal statistical areas. The reported landings of species typical of the warm-water fauna of the Gulf of Guinea that are included in the landings statistics for the statistical area East Central Atlantic Region seem not to reflect the list of species recorded by resource surveys performed in the tropical region. So, I have informally extracted from the FAO regional data those species and groups that are restricted to the warm shelf water of the Gulf of Guinea by accumulating the catch data reported by the individual countries from Guinea-Bissau to the Congo, excluding distant-water catches by other nations. Many subjective judgments had to be made concerning the allocation of individual data sets to named species but, even so, it is clear from Figure 7.2a that the trajectories of the reported landings of the dominant continental shelf demersal species in no way support other evidence that the tropical Gulf of Guinea is one of the most depleted FAO regions. There is a clear conflict between the data used for Figure 7.2a and survey data recently reported from the Ghanaian shelf. The results of routine trawl surveys done over the period 1963–90 showed a reduction in catch rate from 40 kg/h^{1} to only 15 kg/h^{1} over this period from the shallow shelf community (0–30 m) comprising sciaenids, polynemids, drepanids and other demersal species. It would be very surprising if this trend had not continued and if the Ghanaian results were not also applicable to the shelf from, say, Sierra Leone to Nigeria – and yet, if that is the case, how should we explain the data shown in the figure?

Then, how should we interpret a similar analysis (Figure 7.2b) of the landings from the FAO SW Pacific region, encompassing the Indonesian archipelago and coasts of the Gulf of Thailand? Again, this region is often proposed as an example of seriously degraded fisheries resources and is discussed later in this chapter, but the accumulated national statistics suggest a different situation: one of steadily increasing landings in all the principal categories of demersal fish. I can only remark that, as in the case of the data supplied to FAO by China, the annual statistical returns from many nations should be examined even more critically than they have been in the past.

The 18 FAO sea fishing regions into which the statistical data are aggregated are very large, averaging >20 million km^2, and this conceals a great deal of information concerning the actual distribution of fishing fleets and the origin of their catches. Watson and others addressed this problem in 2004 by disaggregating the original data onto a

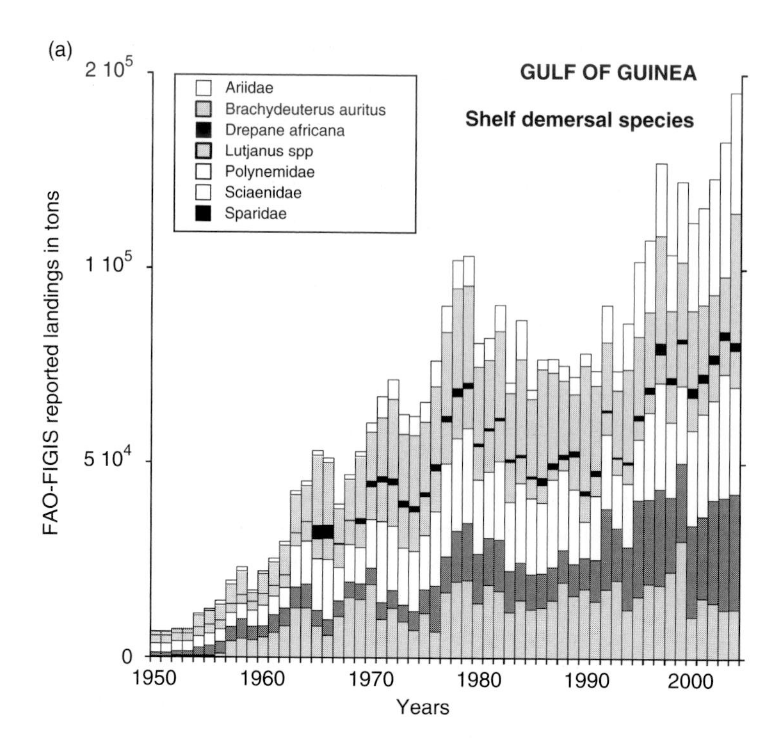
(a)
GULF OF GUINEA
Shelf demersal species
Ariidae
Brachydeuterus auritus
Drepane africana
Lutjanus spp
Polynemidae
Sciaenidae
Sparidae
FAO-FIGIS reported landings in tons
2 10^5
1 10^5
5 10^4
0
1950
1960
1970
1980
1990
2000
Years

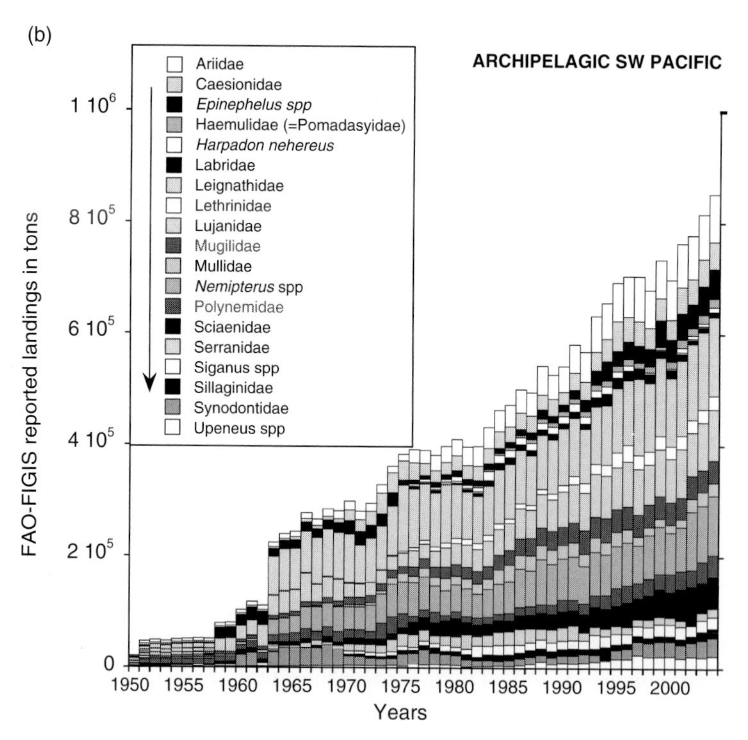
(b)
ARCHIPELAGIC SW PACIFIC
Ariidae
Caesionidae
Epinephelus spp
Haemulidae (=Pomadasyidae)
Harpadon nehereus
Labridae
Leignathidae
Lethrinidae
Lujanidae
Mugilidae
Mullidae
Nemipterus spp
Polynemidae
Sciaenidae
Serranidae
Siganus spp
Sillaginidae
Synodontidae
Upeneus spp
FAO-FIGIS reported landings in tons
1 10^6
8 10^5
6 10^5
4 10^5
2 10^5
0
1950
1955
1960
1965
1970
1975
1980
1985
1990
1995
2000
Years

half-degree grid of cells covering the entire ocean by means of a rule-based approach, with constant reference to ancillary information concerning the biogeographic distribution of fished species, and the areas within which each reporting nation had legal rights to fish.[30] This extraordinary task achieved a data set parallel to that of the FAO data, and for the same period, which can be accessed for any aggregation of the 180,000 spatial cells that represent all ocean areas: these cells range in size from 3000 km^2 at the equator to a tiny area near the poles. From these data, it is possible to plot synthetic quantitative maps of the distribution of catches of individual species or aggregated groups of species. To the extent to which it was possible, species identification problems in the original data were rectified individually during this process.

These products are provided online by the Sea Around Us Project (SAUP) led by Daniel Pauly and located at the University of British Columbia[31]; this project offers graphical and mapped products of sectors of the whole, specified by the user, to visualise regional, taxonomic or temporal data groups. Although the tabular material originates in the FAO capture landings, the output data from the project harmonise this with regional data obtained from ICES, NAFO and other fishery management organisations and also from national sources, such as DFO Canada. Increasingly, this database is used in preference to the original FAO data in recent discussions of the state of marine ecosystems, but I think it would be wise to have the same diffidence concerning the SAUP data as has been suggested would be appropriate for the FAO archives: an FAO staff member once wrote 'The FAO data set can best be used as an indicator of trends and a generator of research hypotheses to be tested by more exact analyses in individual, smaller and more homogeneous fishing areas ...' The harmonised data set from SAUP is now likely to be presented uncritically by other users as representing detailed catch data, finely partitioned over the entire surface of the fished ocean. This, unfortunately, adds unjustified gravitas to the presentation for readers who are unfamiliar with the uncertainties associated with the original data.[32]

Caption for Figure 7.2 (cont.)
(a) Landings of demersal fish from the Gulf of Guinea continental shelf, as reported to FAO by riverine nations, and also by those whose distant-water ships fish in this region; (b) landings from the SW Pacific FAO regional fisheries, 1950–2005 from FAO data for major species groups. Data shown in both figures appear to be at odds with other reports of stock status in these regions.

Turning now to a final difficulty in interpreting the FAO data, or any other long time-series purporting to indicate captures from a population, it is also necessary to remember that these may conceal the consequences of progressive changes in fishing methods or of progressive changes in fishing location within very large statistical regions. This is not a trivial problem, and some of the assumptions commonly made for the extrapolation of fishery data over an entire region or stock may give incorrect results.[33] The problem arises if, as is often the case, the fishery does not exploit the entire region equally: especially in the initial and terminal phases of a fishery, gross errors can be (and have been) made concerning the true state of a population.

True information concerning the evolution of the CPUE relationship is essential information in any stock evaluation and requires the use of stratified sampling statistical procedures. Weighted averages of individual cells or strata must be obtained in such a way that they are not influenced by the numbers of observations, thus placing undue weight on the results from fished cells, which in a new fishery will be concentrated in only a part of the species range. Because, in this case, catch rates rapidly decline, the data will suggest an overall stock status that is worse than it really is: a correct assessment can only be made if information can be obtained concerning population status in unfished areas of the species range. In the early or maturing stages of a fishery, this error may cause what has been termed 'hyperdepletion', when the population appears to have declined to lower abundance levels than has actually occurred.

In the terminal phases of a fishery, the same error of extrapolating CPUE ratios from fished areas to the entire region potentially occupied by the fish species may produce even more serious errors, although in the other sense because the catch and effort data may now suggest that a population is more abundant than is really the case. This well-known phenomenon results from the natural behaviour of some species in response to a reduction in their absolute abundance. McCall has reviewed studies of the phenomenon in insects as well as fish, birds and other vertebrates and has generalised it as the *population theory of density-dependent habitat selection*.[34] As population numbers are progressively reduced, either naturally or by a fishery, the remaining individuals may – so as to maintain a comfortable density of individuals – aggregate into progressively smaller parts of their original range. Because the fishing fleet will aggregate accordingly, in their case to maintain a comfortable CPUE ratio, the aggregation of the target fish may be masked, and can only be revealed by deliberate test-fishing over the entire range that the species occupied prior to fishing. Unfortunately, this does not always occur: in the Newfoundland cod fishery, as was discussed in Chapter 6, reductions in the area occupied by the individual populations accompanied declines in stock biomass and caused significant errors in stock assessment.

UNCERTAINTIES ARISING FROM DISCARDS AND FROM ILLEGAL, UNREGULATED AND UNREPORTED CATCHES

It has become customary to recognise that landing statistics are imprecise, and to distinguish between the landings that are formally recorded by each nation and the remainder of the fish killed by fishing; the latter are habitually grouped as IUU (illegal, unreported and unregulated) catches, this being a catch-all term that refers to several different processes, each of which has different consequences for the accuracy of the catch data: (i) *illegal fishing* refers to practices that contravene any regulation in force, from gear restrictions to seasonal closures – catches from illegal fishing may or may not appear in landing statistics; (ii) *unreported fishing* means what it says, and may or may not result from illegal fishing; (iii) *unregulated fishing* means only that no regulations apply to a fishery, from which landings may or may not be reported, and finally, (iv) *discarding* is considered by some authors to be included in the unreported fishing category, while for others it is to be discussed separately; for convenience, I shall follow the latter course. It is not surprising, therefore, that we lack consensus on the regional and global importance of such diverse processes.

Each of these processes introduces errors into the reported landings that are published annually by FAO, which prints in each annual statistical volume a diagram (see Figure 7.3) to illustrate the difficulties of reaching totals on which we may place confidence and the steps that must be quantified to achieve a true estimate of fishing mortality. Unfortunately, as the FAO text that accompanies the diagram comments rather wryly: 'these steps represent the procedures that should be used, not necessarily those that are actually used, by the entities submitting data to FAO'.

Two recent and comprehensive estimates of the global significance of IUU fishing and of discarding are in agreement that the errors that are introduced into global fisheries statistics are not trivial. The evolution of the losses that have been incurred since the mid-twentieth century have been computed from the database of the Fisheries Center at UBC,[35] and this analysis suggests that IUU mortality and discards,

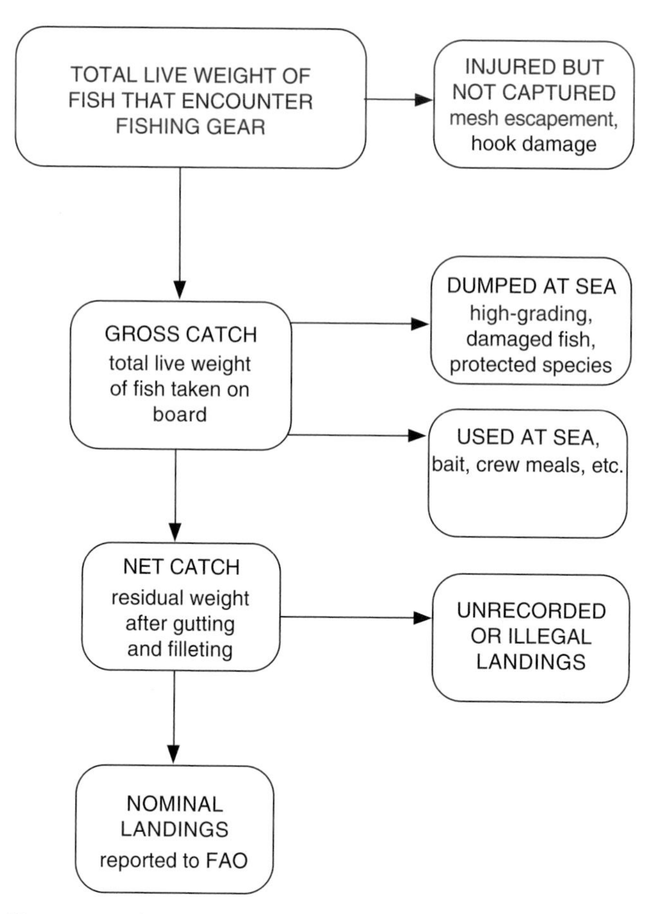

Figure 7.3 Diagram to illustrate the potential pre- and post-catch losses of fish prior to reporting of landings, and therefore the probable bias towards under-reporting of fishery mortality by FAO statistics.

tallied separately, have routinely been equivalent to about half of the total of all legal landings that were reported to FAO in 1970, about 67% in 1980, 62% in 1990 and 48% in 2000. The study suggests that the proportion of discarding to other IUU losses has changed significantly during the same period, building to a maximum in 1990, when discards were approximately similar to other IUU losses, and subsequently reducing to quite a small fraction of total losses.

Despite this apparently progressive reduction in the rate of discarding towards the end of the century, a study commissioned by the UK Department of International Development computed IUU fishing losses to legal trade of between US$2 and US$9 billion in the period

2003–2005, when other studies suggested that discard rates were declining. Of these losses, about US$1.2 billion represented losses in the high seas fisheries: tuna, cod, redfish, orange roughy, and so on.

Case studies of IUU losses in national fisheries in Iceland[36] and Morocco,[37] nations that have very different social, scientific and enforcement regimes, help us understand why and where IUU fishing is most prevalent. Off Iceland, discards appear to have been only 1–11% of all cod catches and 2–20% of haddock until about 1985, after which incentives to discard and misreport became higher but, even then, were less than 14% and 28%, respectively. Off Morocco, proper recording procedures were not put in place until 1974 so that, before that year, landing data were certainly incorrect. In the 1990s, estimates of unreported landings were 47–60% of total landings in the coastal fishery and 47% in the industrial fishery; discards were estimated at 30% of total industrial catches and 16% of those in the coastal fishery. Given the responsiveness of the Icelandic fishery patrols in the Cod Wars with Britain, and the general irresponsibility of foreign industrial fishing off western Africa that will be discussed in the next chapter, these results are not surprising.

Studies of the discard problem have a curious history, which is worth briefly retelling. Three attempts to quantify discards globally are usually considered to be the most useful, and each was published as an FAO Technical Paper or Circular, so that none appears to have been subjected to formal peer review. Neither the 1983 study of Saul Saila, nor that of Alverson and others of 1994 remains on the FAO website, nor, curiously, can they now be ordered from the FAO.[38] In the Foreword of the 2005 study by Kelleher,[39] the authors of the 1994 document withdraw their contribution, which was based on a species-by-species analysis, and urge 'that the 1994 global discard estimates are no longer cited to decry the state of the worlds fisheries ... By-catch and discard problems must be addressed fishery by fishery and we urge that scientists and advocacy groups alike focus on the successes of the past decade rather than on the continued citing of data not applicable to fisheries in this century.' This, at least to me, smacks somewhat of organisational advocacy but, fortunately, their valuable contribution has not been lost to science.

Kelleher's conclusions, which unlike the Alverson study are based on a fishery-by-fishery approach, are upbeat and suggest that the average annual discards from 1992 to 2001 were only a little more than 7 million tons annually, or about one-quarter of the 27 million tons (subsequently adjusted to 20–22 million tons) suggested by the Alverson study, although it is claimed that the two estimates are not directly comparable because they were obtained differently. Kelleher, however, suggests that the apparent reduction in discard rate is real and ongoing and is the result of changes in the fisheries: increased utilisation of catches, improved processing technologies, expanding markets for small fish, more selective fishing gears, introduction of regulations to control by-catch and discarding, and improved enforcement. Although many examples of each of these improvements in fishing procedures are cited, one can perhaps be forgiven for suggesting that Kelleher ignores some other important reasons for the difference in the two estimates; in particular, the collapse of some large fisheries in which discarding had been rife in the 1980s.

I am comforted in this opinion by the recent suggestion of Zeller and Pauly that the Alverson report should be taken seriously because the reduction in the global rate of discarding may result from a global catch rate that is declining more steeply than was previously assumed, rather than from improved management and utilisation practices.[40] Using all estimates of discards since 1975, Zeller and Pauly suggest that interpolation between estimates at 10-year intervals gives a 'most probable' progression of total global catches (recorded landings + IUU landings + discards) that peaked at about 105 million tons in 1989, and then declined progressively by about 2 million tons each year to the end of the century. Kelleher produces no real evidence that improved conservation practices have halted or reversed the observed decline in landings, and other mechanisms are discounted; the report does not discuss the simple fact that when a major fishery collapses – as in the case of the NW Atlantic cod and halibut fisheries in the international and Canadian EEZ zones – then the discards associated with it are no longer tallied into the global total. This is obviously at least part of the reason why the percentage of discards for the NW Atlantic is shown as being significantly lower than in the NE Atlantic.

The FAO report, even if not very convincing as to the effectiveness of remedial measures, is still an excellent source of detailed information on recent discard rates in individual fisheries and regions: the results are chilling. In the North Sea, 20–50% of haddock caught is reported as discarded, as are about 50,000 tons of whiting, while flatfish beam trawlers have discard rates as high as 80%. The French trawler fleet operating in the Celtic Sea discards 33% of its catch, while Spanish multispecies trawlers discard 45%, and so on. More generally, the greatest gear-specific discard rates are associated with shrimp trawls (62.3%), tuna long-lines (28.5%) and benthic dredges (28%), while the smallest rates are associated with midwater trawls (3.4%), and purse-seines used for tuna and small pelagics (5.1% and 1.2%, respectively). Demersal long-lines (7.5%), demersal fish trawls (9.6%) and tuna seines (9.6%) all have moderate discard rates. Consequently, region-specific rates vary significantly: the southwest Pacific, central America and western central Atlantic, where shrimp trawling is extensive, have high regional discard rates (38–68%), while the southeast Atlantic and Pacific, dominated by upwelling ecosystems and fisheries for small pelagics, have very low rates (3–6%). The worst discard rate of all is in the shark-fin fishery – 92.3% of the biomass of the catch goes back over the side, finless!

These dismal figures tend to conceal the fact that some progress has been made, and that in a few fisheries the discard problem is now negligible: in the Southeast Asian and African coastal fisheries there is both increased utilisation of small fish in the markets and also to satisfy an increasing demand for aquaculture food. Parenthetically, it is bizarre that in maritime Nova Scotia, where fishing has been a traditional way of life since the beginning, we are offered frozen fillets of freshwater catfish (*Pangasius hypophthalmus*) from Thailand and of tilapia from Vietnam, fattened in this way.

There has also been some progress in recent decades towards utilising the by-catch, rather than discarding it. In the tropical shrimp fisheries and in some of those targeting high-grade teleosts for northern markets, arrangements are increasingly being made to pick up at sea the small, low-grade fish that would normally be dumped; these are transported to local markets for sale. The use of low-grade fish, invertebrates and offal in the manufacture of feeds for aquaculture is also on the rise. Finally, some thought is being given to the most difficult problem of all – the interaction between quotas, market prices and hold capacity in the industrialised cold-water trawl fisheries – although this problem remains very far from being solved, as shall be discussed later.

An associated problem is the by-catch of organisms other than fish: porpoises in purse-seines, albatrosses on long-line hooks while setting the gear, turtles in tropical benthic fish traps, and so on. Most such problems can be resolved by modifications to fishing gear; for instance, the mortality and discarding of porpoises in the tuna purse-seine fishery was quickly solved in the early days of the fishery as soon as the public became aware of how many were being drowned. Relatively simple modifications to the seine and to fishing procedures were sufficient to release these animals, by backing down briefly after the set is completed, to sink part of the head-rope and allow the dolphins to escape at the surface, while the entrapped tuna remain deeper in the net. In the early 1960s, nearly half a million small cetaceans were drowned annually, but by 1990 this kill had been reduced to about 5000 dolphins annually.

However, the greatest uncertainty in the FAO data set may not be discarding, but rather the unknown extent of unreported landings, resulting from actions by the crew at the instruction of the owners, and also from deliberate misreporting by national fishery authorities: both sources of error are sufficiently serious as to throw much doubt on the real significance of the FAO global data. The FAO itself is in a very delicate position vis-à-vis its member states, and perhaps this is the reason that their major study of the problem, collated by Bray, restricts itself to suggesting the measures that might be taken by regional fisheries bodies with regard to landings by open register states (the flag refuge of many vessels fishing otherwise illegally) and by vessels flagged in non-Party states.[41]

A recent report by the Marine Resources Assessment Group, commissioned by the UK government,[42] on unreported fishing on the high seas points out that problems in individual fisheries vary widely in severity, depending on the effectiveness of the local regulation and incentive levels. For example, unreported fishing on bluefin tuna in the eastern Atlantic is probably <1%, but a figure of 50% has been evoked for the same species in the Mediterranean. The high seas fisheries that have been most seriously targeted by unreported fishing are, according to this report: warm-water tuna worldwide, pelagic sharks worldwide, cod and redfish in the North Atlantic and squid in the South Atlantic and Pacific Oceans. In 2003 the European fleet on the high seas was tasked with 9502 individual infringements – most of which resulted in a penalty – committed by 10% of the fleet, most of which were sailing under the flags of convenience of eight different nations.

A serious level of unreported catches of southern bluefin tuna (SBT) in the Southern Ocean has attracted curiously little attention: an independent investigation[43] of Japanese market statistics undertaken by the governments of Australia and Japan revealed that much larger quantities of southern bluefin were entering wholesale markets in Japan than the total reported catches of that species since the early 1990s, a discrepancy that exceeds 170,000 tons in total (at a rate of up to 14,000 tons per year). This under-reporting appears to have begun when catch limits became constraining for the Japanese long-line fleet; investigations suggest that at least a large part of the illegal catches are taken by vessels fishing legally outside the SBT-regulated areas and misreporting their catches of this species, if at all, as tropical tuna. Aerial surveillance in this ocean is expensive and difficult, and is rarely performed, although one such flight in 1996 over a limited part of the area that had been officially closed by the Japanese fishery agency to SBT fishing revealed the presence of at least 50 long-liners, of which 40 were Japanese.

The Commission for the Conservation of Southern Bluefin Tuna (CCSBT) is well aware of the problem, but to date has been slow to take effective action to bring this illegal fishing to a halt even though, at the 2006 meeting, the Australian commissioner intervened very strongly, as recorded in the minutes: 'The issue before the CCSBT at this meeting [he said] is the illegal catch ... from this fishery of up to 178,000 tonnes of longline-caught SBT. From 1994, when the CCSBT came into being, the overcatch above national allocations has been in the order of 133,000 tonnes of fish ... If this overcatch had not occurred we estimate that the fishery would be 5–6 times larger than it is at present, well on target for our original goal to rebuild this fishery to 1980 levels by 2020. The Commission had taken the right decisions on catch and it has just been the actions of the Japanese long-line industry that has undermined the recovery of the fishery at a significant cost for all members.'

Exactly as you would expect, the Japanese commissioner was having none of this: he shot back that the 'Extremely biased and one-sided information saying that Japan had been engaged in massive over-catch for many years ... is based on estimates with low reliability, and therefore we cannot accept estimates blindly. However, last year, SBT catch by non-SBT registered fishing vessels was detected ... there was SBT catch outside the designated fishing zones ... As a result, we exceeded our allocation of 6,065 tons by about 1,800 tons. This incident indicated that there was a shortcoming within the old [Japanese] SBT management system. Therefore, we initiated a drastic improvement ... immediately after the detection, and we have implemented the following new SBT fisheries management system from April 2006.' Again exactly as you would expect, the Australian Commissioner has been significantly less aggressive when addressing subsequent meetings of the CCSBT: Japan and Australia probably have other fish to fry than southern bluefin.

Obviously, the problem of unreported fishing is not trivial, and is a major factor in our level of uncertainty concerning real catches. In recent years, there has been a great deal of national and international discussion of the problem and how it may be resolved. For instance, FAO has sponsored a series of consultations to develop an 'International plan of action to prevent, deter and eliminate illegal, unreported and unregulated fishing' that was published in 2001. This instrument admonished coastal states to design and implement National Plans of Action for unreported fishing of which the following extract gives a flavour: 'States should develop and implement, as soon as possible but not later than three years after the adoption of the IPOA, national plans of action to further achieve the objectives of the IPOA and give full effect to its provisions as an integral part of their fisheries management programmes and budgets. These plans should also include, as appropriate, actions to implement initiatives adopted by relevant regional fisheries management organisations to prevent, deter and eliminate unreported fishing. In doing so, States should encourage the full participation and engagement of all interested stakeholders, including industry, fishing communities and non-governmental organisations.' Whether such formal admonitions can solve this kind of problem remains to be seen: certainly, it cannot be resolved without the determined cooperation of those states whose vessels dominate the practice. The economic incentives and social constraints on IUU fishing will be discussed in Chapter 8.

HOW MUCH CONFIDENCE CAN WE HAVE IN ASSESSMENTS OF GLOBAL STOCK DEPLETION LEVELS?

One of the tasks of the UN Food and Agricultural Organisation is to publish regular reviews of the state of global fish populations, formally listing those that it considers to be overexploited or depleted; at present, these reviews are published biennially. By the mid 1990s, FAO reported that the number of individual fisheries with a downward trend was beginning to cause serious concern. Their 1996 report, entitled 'Chronicles of Marine Fishery Landings' was unequivocal, and went much further than the 1971 study noted above: 'The analysis of the shape of the development trajectory of the different clusters of species supporting the world fishery clearly demonstrate the rapid increase in fishing pressure on the top 200 resources of the world, the result of which is a gradual increase of the estimated number of stocks requiring management from almost none in 1950 to over 60% in 1994. This underlines the urgent need for effective measures to control and reduce fishing capacity and effort.' Unlike the pelagic fisheries, all regional catches of demersal fish had already peaked, and landings were significantly reduced by 1995 in several regions: to around 39%

and 32% of peak demersal landings in NW and SE Atlantic regions, respectively, to about 65% in the EC Atlantic and to 80% in both NE Atlantic and NW Pacific. Towards the end of the century, nobody could ignore the fact that the global fishing industry had its head against the ceiling.

Because of the insufficient and incomplete coverage by the formal stock assessments, it is inevitable (as already noted) that global evaluations of the status of fish populations should preferentially be based on the historical trajectory of landings recorded for each species in the FAO statistical database. Most evaluations assume that the level of captures bears some direct relationship to the relative size of the stock biomass remaining in the sea, although FAO officials themselves are quick to point out that the information required to assess the state of global fish populations in any ideal sense is not included in the data sets distributed by their organisation. Caddy and Garibaldi reminded us recently that stock status cannot be quantified without access to estimates of how fishing and natural mortality rates have changed over time, as well as serial data on population size.[44] It is only in relatively few regions that such information is available from the work of national or regional fisheries organisations.

Use has been made of indices to characterise the state of each population, derived from the progressive evolution of the catches and associated with other information. Most of the indices, or categories, that have been used in this way are frankly subjective, as in the case for the most widely used system, that of FAO itself. This categorisation of stock status is based 'on the best information available, which may include the results of peer-reviewed published reports as well as the analysis of qualitative data and information whose reliability may vary from one region to another as well as between stocks or groups of the same or of different species within the same area'. The categories provide useful and believable indications of stock status, but are neither quantitative nor fully objective: in practice, these categories prove to be overprecise, and the tabulated data offered by FAO contain many cases where a compromise is suggested between two adjacent categories. They are, nevertheless, the best we have to make judgements at global scale.

> ? – *Not known or uncertain*. Not sufficient information is available to make a judgement.
>
> U – *underexploited, undeveloped or new fishery*. Believed to have a significant potential for expansion in total production.

M – moderately exploited, exploited with a low level of fishing effort. Believed to have some limited potential for expansion in total production.

F – fully exploited. The fishery is operating at or close to an optimal yield level, with no expected room for further expansion.

O – overexploited. The fishery is being exploited at above a level, which is believed to be sustainable in the long term, with no potential room for further expansion and a higher risk of stock depletion/collapse.

D – depleted. Catches are well below historical levels, irrespective of the amount of fishing effort exerted.

R – recovering. Catches are again increasing after having been depleted or a collapsed from a previous high.

The FAO *Review of the status of world marine fishery resources* of 2005 suggests that the level of uncertainty concerning stock status varies markedly between the 17 FAO statistical regions that cover the global ocean: for the NW Atlantic, of 25 fish species selected for listing in the accompanying tables, only 2 were categorised as having a *not known or uncertain* status, while (at the other extreme) for the W Indian Ocean, of 29 species of fish selected for listing, for only 6 could any status be inferred from the available data. Some important species are not listed in the data tables, leading to further regional difficulties: individual stocks of Atlantic salmon, for instance, are properly indicated as being either *fully exploited* or *depleted* in the NE Atlantic, but the species is not included in the NW Atlantic tabulation, although most stocks in that region are also depleted. I do not point this out with any intention of criticising an admirable work, but rather to emphasise the fragile nature of the information on which any global statement – even the most categorical – must have been based.

Although not alarmist, the general conclusions are not encouraging: the latest publication (that for 2008, which presents data to 2006) reports that 52% were again qualified as *fully exploited*, 18% as *moderately exploited*, while only 1% of stocks were thought to be *underexploited*. The fraction (28%) of stocks qualified as *overexploited* (19%), *depleted* (8%) or *recovering* (1%) had changed little since the previous report.

Table 7.1 shows the status of the 256 populations of fish recorded in the 2005 tabulations, of which only 3 were thought to be unexploited: silver hake in the NW Atlantic, anchovy in the SW Atlantic and myctophids in the Southern Ocean. The FAO reviewers point out that of the 10 species from whose stocks presently the largest catches are yielded globally, together accounting for 30% of all landings, 7 must

Table 7.1. *Estimation by FAO of the status of 256 stocks of sea fish selected by them for evaluation in 2005*

FAO fisheries regions	NWA	NEA	WCA	ECA	M/BS	SWA	SEA	WI	EI	NEP	NWP	WCP	ECP	SEP	SO	N
Uncertain status	1												1		1	3
Uncertain or moderately exploited						1										1
Moderately exploited				1	2	1		1	4	1		8	1			19
Moderately–fully exploited					6	5	3		7			13	4	1		39
Fully exploited	15	9	3	13	3	1	2	3	12	13	6	5		1		86
Fully–overexploited		6	6		2	1	1	2	5		2			7		32
Overexploited		2	1	6	3	1	2									15
Overexploited–depleted		2				1								1	1	5
Depleted	2				5		2								3	12
Recovering														1		1
Unknown status	1	1	4	13				23		1						43
N	19	20	14	33	21	11	10	29	28	15	8	26	6	11	5	256

be categorised as fully or overexploited: *Engraulis ringens* in the SE Pacific was recovering after a previous decline (to 1.7 million tons in 1998) and *Micromesistius poutassou* of the NE Atlantic was overexploited, while *Theragra chalcogramma* of the NE Pacific, *Mallotus villosus* and *Clupea harengus* of the N Atlantic, *Scomber japonicus* of the SE Pacific and skipjack tuna in the Pacific and Indian Oceans were all judged to be fully exploited. Very little hope could be offered of further expansion of these, some of the greatest of the sea fisheries; there is, unfortunately, clear support for the evidence presented in this table in the formal stock assessments of that doyen of international fisheries organisations, the ICES: the status of stocks of the 8 most important of the high-grade fish species in the ICES area at the turn of the century was an unequivocal message of depletion: 11 individual stocks of cod from Greenland to the North Sea, 2 of Greenland halibut, 5 of haddock, 2 of hake, 2 of saithe and 1 each of anglerfish, whiting and redfish were all considered to be *outside safe biological limits*, while only 2 stocks, both of saithe, were considered to be in good condition following the recruitment of excellent year-classes.[45]

The FAO review also provides some appreciation of the progress of the status of stocks since record-keeping began at the global scale in 1950. Using a somewhat different scaling than is discussed above, the review suggests that the *undeveloped* category formed about 65% in 1950, but disappeared from the statistics soon after 1975; similarly, a *senescent* category of catches, consistently falling below their historical maxima, formed <5% in 1950 but included about 30% of all stock items by 1975, since when it has fluctuated between 20 and 30%. The historical analysis of catch trends does, however, suggest that there has been some progressive improvement in the management of large predatory species, at trophic level >4; while the proportion of populations allocated to the senescent or over-exploited category stabilised after the 1970s, there was a small but significant increase in the relative number of populations judged to be recovering from depletion. Unfortunately, for pelagic and oceanic species the percentage of depleted stocks continued to increase – even if only moderately – during the latter decades of the twentieth century.

A more studied approach to the problem of quantifying depleted stocks was used recently by Garibaldi and Caddy,[46] who suggested that the trajectory of the catches of each stock item should be measured against three quantitative criteria for depletion: (i) that a negative trend should have occurred in the most recent five years, (ii) that a long-term of decline of catches from a historical high should have occurred, and (iii) that this decline exceeded 20% of the peak value. As they explain, although a criterion of 10% of unfished population size ($B_{0.1}$) is a widely accepted limit reference point, this would have been too restrictive given the other criteria chosen. This study provides the most comprehensive and balanced view yet of the state of the stocks on which the global fisheries are based: it is hard to see how it could be bettered, yet – as I shall discuss below – it is unable to tell us everything we might want to know.

Restricting their analysis to the years since 1970, and to species that have yielded at least 100,000 tons annually, and thus to 660 items, Garibaldi and Caddy found only 62, or 9.4%, that matched all 3 criteria and could be considered as fully depleted. However, as they pointed out, the numbers of items matching all 3 criteria would have been greater if the analysis had been retrospective to 1950. Further, 53% of the 660 items matched the first criterion by exhibiting a negative slope over the final five years, while almost half (44%) had declined at a rate exceeding 5% per year since their respective historical high.

In this analysis, three FAO regions stand out as having suffered the greatest damage from fishing: the NW Atlantic (23% depleted of 43 stock items), the EC Pacific (17.4% of 23) and the SW Pacific (19% of 26) although, otherwise, these regions and their fishing histories have little in common. The stand of the NW Atlantic at the head of the depleted regions reflect major losses of apparently well-managed gadoids (*Gadus*, *Pollachius*, *Merluccius*, *Brosme*), small pelagics (*Alosa*, *Mallotus*) and other species which had together supported a maximum catch exceeding 2×10^6 tons annually in earlier years. In the region having the next greatest percentage of depleted species, the SW Pacific, just one demersal fish (*Rexea*) and three pelagic species (*Thunnus albacares*, *T. obesus* and *Trachurus trachurus*) have been depleted; these had produced a maximal catch in previous years of only 14×10^3 tons. Likewise, in the third region with a high percentage of depleted populations, the EC Pacific, the consequences are equally trivial globally: one bonito, a marlin and one species of tuna meet the criteria, and had a potential of only about 35×10^3 tons together.

Yet another detailed dissection of the global landings data by Mullon *et al.* confirms the extent of the apparent collapse of fisheries.[47] In this case, a very critical examination of input data resulted in removal of possibly spurious national submissions by ex-USSR nations, Yugoslavia, South Africa and Namibia, because they include series that are best interpreted as the result of administrative decisions that might mimic population collapses, and also data from China and from distant-water fleets subject to fishing rights agreements. Aggregated data identified only above generic level, series having missing data within the previous 20 years and a few with extremely low catches (<500 tons annually) were also excluded. This process resulted in about 1500 data series from national species-specific fisheries, most dating back to 1950. Using the pattern of changes in landings after reaching their maximum as the criteria for collapse revealed that 366 (or 24%) of fisheries thus defined had already collapsed. The list of collapsed taxa is not dominated by one or a few groups; the percentage of collapsed fisheries ranges only from 19 to 33% between the 9 FAO species-groups. The study suggested that the rate of collapses had been approximately stable during the period 1975–2000, with about 60–70 collapses occurring in each 5-year period.

This is certainly a less alarmist result than some others, but the particular interest of this study is not the result itself but rather the authors' analysis of the small number of repetitive patterns of collapse that it reveals, and what these patterns suggest concerning the causes and dynamics of collapse. I shall be returning to this aspect of the study in the next chapter.

We are fortunate that the Asia-Pacific Fishery Commission (APFC) has recently produced a very detailed analysis of stock status in the very complex and interesting region from the Yellow Sea to the Arabian Sea, including the archipelagic region of the SW Pacific Ocean.[48] Of the 10 major fishing nations, only 5 (China, Japan, India, Malaysia and South Korea) perform formal single-species stock assessments, while the others (Indonesia, Philippines, Myanmar, Thailand and Vietnam) perform multispecies assessments based on aggregated catch and effort data. All perform, or have performed, resource surveys and all compute specific or generic reference points, although only six nations use these in the management of some species.

The results of the regional trawl surveys that have covered a large part of the entire Asian sub-region are unequivocal. In those regions where surveys had been carried out over an extended period, the standing stock of demersal fish (tons/km^2) has declined by as much as 30–40% of its original value, the most serious declines being in the Gulf of Thailand (by 86%), Manila Bay (by 65%) and also along the east coast of Malaysia. The great majority of the 427 stocks analysed in these surveys show signs of depletion, and a marked change in dominance has been observed from large predatory fish (bream, goatfish, snapper, groupers, snappers, rays, sharks) to small and less desirable species, at lower trophic levels; this region was, of course, where the fishing down the food web syndrome was first evoked.

This APFC investigation was partitioned between 15 LMEs and the fisheries in each were categorised:

(i) offshore deep-water systems dominated by pelagic fish (e.g. Kuroshio and NW Pacific);
(ii) Heavily fished coastal systems that display sequential depletion of species groups, and 'fishing down the food chain' (e.g. Yellow Sea, Gulf of Thailand and NW Australia);
(iii) coastal ecosystems still showing increase in reported catches (e.g. Bay of Bengal and South China Sea); and
(iv) fisheries managed under tight control of access rights (e.g. SE Australia and New Zealand).

In each region, with the exceptions of the Bay of Bengal and the New Zealand shelf, demersal catches are recorded as having peaked in the past, most usually during the 1970s. This reported pattern in no way corresponds to the statistical data for the entire FAO West-Central Pacific Region that includes the Indonesian archipelago and the coastal regions of SE Asia. This anomaly occurs, where it would be least expected, in plots of coastal demersal species, in which the catches of the principal genera increase continuously since the 1950s as shown in Figure 5.3: this appears to support some discordance between landings reported to FAO and independent survey data.

A less formal approach to the evaluation of global resource status was introduced by Hilborn, who examined the relationship between present landings and maximum landings in the past, in suitably smoothed data, excluding only Chinese landings.[49] No formal measure of collapse was suggested, but the results are far less alarming than from the studies so far discussed: in only 14.5% of 495 fisheries were current landings less than 20% of maximal landings, while for 32% of fisheries, present landings were at least 90% of the maximum. Viewed this way, Indonesia, India and Thailand were the nations whose fisheries are in the best shape, while Canada, Japan and South Korea are those whose fisheries are most depleted.

Personally doubtful that this ranking tells us anything useful about global resources, I tested the technique myself by selecting from the 1950–2000 FAO landing statistics all national entries that were identified to a single species, and then partitioned these by the FAO fishing regions. This procedure gave me 572 species–region data that described current landings as a percentage of maximal landings (Table 7.2) and a very different overall result, with 385 fisheries in what would be described in other studies as depleted (present landings <40%

Table 7.2. *Reported regional landings of 572 species of sea fish in the year 2000, expressed as a percentage of their reported maximal landings during previous 50 years*

	>80%	80–60%	60–40%	40–20%	20–0%	N species	N > 60%	N < 20%	% > 60%	% < 20%
Atlantic, Northwest	2	2	1	10	21	36	4	31	11.1	**86.1**
Atlantic, Northeast	10	10	14	16	13	63	20	29	31.7	46.0
Atlantic, Eastern Central	2	7	10	9	16	44	9	25	20.5	**56.8**
Atlantic, Western Central	2	11	5	8	8	34	13	16	38.2	47.1
Mediterranean and Black Sea	5	5	9	5	7	31	10	12	32.3	38.7
Atlantic, Southwest	9	4	7	5	14	39	13	19	33.3	48.7
Atlantic, Southeast	2	3	6	6	22	39	5	28	12.8	**71.8**
Pacific, Northwest	5	6	13	12	26	62	11	38	17.7	61.3
Pacific, Northeast	4	4	3	4	7	22	8	11	36.4	50.0
Pacific, Eastern Central	6	3		4	11	24	9	15	37.5	62.5
Pacific, Western Central	17	4	6	5	4	36	21	9	**58.3**	25.0
Pacific, Southeast	7	2	6	7	10	32	9	17	28.1	53.1
Pacific, Southwest	10	6	4	9	7	36	16	16	44.4	44.4
Indian Ocean, Western	13	5	5	4	1	28	18	5	**64.3**	17.9
Indian Ocean, Eastern	17	6	7	3	4	37	23	7	**62.2**	18.9
TOTAL (species)	*111*	*78*	*96*	*107*	*171*	563				

of maximum) and 190 fisheries (>60% of maximum) in what we might call a satisfactory state. Looked at regionally, the NW Atlantic stands out as being heavily depleted (86% of fisheries having <20% of maximum), dominated by losses in 8 gadoid, 4 flatfish and other demersal groups. The West-Central Pacific and the Indian Ocean are shown as being in the best shape (>60% of fisheries having >60% of maximum landings), and species in the FAO clupeids, miscellaneous pelagic, miscellaneous demersal and tuna groups are the most successful. These two results confirm those of Garibaldi and Caddy (noted above) concerning individual regions. I can find no significant difference in relative depletion (<20% of maximum catch) among the major FAO species groups: tuna, etc. −48.6%, miscellaneous pelagics −41.2%, miscellaneous demersal −41.6%, herrings, sardines, etc. −45.1%, flatfish −44.6% and gadoids −40.5%. This finding conceals the fact that some groups of fish are relatively intolerant of fishing mortality, as will be discussed later: significant information is lost when aggregating species having as diverse ecological characteristics as those that are included in the FAO category miscellaneous demersal fish.

I offer these simple results for no other reason than to illustrate the ease of generating contradiction, personally having very little confidence in the results obtained by analysis of the trajectories of landings reported to the FAO. If I had to choose one on which to hang my hat, I believe it would be the analysis of Mullon *et al.*, referred to above, because of the very careful and objective manner in which data sets that probably included bias – for a variety of carefully thought-out reasons – had been eliminated. I shall return to this study in the next chapter.

The potential of ecological modelling for the estimation of the effect of fishing on stocks has been investigated by Christensen, and others at the Sea Around Us Project (SAUP) at UBC Fisheries Center.[50] This study was based on an integration of 23 ecosystem models that were constructed with the well-known Ecosim with Ecopath (EwE) software, each model representing conditions within one of 15 spatial realms (each comprising a variable number of 0.5×0.5 degree cells) representative of the North Atlantic from Lancaster Sound to the coast of Morocco, for a given year or short period between 1980 and 1998. Catch data from FAO and the ICES Statlant database of large fish at trophic level >3.75 were used with multiple linear regression to obtain the evolution of abundance of these and of species at lower trophic level in each spatial cell.[51] Relevant chlorophyll and 10-m temperature fields from satellite imagery and NOAA data banks were treated as unchanging between years. The output of each model describes the abundance of fish species as a function of time, primary production, depth, temperature, latitude, ice cover and catch composition within one of the 15 areas and also, to the extent possible, the numbers of fishing vessels active there from 1950 to 1999.

A general linear model, based on 18,000-odd half-degree cells, was derived to obtain the abundance of high-trophic level species in each of the specified regions for the years 1900, 1950, 1975 and 1999, and these were mapped both as abundance and catches within each region. Also obtained at the same spatial scale was a measure of fishing intensity, based on indicated biomass and catches: thus, over the period since 1950, while biomass declined linearly by about two-thirds, and catches peaked around 1970, the level of fishing intensity has remained consistently high as catches and biomass declined. Declines of individual species have been more complex than that of the overall biomass of all species, each of which exhibits a more or less complex progressive collapse. The overall fishing mortality of 35 populations progressed from about 0.3/year in 1950 to a fluctuating plateau at 0.5–0.6/year since 1975. Fishing intensity followed a similar trajectory, doubling between 1950 and 1975.

'The results indicate', report the authors, 'that the biomass of high-trophic level fish has declined by two-thirds during the 50-year period' since 1950 and suggest a biomass decline by a factor of nine during the twentieth century. The authors warn us that the regression presented does not serve to explain what caused the changes in biomass and that the relationship between fishing pressure and biomass decline is merely correlative; they suggest that there are no plausible environmental mechanisms that might have caused the observed effects. Although I am no great fan of ecological modelling, I am personally confident that these results do reflect the general consequences of industrial fishing, at least in the North Atlantic. To support this statement, I note that the analysis of the catch statistics of Mauritania discussed above (which will serve as a proxy for Christenson's Moroccan region) suggests that a depression of biomass of demersal populations – mostly sparids – by 25% occurred in the last 20 years of the last century, and has been associated with 'a huge increase in effort'.

Studies and evaluations of the status of fisheries commonly proceed solely from estimates of the status of the fish populations themselves, as in the examples discussed above, although the real situation is greatly more complex than that. It would be useful if we could accommodate other information concerning the fishery, such as the level of capitalisation of the industry exploiting the population relative to a rational level, the quality of the management regulations applied and the level of compliance of the fishery with them, the quality of the scientific management assessments, the level of unreported and illegal fishing, conflicts between fishing and offshore extraction of mineral resources – the potential list is almost endless. I am struck that the studies that I have discussed in this chapter all ignore what I have suggested is one of the crucial criteria of the viability of a fished population: the extent to which its present age-class distribution departs from the pristine state (see Chapter 3).

To extend the list of useful criteria for the relative health of a population might be difficult to achieve in models such as that of Christenson discussed above, but if ordination of fisheries along a scale from 'good' to 'bad' is useful, then a suitable tool is to hand that is capable of incorporating any and all sources and types of information. This is the Rapfish technique that was developed at UBC in collaboration with the FAO in Rome.[52] This 'employs simple, easily-scored attributes to provide a rapid, cost-effective and multi-disciplinary appraisal of the status of a fishery, in terms of sustainability' and achieves this by the ordination of sets of attributes, which are bounded by reference points representing the best and worst possible fisheries. Attributes are defined for a series of fields – ecological, technological, economic, social and ethical – which are treated as fields of components of sustainability. Thus, in the ecological field the following attributes are among those that might be appropriate, scored as indicated – (i) exploitation status (0–3, using the FAO scale), (ii) recruitment variability (0–2, low to high), (iii) change in trophic level with time (0–2, increasing or decreasing?), and (iv) range collapse (0–2, none to rapid).

The ordination resulting from Rapfish is displayed as a two-dimensional field on which are scattered the points representing the integrated results for each fishery, scored from good to bad; test ordinations, for example, of Canadian east and west coast fisheries successfully point the finger at the east coast fisheries that have collapsed. It is suggested that such techniques may supplement the use of standard single-species reference points in complex multispecies fisheries situations. Although this non-parametric method seems promising enough, it has not been picked up widely and I have seen no analyses that will help us in this chapter, although Chuenpagdee and Alder[53] use three measures (Rapfish criteria of sustainability, FAO Code of Conduct criteria and national compliance with international regulations) to rank national fisheries in the North Atlantic (see Chapter 8). I judge that a formal ranking of a dozen fishing nations with Greenland and Faeroes at the top and Canada and the Netherlands at the bottom doesn't really tell us as much about the 'sustainability ranking of North Atlantic fisheries' as is claimed.

CONSEQUENCES OF THE DIFFERENTIAL SENSITIVITY OF MARINE BIOTA TO FISHING

The discussion thus far has been concerned only with attempts to reach a quantitative assessment of stock status, either regionally or in all oceans, and these have rarely attempted to invoke the strong differences in the biology of the very diverse range of biota targeted by individual and often specialised fisheries. As has already been discussed in earlier chapters, the growth, reproductive and habitat characteristics of some groups of fish and invertebrates make them unusually sensitive to additional mortality induced by fishing; the inverse is, of course also true – some species have characteristics that make them especially resistant to fishing. Of the studies touched on above, only the analysis of the suggested rapid decline in shark populations was directed specifically at a taxonomic group, the elasmobranchs, which have reproductive characteristics that we must expect would render them unusually sensitive to fishing pressure. These characteristics have been generalised many times as having long intervals between successful year-classes and unusually slow inherent population growth – those species lying at one extreme of the r- /K-selected series of life-history characteristics: extreme longevity, late maturation, slow growth and low fecundity.

Among the most extreme examples of such organisms in the sea are, of course, the great whales, and these have responded to whaling exactly as would have been predictable if anybody had thought seriously about it in the early years: each species in the Southern Ocean was reduced to commercial extinction in just a few decades of industrial whaling after World War II. Blue whales were essentially gone by 1955, humpbacks and fin whales by 1965, sperm and sei whales by 1980; only minke whales were being killed at anywhere near the maximum rate that had been previously achieved (around 8000 animals per year) when commercial whaling ceased in 1986 after the IWC set catch limits for all species to zero: the intensive period of catches for each species lasted no more than 25–30 years. Yet, from the 1930s to the 1960s, Southern Hemisphere whaling was more valuable, in terms of extraction of useful products – principally of whale oil – than any of the great fisheries of that period.

This pattern of the exploitation of whale populations is as close to mining a non-renewable resource as makes no difference; the pattern of population growth at present estimated by the IWC[54] is from 3 to 12% annually for those species for which estimates have been made: blue, grey and humpback whales. The present population estimates of the IWC compared with maximum kill rate in the past are illuminating: fin whale 28,000 (*c.* 30,000 killed annually in 1955); blue whale 2300 (*c.* 35,000 in 1930); southern hemisphere humpback whale 42,000 (*c.* 16,000 in 1960). You will be aware of the efforts of the Japanese whale scientists to convince public opinion that commercial whaling is viable and should again be permitted by the IWC, and also perhaps of the Norwegian argument that minkes should be culled in the Barents Sea to protect fish populations and to allow fish populations to rebuild after overfishing. I hope that you will also be aware of the demonstration by Sidney Holt, Peter Corcoran and others of the extent to which these arguments, and the models of fish consumption by whales on which they are based, are fallacious, because the desired result is obtained by selection of data.[55]

However that may be, it is often claimed that the past pattern of exploitation of whales is currently being repeated among some species of fish that have similar population characteristics as marine mammals: great longevity, slow reproduction and – what is worse – very infrequent and unpredictable recruitment events. The large elasmobranchs, like most large predators, are strongly K-selected species and are very intolerant of additional mortality imposed by fishing, and they have accordingly attracted much attention – especially in the case of the indecent fishery for their fins, an essential ingredient in an oriental soup. But perhaps even more vulnerable than these are the teleosts of deep offshore banks and of the continental slope that were discussed briefly in Chapter 3; stocks of these fish may be very rapidly reduced – in the space of 5–10 years – if they are targeted by a modern deep-water fishery, as many have been in recent years.[56] Two distinct groups of fish are implicated in these fisheries, those aggregated on deep banks and seamounts are mostly gadoids (*Molva*, *Brosme*) and flatfish, and those of the open sea-floor of the upper slope are diverse, strong-swimming teleosts (e.g. *Hoplostethus*, *Beryx*, *Dissostichus*, *Pseudopentaceros* and *Sebastes*).

Those species targeted by these fisheries on deep banks and seamounts generally have the general characteristics of the rockfish that were discussed in Chapter 3, and this accounts for the response of their stocks to fishing, being extreme examples of K-selected species. They have great longevity, their growth rate flattens out after maturity and they have very low fecundity with a 'bestowal' maternal strategy (see Chapter 2) and relatively large eggs. Within size-classes, there is an unusually wide range of age-classes, so that the very loose correlation between age and length leads to population frequency distributions that often show two modes: a dominant adult mode and the smaller recruitment mode that is characteristic of K-selected populations. Such a bimodal pattern of size distribution appears to occur characteristically in organisms in extremely steady-state, bounded environments like arctic lakes (arctic charr) or atolls (giant clams), and is associated with large accumulation of biomass and low levels of recruitment.[57] The environment of many population of the species we are concerned with here is also bounded vertically, on individual seamounts between the surface and the low oxygen zone that often occurs at depths around 800 m; both the benthos and fish of seamounts depend on the passive drift of diel-migrant plankton and mesonekton past the feature for their nourishment, so that energy-flow into the system is very limited.

These fish have great longevity, and individuals of several genera (*Sebastes*, *Hoplostethus*, *Pseudocittus*, *Allocytus*) have been aged at >100 years. One of the two cryptic species within *Sebastes aleutianus* was reliably aged at 205 years, and in the same sample as this particular individual, several other specimens were aged at between 100 and 160 years.[58] However, other species of these and other genera of deep-water fish may have growth curves and age–length frequency distributions more typically teleostean: *Hoplostethus mediterraneus* and *Antimora* spp., for example.

I have the impression that, of all these deep-water fisheries, the longest established and most intelligently managed is for the enormous variety of rockfish (*Sebastes* spp.) in the NE Pacific from California to Alaska and also, to a lesser extent, the redfish of the NW Atlantic, where the fishery that had been viable for half a century suddenly collapsed in recent years. Even if there has been some serial depletion of slope species, as in the case of *S. alutus* off Washington–Oregon, this progression is quite unlike the situation of, for example, the orange roughy populations of the SW Pacific, where the fishery has caused the progressive depletion of stocks of individual seamounts. On the Chatham Rise, a stock reduction of 80% occurred in the first 15 years of the fishery; age structure remained unchanged in the diminishing population, no significant recruitment having occurred! The Russian fishery for *Pseudopentaceros wheeleri* on seamounts in the central N Pacific extracted as much as 200,000 tons annually and exhausted the populations in about one decade, 1967–77. By 1982, the populations were commercially extinct. In the cold Southern Ocean, two species of toothfish (*Dissostishus* spp.) appear to be going the same way; high longevity, late maturation, a very high price on the Japanese sashimi market and very weak surveillance at sea off South Georgia has led in just a very few years to a rapid depletion of the stocks: curiously, the Marine Stewardship Council certifies the deep long-line fishery on the South Georgia population as sustainable.

So, it is not surprising that the populations of deep demersal habitats should have become a popular model for what ails industrial fishing and its propensity to seek fresh resources for exploitation, so that any suggestion that perhaps such species as orange roughy might be fished sustainably is likely to be treated with derision. Yet a careful analysis of the dynamics of six populations of this species in deep water off New Zealand yields a surprising result, what Hilborn describes as a 'beautiful illustration of the divide between perception and objectivity'.[59] Although the biomass of each of these populations was reduced to at least 20% of the virgin population, it has been calculated that <10% of potential yield is currently lost to fishing beyond MSY and that, in fact, each fishery appears to be managed close to its economic optimum.[60] I can offer no informed opinion on the future of populations of deep-living, slow-growing species when they are targeted by a modern fishery, although – as is clear from what I have written – I am very unhopeful and will be surprised if the popular opinion is incorrect. But informed scepticism of popular opinion is absolutely essential in these days of mass communication, when informational cascades can be generated only too easily, and subsequently render criticism both difficult and unpopular. It is for this reason that I welcome Hilborn's intervention.

I cannot pretend that this chapter has offered a satisfactory account of what actually is the state of global fish populations in our new century, and I am acutely aware that this is not a satisfactory result. Rather, I have deliberately insisted on the unsatisfactory nature of the information on which such an assessment must be based, because I believe that it is more important to achieve a wider understanding of this fact than to suggest that we are more certain of the state of global stocks than we have any right to be. I am also quite aware that no journalist, and perhaps not many readers, will find this a satisfactory result but, as the recently lamented Walter Cronkite would have said in signing off, 'that's the way it is' at this time in the history of fisheries – and the way it has to be, considering our present state of knowledge of a most extraordinarily complex and opaque system.

ENDNOTES

1. Moiseev, P.A. (1971) Present fish productivity and bioproduction potential of the world's oceans. *Sci. Marina* **59**, 565–9.
2. Ryther, J.H. (1969) Photosynthesis and fish production in the sea. *Science* **166**, 72–6.
3. Alverson, F.G., *et al.* (1970) How much food from the sea? *Science* **168**, 503–05.
4. The Department of Fisheries of the UN Food and Agriculture Organisation is located with the parent body in Rome.
5. Watson, R. and D. Pauly. (2001) Systematic distortions in world fisheries catch trends. *Nature* **414**, 534–6.
6. Hilborn, R. (2006) Faith-based fisheries. *Fisheries* **31**, 554–5, *and see also* Hilborn, R. (2007) Moving to sustainability by learning from successful fisheries. *Ambio* **36**, 296–303.
7. Pauly, D. (2009) Beyond duplicity and ignorance in global fisheries. *Scientia Marina* **73**, 215–24.
8. Polacheck, T. (2006) Tuna longline catch rates in the Indian Ocean: did industrial fishing result in a 90% rapid decline in the abundance of large predatory fish? *Marine Policy* **30**, 470–82.
9. *Le Monde*, 19–20 April 2009.
10. Worm, B., *et al.* (2006) Impacts of biodiversity loss on ocean ecosystem services. *Science* **314**, 787–90.
11. Longhurst, A.R. (2007) Doubt and certainty in fisheries science: are we really headed for a global collapse of fish stocks? *Fish. Res.* **86**, 1–5.
12. Andrew Sugden, *in litt.*
13. Hilborn, R. (2007) Moving to sustainability by learning from successful fisheries. *Ambio* **36**, 296–303.
14. Longhurst (2007), op. cit.
15. Myers, R. and B. Worm. (2003) Rapid worldwide depletion of predatory fish communities. *Nature* **423**, 280–3.
16. Polachuk, T. (2006), op. cit.
17. Hampton, J., *et al.* (2005) Decline of Pacific tuna populations exaggerated? *Nature* **434**, E1–E2.
18. Presentation to an AAAS meeting in February 2007 '*Confronting misleading, exaggerated, and false information*'.
19. See, for example, Jackson, J. (2008) Ecological extinction and evolution in the brave new ocean. *Proc. Natl Acad. Sci.* **105**, E1–E2.
20. Baum, H.K., *et al.* (2003) Collapse and conservation of shark populations in the northwest Atlantic. *Science* **299**, 389–92.
21. Baum, H.K. and R. Myers. (2004) Shifting baselines and the decline of pelagic sharks in the Gulf of Mexico. *Ecol. Lett.* **7**, 135–45.
22. Burgess, G.H., *et al.* (2005) Is the collapse of shark populations in the NW Atlantic Ocean and Gulf of Mexico real? *Fisheries* **30**, 19–31.
23. Worm, B., *et al.* (2009) Rebuilding global fisheries. *Science* **325**, 578–85.
24. Polacheck, op. cit.
25. North Atlantic Fisheries Organisation, InterAmerican Tropical Tuna Commission and International Commission for the Conservation of Atlantic Tuna.
26. Watson, R. and D. Pauly. (2001), op. cit., and see also Kwong, L. (1997) 'The Political Economy of Corruption in China' (ME Sharpe, New York. pp 75).
27. Watson, R. and D. Pauly. (2001), op. cit.
28. Gascuel, D. (2007) Lessons from a reconstruction of catch time series for Mauritania. *SAUP Newletter* **39**, 1–4.
29. See Table A1 in the FAO annual statistical yearbook series.

30. Watson, R., *et al.* (2004) Mapping global fisheries: sharpening our focus. *Fish. Fish.* **5**, 168–77.
31. http://www.seaaroundus.org/
32. A case in point is the use of such a map in the paper by Worm, B., *et al.* (2009) quoted above.
33. Walters, C. (2003) Folly and fantasy in the analysis of spatial catch rate data. *Can. J. Fish. Aquat. Sci.* **60**, 1433–6.
34. MacCall, A. D. (1991) *Dynamic Geography of Marine Fish Populations*. Seattle, WA: University of Washington Press.
35. Pauly, D., *et al.* (2002) Towards sustainability in world fisheries. *Nature* **418**, 689–95.
36. Forest, R., *et al.* (2001) Estimating illegal and unreported catches from marine ecosystems: two case studies. *Sea Around Us: North Atlantic*, 81–93.
37. Baddyr, M. and S. Guénette. (2001) The fisheries of the Atlantic coast of Morocco, 1950–1997. *Fish. Cent Res. Rep.* **9**, 191–205.
38. Saila, S. B. (1983) Importance and assessment of discards in commercial fisheries. *FAO Fish. Circ.* **C765** and Alverson, D. L., *et al.* (1994) A global assessment of fisheries by-catch and discards. *FAO Fish. Tech. Paper* **T339**.
39. Kelleher, K. (2005) Discards in the world's marine fisheries: an update. *FAO Fish. Tech. Paper* **470**.
40. Zeller, D. and D. Pauly. (2005) Good news, bad news: global fisheries discards are declining, but so are total catches. *Fish. Fish.* **6**, 156–9.
41. Bray, K. (2000) A global review of IUU fishing. Background paper for Expert Consultation on IUU Fishing, Sydney, Australia, 2000.
42. Marine Resources Assessment Group (2005) *IUU Fishing on the High Seas* (Final Report).
43. Curiously, this report is confidential under the rules of the CCSBT, but information on unreported catches is available in reports of the Scientific Committee and the Stock Assessment Group of the CCSBT; however, the problem is fully discussed by Polacheck, T. and C. Davies. (2008) Considerations of the implications of large unreported catches of SBT for assessment of tropical tunas and the need for independent verification of statistics. *CSIRO Marine and Atmospheric Research Paper* **23**.
44. Caddy, J. F. and L. Garibaldi. (2000) Apparent changes in the trophic composition of world marine harvests: the perspective from the FAO capture database. *Ocean & Coastal Mngt.* **43**, 615–55.
45. See report of ICES Adv. Ctee. Fish. Mngt. for 2001.
46. Garibaldi, L. and J. F. Caddy. (2004) Depleted marine resources: an approach to quantification based on the FAO capture database. *FAO Fish. Circ.* **1011**, 1–32.
47. Mullon, C., *et al.* (2006) The dynamics of collapse in world fisheries. *Fish Fish.* **6**, 111–20.
48. Lungren, R., *et al.* (2006) Status and potential of fish and aquaculture in Asia and the Pacific (Asia-Pacific Fishery Commission). *FAO RAP Publication* **2006/22.**
49. Hilborn, R., *et al.* (2003) The state of the world's fisheries. *Ann. Rev. Envir. Res.* **28**, 1–15.
50. Cristensen, V., *et al.* (2003) Hundred-year decline of North Atlantic predatory fishes. *Fish Fish.* **4**, 1–24.
51. These include halibut, flounder, turbot, saithe, cod, hake, whiting, horse mackerel, mackerel, bluefish, bluefin tuna, striped bass, snappers, salmon, rays, skates, sturgeon, and so on – largely, those fish taken for human consumption.
52. Pitcher, T. J. (1999) Rapfish, a rapid appraisal technique for fisheries, and its application to a Code of Conduct for Responsible Fisheries. *FAO Fisheries Circular* **497**, 52 pp.

53. Chuenpagdee, R. and J. Alder. (2001) A sustainability ranking of North Atlantic fisheries. In: T.J. Pitcher, U.R. Sumaila and D. Pauly (Eds.). Fisheries Impacts on North Atlantic Ecosystems: Evaluation and Policy Exploration. *Fisheries Centre Research Reports*, **9**(5), 93 pp.
54. International Whaling Commission, see http://www.iwcoffice.org/index.htm.
55. See, for example, Holt, S.J. (2007) So, Farewell, then, Fishes and Whales. Presentation to *Pacem in Maribus 23*, November 2007.
56. Koslow, J.A., *et al.* (2000) Continental slope and deep-sea fisheries: implications for a fragile ecosystem. *ICES J. Mar. Sci.* **57**, 548–52.
57. Gauldie, R.W. (1989) K-selection characteristics of orange roughy (*Hoplosthesus atlanticus*) in New Zealand waters. *J. Appl. Ichthyol.* **5**, 127–40.
58. Kristen Munk (2000), in press.
59. Hilborn, R. (2007), op. cit.
60. Hilborn, R., J. Annala and D.S. Holland. (2006) The cost of overfishing and management strategies on slow-growing fish: orange roughy (*Hoplostethus atlanticus*) off New Zealand. *Can. J. Fish. Aquat. Sci.* **63**, 2149–53.

8

The mechanics of population collapse

'The amount of uncertainty in an assessment is directly proportional to the amount of scientific rigor applied. The better you get at describing uncertainty, the worse you will be at providing useful advice.'

Peter Koeller, 2003[1]

The collapse of a fishery is usually interpreted as being the result of a simple negative balance between fishing mortality and the population growth of the target species: in short, as a result of 'overfishing'. It is often suggested that this occurs because scientific management depends on inadequate models, or on incorrect assumptions concerning acceptable levels of mortality, with the consequence that permissible catch levels may be set too high. But this ignores some very important aspects of the environment in which management decisions are made. For a start, it ignores the simple 'garbage in–garbage out' effect due to incorrect or insufficient stock assessment data, and also ignores the likelihood that, even if the scientific advice is correct, it may be ignored or manipulated by the regulators.

It also ignores the simple fact that the dynamics of a fishery are very complex. Population collapses cannot necessarily be understood by reference to the performance of single-species analyses alone, even if these are the central subject matter of fishery science, which has concentrated too much of its attention on the relationship between fish population growth and fishing mortality and too little on the internal dynamics and stresses of the fishing industry. These are equally complex and – in many ways – more opaque and difficult to study, yet, by fair means and foul, the fishing industry does have a strong influence on the actual pattern and intensity of fishing that occurs at sea, and this may be quite different from what has been recommended by the scientists to the regulators.

It is easy to forget that the fishing industry is an industry like any other and that all its sectors – from small-boat fishermen to multinational corporations – have the same profit motive and the same debt pressures, but they do not all place the same value on continuity. Indeed, conventional discounting models of fishing economics, associated with cost–benefit analysis, suggest that aggressive rather than conservative harvest strategies should be followed to maximise rents.[2] Without inter-generational discounting, and using market interest rates, it may be more economic under conventional valuation to harvest populations to collapse rather than to sustain them.

One doesn't necessarily have to take theoretical economic models very seriously – considering their failure to predict the events of 2008 – but the whaling industry, it will be recalled, deliberately chose to mine the Southern Ocean resources, over a sufficiently short period that the natural reproduction rate of the great whales scarcely slowed their population decline: the industry then sold out and moved its capital to other endeavours. For the Christian Salvesen enterprise, which dominated whaling in the decades after World War II, this was but one component of a shipping-based company and formed an important part of their business for only a few decades. This was at a time when food oils were scarce and much sought after, but when it became clear that the resource was approaching depletion, the Salveson whaling fleet was liquidated. I would be very surprised to find that the same strategy was not in the minds of some distant-water fishing operations today, because it is a very common pattern in commerce and industry.

So, to understand fully the causes of the observed decline and collapse of a fishery, we need information not only on the resource but also on the changing economic incentives of the industry, and on any political influences on its operations. We must also understand the structure and functioning of the marine ecosystem of which the collapsing target species is a component, so we need access to current information on the state of the ecosystem as it responds to changes in wind stress, solar radiation and ocean circulation. Such an ideal is rarely if ever achieved because, although the relevant natural phenomena may be relatively accessible, it is difficult also to include in the same analysis the relevant complexities of business finance, national and international politics, market demand, the sociology of fishing communities and the evolution of management procedures.

The fishing industry is susceptible to many external influences over which it has no control, from states of war to changes in environmental forcing on its resource, and these must be reflected in the patterns of collapse that we observe. Not even the immediate closure of the fishery will, in some cases, prevent the progressive erosion and eventual collapse of the resource; this lesson should have been learned when unusually cold water in the slope habitat off New England brought disaster in 1882 to the fishery for tilefish (*Lopholatilus chamaeleonticeps*). Almost the entire population died and floated to the sea surface; population re-growth began only 15 years later, eventually permitting a fishery to be re-established, but consumer preferences had changed by then and the fishery once more failed through lack of demand. Because the collapse of the Grand Banks cod was mainly caused by very heavy fishing, it is easy to forget that American plaice (*Hippoglossoides platessoides*) in the same region, and almost during the same years, also collapsed, but for a different reason. This species was very lightly fished (<2% exploitation rate, 1970–90) but recruitment, individual growth rate and population biomass all progressively dwindled; the remnant population finally avoided a cooling trend in surface layers by retreating to deeper and warmer water as discussed in Chapter 5.[3]

Definitions of fishery collapse have varied rather widely, from the simple reduction to 10% of maximum historical landings used by Worm and his co-authors[4] to the more complex formulations of the FAO staff, for whom collapse was the terminal state in the sequence: *undeveloped, developing, mature* and *senescent* (see Chapter 7). For the present purposes, I find the analysis of Mullon and others to be very useful because of its careful attention to the dynamics of collapse[5]: this study was (as already noted) based on about 1500 subsets of the FAO data, carefully selected to exclude very small populations, those insufficiently specified or from doubtful sources, and those subject to distant-water fishing agreements. Categorisation of the pattern of collapse was achieved by means of a k-means clustering model fitted to each of the 366 data sets, or about 25% of the whole, that matched the definition of a collapsed population. The rate of collapse (12–15 stocks annually) was observed to have increased marginally from 1975 to the end of the century; salmonid fisheries have collapsed more frequently than others (33%), followed by gadoids (31%) and large pelagic fish (28%). It should be noted that this tells us nothing about why populations have collapsed – in some cases, it may be that the market for the product has collapsed: this illustrates the significant danger that macro-ecological methods will mislead as to the nature of processes observed.

However that may be, a k-means analysis produced nine clusters of collapsing fisheries that could be aggregated into three general patterns: (i) a rapid terminal crash following a period of erratic pattern of landings (45%), (ii) a plateau-shaped trajectory, with a terminal collapse after lengthy period of sustained landings (21%), and (iii) a relatively smooth downward trend of landings leading to a terminal collapse of the fishery (33%). There is some correspondence between the pattern of collapse and the characteristics of the fishery itself; large fisheries, based on a single and abundant species, tended to follow plateau-shaped or smooth pattern trajectories towards collapse, while small fisheries tended to have an erratic pattern of variable landings prior to eventual collapse. The major clupeid fisheries fell into the former cluster, while many fisheries for minor pelagic species were included in the latter.

Mullon suggests that the plateau-shaped trajectories having a rapid terminal collapse demonstrates that a depensation mechanism in the reproduction of the population operates below a certain population size, so that such a non-linear relationship between population size and CPUE is termed hyperstability.[6] This pattern of collapse is often also associated with a contraction of the area occupied by the stock, even as its total population is reduced in size; such an effect may be widespread and has been detected in organisms as diverse as northern cod on the Grand Banks and banana prawns (*Penaeus merguensis*) in the Gulf of Carpentaria.[7] Moreover, if we can generalise from the behaviour of these latter, such a fishery-induced range contraction into a stable hot-spot may become a permanent feature of the ecology of a heavily fished species: in this case, <85% of the standing stock of prawns has been removed annually by the fishery.

In such cases, progressively increasing effort or progressive evolution of gear efficiency (the typical course of many fisheries) maintains landings at a level to which we become accustomed, even as it progressively reduces population biomass until it is sufficiently attenuated for depensatory effects to occur: typically, this will involve persistent recruitment failure due, in some instances, to the production of insufficient numbers of planktonic larvae to swamp predator demand. Other effects, such as the dispersal of the remnant adult individuals, have also been postulated and, it goes without saying, management by controlling fishing effort may be ineffectual.

Plateau-shaped landing trajectories are very easily misinterpreted and, in at least some such cases, it has been wrongly assumed that the pattern occurs because fishing effort has stabilised and is in equilibrium with population growth; in some cases this interpretation has been preferred even when stock assessment surveys have appeared to reveal a progressively declining biomass.[8] Mullon and his co-authors suggest that 'a stable level of catch over several years is shown to conceal the risk of sudden collapse. This jeopardizes the common assumption that considers the stability of catch as a goal for fisheries sustainability.'

Although these repetitive patterns of collapse are informative, each is the result of interaction between so many different processes that it would not be easy to describe what has caused the characteristic trajectory of any individual collapse. That task is not attempted here, and the remainder of this chapter is no more than a review of some of the processes that have been observed to cause populations to fail, even in scientifically regulated fisheries: defining closed areas and seasons, establishing gear restrictions, and setting quotas for catch and by-catch do not guarantee successful management and a sustainable fishery. What is not discussed in this chapter is the very real possibility that permissible catches, even when determined by accepted scientific methods, may be set sufficiently high as to ensure that a population shall collapse, even in the absence of any of the problems discussed below.

THE MORASS OF UNRELIABLE OR INCORRECT STOCK ASSESSMENT DATA

The performance of any system of stock management, no matter how it is structured or what tools it uses, is no better than the relevance and quality of the input data, and how well they are interpreted. The failure of Canadian officials to interpret correctly the data that were used in the management of the northern cod has become a prime example of this problem because of the extraordinary consequences of their failure, but it was by no means unique.

Both quantity and quality of data are constrained by the costs of obtaining information at sea and processing it ashore; these are not negligible and are habitually borne by national or provincial governments, as a public good, rather than by the industry.[9] In terms of direct return on the investment in management, the cost/benefit ratio to governments is probably negative; nevertheless, both the benefits and the costs are borne by a wide group – the public – and are politically acceptable because both jobs and the needs of the consumer are supported. Indeed, public expenditure is essential if the industry is to have any chance of being sustainable in the long-term, because an unregulated and over-capitalised fishery will almost inevitably follow a devil-take-the-hindmost trajectory resembling the pattern of mining or whaling.

Public purses are not bottomless and the resources applied to management and data collection are commonly quite inadequate for the tasks to be performed: the sea is a very big place. Canada, a relatively rich and maritime nation with coasts on three oceans, has only three fishery research vessels (50–60 m stern trawlers) to monitor and investigate the resources on the extensive shelf, including much of the Grand Banks, from the demersal fisheries of Nova Scotia (45°N) to the arctic shrimp of Baffin Bay (75°N), a distance of about 5000 km. Two small vessels cover the Gulf of St. Lawrence (of similar dimensions to the southern North Sea) and one is allocated to work in the Bay of Fundy and Gulf of Maine.

Given that passage time from the home port to the fishing grounds is not trivial in these seas, the number of ship-days available for work at sea off eastern Canada is very limited. Even with the use of survey tools as effective as modern multibeam echo sounders, capable of species identification over a wide swath on each side of the survey track, the percentage of the habitat of a population that can realistically be surveyed is trivial. Conclusions based on the assumption of uniform distribution of the population may be totally false if the species concerned forms social aggregations (like cod). As a Canadian scientist commented recently 'during long steams on groundfish surveys I worried that the amount of fish caught at the last station plus the amount we would catch at the next, divided by two, was not as representative of the hundreds of square kilometres we were steaming across as the models assumed. And I doubted the models more than the data.'[10] Unfortunately, just as the sophistication and precision of simulation models give their output an authority that may be far beyond what they really deserve, so the cost and difficulties of work at sea may validate assessment data in the eyes of the user.[11]

As elsewhere, Canadian technicians routinely sample catches aboard domestic fishing vessels and international observers are supposed to be carried aboard all vessels from NAFO nations when fishing in the region. However, costs are a significant and even prohibitive constraint on data collection in this way. Although a single technician can sample all the landings of a large number of trawlers as they discharge at the fish dock and obtain data representing several weeks fishing from several different grounds, these data will give incorrect information concerning real catches because discards at sea will remain unrecorded. This problem can be resolved by sending the same technician to sea for the same length of time, but in this case accurate information will be obtained concerning catches of a single fishing vessel fishing in one region. Combining these two types of data sets does not necessarily lead to improved advice.[12]

Good management requires information not only about the present status of the resource, but also background understanding of the general biology of each species and the place it occupies in the regional ecosystem; such information is not obtained by routine stock assessments, or by research on interactions between fish and gear, although it is these that are prescribed in management manuals. Yet management decisions must be made even in the absence of ideal information. It is hard to overemphasise the extent of the gap between the assumptions often used in establishing allowable catches and reality, for which the NW Atlantic cod fisheries once again offer an excellent example. It appears that scientists, in the years preceding the population collapse, failed to distinguish the six individual stocks that together comprised the resource, each having individual characteristics, distributions and migration patterns. This failure was all the more remarkable because a senior DFO fishery scientist in the Atlantic region is the author of a volume in which he discussed the importance of treating individual populations individually in management![13]

The equally essential investigation and monitoring of the oceanography and ecosystems of the Atlantic–Arctic region of Canada is performed aboard a single large (and ageing) oceanographic research vessel or on coastguard ships. Primary oceanographic research, of course, may be done by ships of other nations, as in the case of studies of Baffin Bay that extended over several months in 2007 by a ship from the University of Washington on which Canadian scientists participated as guests.

I have discussed this region only because it is the one with which I am most familiar, but I believe that it is closer to the reality of most fishery regions than the much better-endowed shelves of northern Europe, where many national research fleets cooperate within the framework of ICES. In the less-than-ideal situation in the NW Atlantic, it would be naïve to expect that the data available for stock management would fulfil the textbook ideal, and this is certainly far from being the case: problems arising from imprecision and lack of data are widespread. Such facts are not made much of in the open literature or in textbooks, and perhaps the best source for information is the extraordinarily extensive library of informal publications of ICES, whose structure of working parties and consultations covers every imaginable aspect of fishery management.

The results of these are not always as hoped for, and one readily locates items such as the Study Group on Baltic Cod Age Reading that was organised in 1972 to solve the problem of the differing interpretation of scales and otoliths of this species, which is the object of an international fishery: it is reported that scale readers from western nations have consistently allocated individuals to younger ages than their counterparts in eastern nations.[14] The critical issue is whether errors introduce bias or noise into subsequent computations of, for example, fishing mortality or spawning population biomass: this is what has been called the 'precisely wrong or vaguely right' problem. At least in the case of fishing mortality, age reading from scales or otoliths risks introducing bias because these techniques commonly under-estimate the true ages.[15] Although this conclusion was based on a study of only regional scope, it may indicate the existence of a general malaise, for there appears to be little uniformity or cross-calibration of techniques between different fisheries laboratories and even within some single teams of readers.

Almost 30 years after it was set up, the Study Group on Baltic cod reported that 'after many successive meetings there are still systematic differences between age readers'[16]: unfortunately, neither of the competing interpretations is falsifiable. Disregarding such difficulties, most age-based stock assessments and models these days are based on an assumption of zero errors and, as Beamish and McFarlane noted, validation of operational scale reading appears to have been long forgotten.[17]

This is but one example of a constellation of such issues ranging from complex assumptions habitually used for fishing mortality or recruitment, to apparently simpler issues, such as trends in individual weight with increasing age. The difficulty of reaching correct quantifications appears to be enhanced in heavily fished populations that have truncated age profiles. Unidentified trends in these variables will influence the quality of the final computations that are made with assessment data obtained from year to year in an unpredictable way.

Where management advice is formulated internationally, as in the ICES region, it is critical that standards of precision and accuracy of data should be agreed internationally. However, as we might expect, the techniques used to process data to reach numbers of fish at age *N*, the total catch weight and individual mean weights in the catch are determined differently in each national laboratory.[18] Nevertheless, success can be achieved: in the large Dutch flatfish fisheries in the North Sea, uncertainty concerning mean weight of catch is low (*c*. 5%) and concerning numbers of fish is moderate (*c*. 25%); in other fisheries, the situation is much worse and precision may be very poor indeed.[19]

Retrospective analysis of the historical performance of stock assessment models suggests that the most important sources of error in estimates of allowable catches, apart from model assumptions, are changes in serial catch data and the quality of abundance indices. In general, the relationship between catch forecasts and realisations are weak. There was a tendency during the 1990s to over-estimate population size of some of the major species in the North Sea (cod, haddock, plaice, sole): bias in estimates of fishing mortality was usually opposite to bias in estimates of spawning stock biomass.[20]

THE TRICKY TRANSLATION OF SCIENTIFIC ADVICE INTO MANAGEMENT DECISION

Because common wisdom holds that the collapse of fisheries is usually the result of excessive catches and a lack of effective regulation by those responsible for management of the resource, it may be useful to consider how scientific advice concerning safe levels and patterns of fishing is delivered: however excellent the quality of the original advice, its utility is readily degraded during its transmission to the decision-makers. Parenthetically, it may be worth recalling that this nebulous group of people, constantly evoked in the fisheries literature as requiring scientific advice, may be very hard to define in any concrete instance; this is a problem that will need some clarification, but it is not simple because decision-making often occurs behind closed doors. In national jurisdictions, as in Canada, the fisheries minister usually has complete discretion in making management decisions,[21] and he or she is habitually presented with a variety of options, formulated by senior departmental and ministerial staff who will have analysed such issues as 'What might be the scientific concerns if the TAC was increased?' or 'What might be the social/political consequences if the TAC is NOT increased?' The periods over which such concerns are extrapolated are probably rather short, of the order of 1–3 years, responding to parliamentary rather than ecological time scales.[22]

Irrespective of how the consequences of each option are derived, what must be evaluated are the consequences of the interactions between a stressed natural ecosystem and an industry that may be equally stressed. Inevitably, the information content of the original scientific information must be degraded by the search for confident advice to be delivered to the minister. Knowledge of the natural world, except in the case of physics, usually permits a degree of interpretive flexibility so that alternative conclusions may be drawn from the same data or model: because many of the relevant variables are largely uncontrollable, and only partially observable, understandable and predictable, all the problems of predicting the performance of complex, open ecosystems discussed in earlier chapters must be acknowledged.

In an ideal situation, management advice developed by technical groups would be unequivocal and permit no freedom of interpretation; fisheries ministers must inevitably deal with uncertainty, exaggeration and changing positions from the industry for whose well-being they are responsible, and do not appreciate also having to make their own interpretation of the state of the resource, based on equivocal evidence presented to them. At least in the recent past in Canada, it was understood by senior administrators that the advice on catch levels taken to the minister must be as clear and as unequivocal as possible. The consequences of the wide spectrum of advice received by the person or body responsible for taking decisions in fisheries jurisdictions have been well reviewed recently by Rosenberg in his essay on 'Fishing for Certainty'.[23] He suggests that the chorus of advice from those interested parties who try to emphasise how little we know about a fish population may drown out the voices of those who intervene, more realistically, with emphasis on what we actually do know, and who therefore propose rational restrictions on its exploitation.

The ideal, of course, would be to draft options having quantified consequences, so it is not surprising that fishery management science has opened the Pandora's box of formal decision analysis in an attempt to introduce these techniques into fisheries science – both in a-posteriori analysis of decisions already taken, and also in drafting operational management options. Much of the initiative for the translation of Bayesian and other forms of decision analysis to fisheries management science has come from the universities, although the Revised Management System of the International Whaling Commission has included a set of robust procedures that are thought to be relatively insensitive to the use of incorrect assumptions: they are yet to be tested, of course, and one may hope that they will never actually be needed.

There is very little agreement on how risks and uncertainty in fisheries management can be quantified and the discussions (and associated terminology) concerning decision analysis have assumed something of a theological air. While suggesting that risk and uncertainty are difficult to define, and that the distinction between them does not appear to have been found useful by many authors, Francis and Shotton nevertheless identify six kinds of uncertainty (*process, observational, model, estimation, implementation* and *institutional*) and discuss the difference between reducible and irreducible uncertainties, to say nothing of the distinction between probability and severity of risks.[24]

Very numerous outcomes are typically produced by Monte Carlo analysis of a system having as great a number of uncertainties as a fishery, and such a result may not be very conducive to the selection of the most suitable option in the real world. In practice, the application of such techniques to decision-making in the setting of annual catch limits has been of limited utility, perhaps because the analysis does not (cannot?) include all the issues of importance to the person or group who must take the final decisions: it goes without saying that this group will not comprise the assessment scientists alone, even in the unlikely event that it includes them.

A system of risk management has been tested in the management of the single-species fishery for Cape anchovy (*Engraulis capensis*); the experimental procedure, based on formal decision analysis, automated the delivery of annual permissible catches by specifying precisely the data to be obtained and the algorithm to be used, so as to find a satisfactory trade-off between high average annual catches and between-year population variability.[25] At the outset, the managers accepted that this system might require temporary, one-year closures of the fishery because recruitment is rather variable in this species. But in practice, they subsequently accepted the recommendation of the model in only 4 of 9 years of the trial, deciding in the other years that the catch limits were too low to permit the industry (or the politicians?) to survive. It would be idealistic to believe that this would not occur everywhere, because what most political decision-makers have in mind is the multi-year election cycle, and the support of their friends in industry, rather than any natural cycle in the oceans. It is unlikely that any Bayesian or Monte Carlo model can resolve the pressures that a fisheries minister faces each year: voting fishermen are likely to demonstrate angrily, unhappy consumers may express their opinion at the polls, and financial institutions know how to apply political pressure where it is most painful.

But to return to the organisational structures in which options are discussed and decisions taken, although this is not the place for an extensive review, because there are very many – and very rapidly evolving – national mechanisms in place and a small sampler will suffice to draw attention to the diversity of practices.

The management process is unusually transparent in the United States, where members of the Regional Fisheries Councils are drawn from State fishery departments, the industry, the federal government and so on. Recommendations concerning the management of each fishery are reported to the US Secretary of Commerce for action, and the formal minutes of their debates are available to the public. They make fascinating reading. For the 2009 fishing season, for instance, we can find in the proceedings of the New England Fishery Council that 'Mr. Preble moved and Ms. McGee seconded: that the Council recommend that the Eastern Atlantic bluefin tuna fishery be indefinitely suspended' and then that 'The motion was perfected to read: that the Council recommend that the Eastern Atlantic bluefin tuna fishery be suspended until mandatory compliance measures are instituted by ICCAT nations'; finally, we learn that 'The motion, as perfected, was carried on a show of hands.' This is an admirably open way of establishing protocol for delegates to the annual meeting of the regulatory body, although there must be a great deal of unrecorded argument and arm-twisting between the public sessions.

On the other side of the Atlantic, things are neither so simple nor so transparent, except in the press when disagreement erupts – as it frequently does. Twenty-one nations, members of the European Union, share the coastline from the Baltic to the Black Sea, and fishing vessels from all member states may fish within these contiguous seas, their activity being regulated by an organisational cascade of advisory bodies. The EU Standing Committee on Fisheries comprises 60-odd EU parliamentarians from 17 nations who adhere individually to 9 different political groups[26] within the parliament, while executive powers for management are delegated to a Directorate-General for Fisheries of the EU Commission, which reports to the EU Commissioner for Maritime Affairs and Fisheries, presently Joe Borg from Malta.

The Directorate-General for Fisheries is supported by a series of expert committees: an Advisory Committee on Fisheries and Aquaculture (21 individuals representing industrial sectors), a Scientific, Technical and Economic Committee (scientists appointed for three-year periods) and so on. The Commissioner submits proposals for management actions and quotas to the end-of-the-year meeting of a Council of Ministers, comprising those responsible for agriculture and fisheries in each member nation; presidency of the meeting rotates every six months conforming to the rotation of the EU President so that, as I write, it is headed by the agriculture minister of the Czech Republic. Meetings of none of these bodies are open to the public and only very summary agendas and reports are published.

A Common Fisheries Policy (CFP) of the EU has guided the management of stocks and support of the industry for many years, but has been characterised by a perpetual state of evolution: to the 2002 reform of the CFP is now added – in March 2009 – a Green Paper detailing additional reform required and which will be debated in the Fisheries Committee in coming months. This is to be a substantial rewriting of policy, since it is generally acknowledged that the CFP has failed: 'We are questioning even the fundamentals of the current policy. We are not just looking for another reform – it is time to design a modern, simple and sustainable system for managing fisheries in the EU', wrote Joe Borg in introducing his paper, which was intended to encourage public debate in the coming 12 months concerning fleet modernisation and capacity, protection of small-scale fleets, decision-making mechanisms and the need to return greater responsibility to individual nations, developing a culture of compliance and resolving the problem of discards, improving long-term fishing plans and the protection of MSY, and so on. Naturally, the national fisheries press is buzzing with recommendations to repatriate the authority for fisheries management from the EU to each State.

The EU has no operational capability, so all technical services at sea or in the Commission are provided by the staffs of national fisheries laboratories, who also have to provide the expertise required by the international fishery commissions, particularly ICES which acts as a forum in which national fisheries laboratories pool their resources to achieve consensus. Here, they collectively review national stock assessment data and are responsible for the generation of management advice to be transmitted to the EU Fisheries Commissioner, who proposes fishing plans for species within EU seas; these plans must be agreed collectively at an annual meeting of the fisheries ministers of member nations, each of whom must balance domestic political pressures against scientific recommendations. But it is inevitable that EU delegates to international fisheries management bodies will interpret EU policy in the light of political and economic imperatives at home in their own nations.

This aspect of the meetings emerges only in the press, and only if a member wishes to make a point: 'The Commission have produced a number of proposals, including reductions in eighteen fish stocks of key interest to Ireland and cuts in the quotas available to Irish fishermen by up to 35%. Needless to say this could have serious negative consequences for the Irish fishing industry and our vulnerable coastal communities. I am holding a series of bi-laterals with my counterparts across Europe as the day goes on and even deep into the night if that is what's required . . . I am determined not to give up until a fair deal is achieved for Ireland', said the Irish Minister during the 2006 quota-setting meetings of the Council. Doubtless, ministers of other nations were making equivalent statements destined for their own constituents.

Prior to the 1990s, ICES undertook to define management objectives in each fishery and it was assumed that scientists had a central role in management; this seems now a very naïve attitude that neglects the realities of national administrative procedures. The pattern of research administration in government fisheries departments leads to what Finlayson[27] has described as an inbuilt ambivalence between the culture, traditions and goals of scientists and those of the senior administrators who are responsible for advising the minister. He or she, rather than the scientists, must make all policy decisions affecting the fishery, as would be the case for any other industry subject to government controls that are imposed for the public good. This may lead to a situation in which the political and corporate commitment to a stated policy may be so powerful as to overwhelm objective judgement of the condition of the populations, as appears to have occurred in the Canadian DFO in the 1980s concerning the departmental mantra of rebuilding the northern cod stock.

The coming of age of the EU Fishery Commission required a clarification of the role of national scientists in their affairs; this, it seems now generally agreed everywhere, should be that stock assessment scientists are not asked to make specific recommendations to management, but rather should advise managers of the consequences of alternative strategies.[28] So, since the early 1990s, ICES does just that, although it still reserves the right to raise red flags, and recommend specific actions, when populations are at, or below, biologically acceptable levels and hence liable to total collapse.

Unfortunately, as in regional fisheries management organisations, the EU Fishery Commissioner (who, in such places, heads the European delegation) operates under pressure from delegates whose nations have special interests in the fishery. In the case of the eastern Atlantic bluefin tuna quota for 2009 that was discussed at ICCAT the previous November, among the EU delegates were those representing the interests of nations with large fleets of tuna seiners: France with 600 boats, Italy with 200, as well as Libya, Malta and other Mediterranean nations also having significant fleets. Not surprisingly, it was widely reported that the meetings in Marrakech were tense, that the Libyan delegation had walked out, and that EU delegates had threatened some smaller member nations with trade sanctions unless they repudiated a proposition that was intended to support lower catch limits and longer closed seasons.[29] The 2009 catch was finally set at 22,000 tons, despite the scientific advice that it should be between 8000 and 15,000 tons. After the meeting, the EU Fishery Commission pronounced itself very satisfied with the outcome and emphasised that ICCAT had agreed that quotas would be progressively reduced in future years, that fishing seasons will be shortened and capacities reduced and, finally, that restrictions would be placed on trans-shipment of catches at sea.

Although the EU Commissioner expressed satisfaction at these results – as was politically necessary for him to do – the outcome must have been very far from satisfying the US Regional Fishery Commissioners who, as discussed above, had decided to 'go with the science' when voting to recommend outright closure of the fishery in the eastern Atlantic. Perhaps they had in mind, when they voted on closure of the fishery, the strength of the reaction of European politicians to EU Fishery Commissioner Borg's decision to close the Mediterranean fishery the previous summer, 15 days before the end of the fishing season, when it became clear that the annual quotas were already exceeded. Voices had been raised in the Commission and 'Je confirme que je ne suis pas convaincu'[30] declared Michel Barnier, French fisheries minister, who wanted to see Borg's proof that the quotas allocated to French seiners had already been filled. Borg (a Maltese, and a tough negotiator) retorted that, on the contrary, he had excellent data, and Barnier was left to regret that to re-open the fishery would require a majority vote of the 27 EU members, which would be hard to get. In fact, he was in a very weak position, because French seiners had overfished their quota by a factor of two during the previous season. Nevertheless, 'Nous avons affaire à des bolcheviques ... pire que les années 1940',[31] suggested Mourad Kahul, president of the Mediterranean tuna fishing association.[32]

Further to the east, the management of fisheries in the Russian Federation – which is one of the most important fishing nations – are also rather opaque and apparently in equally rapid transition. Here, the State Commission for Fisheries (the lineal descendant of the Ministry of Fisheries of the USSR) delegates regulatory authority to five regional bodies, each controlling a fisheries basin of which the northern and Pacific are the most important. The State Commission has not had a comfortable ride, because its authority to legislate has been challenged in recent decades by several other federal ministries, especially Economic Development. As Hønneland remarked recently,[33] the relationship between federal and regional authorities has not yet found its final form and, meanwhile, corruption is endemic.

The pattern of management in Japan is an interesting anomaly because in addition to national and prefectural licensing systems, coastal fisheries are regulated with a rights-based management system: the whole system implements the 'Law Regarding Conservation and Management of Marine Living Resources' which provides for total allowable catches or effort levels. The rights-based component of the system is based primarily on Joint Fisheries Rights (*Kyodo-gyogyoken*) afforded to cooperatives that assume responsibility for management and sustainability of their resource: there are also complex constraints and obligations, as the fact that two-thirds of their members must spend at least 90 days a year at sea in the designated area. A Demarcated Fishery Right (*Kukaku-gyogyoken*) gives a cooperative the right to engage in coastal aquaculture of invertebrates or fish: astonishingly, >10,000 such licences were in effect recently[34]: I shall have more to say on this phenomenon in Chapter 10.

It is also important to note that Japanese fishery managers are unusually aware of and sensitive to the importance of environmental changes in the ocean around the Japanese islands, and the consequences of these for fisheries management. Japan, of course, lies athwart the Oyashio–Kuroshio front and is consequently exposed to changing dominance of cold northern water and warm water from the south that is associated with changes in the Aleutian Low Pressure Index; regime shifts occur between anchovy–squid fisheries in periods of Oyashio dominance and of sardine–chub mackerel fisheries when warm Kuroshio flow dominates (see Chapter 5). Such switches cause major problems for the management of the industry; prior to the 1988–9 switch from anchovy to sardine dominance the anchovy fleet had been built up very strongly because the Peruvian source of fish-meal was failing in those years.[35] The result was excess capacity during the next decade that was difficult to manage for industry and also for fishery regulators: the longevity of fishing boats is a problem that is frequently encountered.

LEVELS OF NATIONAL COMPLIANCE WITH TREATIES AND AGREEMENTS

Until one learns otherwise, it is natural to suppose that when a national government signs onto one of the host of Conventions and Agreements that proliferated during the twentieth century, it intends to implement the requirements of that instrument. Because the great majority of those that are applicable to fisheries are drafted not only to allocate access to resources in the extended economic zones (EEZs) of the signatories, but also to ensure their sustainability, one might reasonably hope that some progress had also been made in that direction.

Unfortunately, neither hope has been realised: compliance is not universal, and a high proportion of fisheries falling under these instruments is not well managed.

A recent study has identified about 20 international instruments directly relevant to fish stock management in the North Atlantic region (e.g. Common Fishery Policy of the EU), 15 of limited relevance (e.g. Convention on the Continental Shelf) and 9 that are now inactive or superseded.[36] Using a simple three-level estimator[37] for national compliance with criteria that represent the provisions for fishery management in the 20 most relevant instruments, the study reaches the surprising conclusion that overall compliance was no better than 58%. The range of compliance for instruments ranges from 33 to 81% and for the overall performance of nations from 27.4 to 69.6%.

There is little or no correspondence between the importance of the provisions of an instrument and national compliance with it. Consider, for instance, the instrument that is abbreviated as the UNCLOS Compliance Agreement, which is intended to resolve some problems associated with fishing on the high seas. The parent instrument, the UN Law of the Sea Convention, has been ratified by a high proportion of maritime nations and compliance with it is high, almost 80%; but although the Compliance Agreement suffers the lowest compliance rate of any of the 20 studied instruments, it covers some issues essential to the control of fishing on the high seas, starting with the reflagging problem. 'Flag states', it stipulates, shall 'take the necessary measures to ensure that fishing vessels entitled to fly its flag do not undertake any activity that undermines the effectiveness of international conservation and management measures' and 'Parties shall cooperate with Developing Countries' at all levels. Even though this essential measure was drafted in 1993 and is lodged at the UN, it is not yet in force because only 18 states of the necessary 25 have ratified it. Of these 18 states, 13 have North Atlantic coasts and, of those that have ratified, only Denmark and Norway have taken any steps towards implementation of the Agreement. There is therefore no control, or possibility of control, over the massive, perfectly legal but perfectly destructive use of flags of convenience by distant-water fishing operators. Many flag-of-convenience states are notorious for their almost complete lack of interest in how their flag vessels operate or their standards of safety or hygiene. More than half of the foreign-flag distant-water fishing vessels are registered in Belize, Honduras and Panama: these ships are among the most notorious perpetrators of IUU fishing. Similarly, it is difficult to rationalise the level of compliance of nations with the voluntary FAO Code of

Conduct for Responsible Fisheries, ranked according to indicators of rational central government, gross national product, or degree of economic development: the 22 nations with the lowest level of compliance are responsible for almost half of all catches![38]

One may be forgiven, I think, for being somewhat sceptical of initiatives like the Rome Declaration on Responsible Fisheries, made by assembled fisheries ministers at FAO in 1999 to the effect that they 'will develop a global plan of action to deal effectively with all forms of illegal, unregulated and unreported fishing including by fishing vessels flying flags of convenience'. Concern for this and related problems of foreign fishing in national jurisdictions had been addressed formally at a conference in Cancun held in association with the 1992 UN Conference on Sustainable Development. IUU fishing – as a portfolio of concerns – was not recognised until 1997. The association between IUU fishing and the flag State issue was early recognised, and the International Maritime Organisation had been invited to develop relevant measures to regulate this practice in relation to fishing.

Finally, in 2001, the Committee of Fisheries of FAO approved an International Plan of Action, the IPOA-IUU, that was intended to address these problems.[39] The scope of this instrument is worth noting, because of the scale of its ambition: 'The IPOA-IUU is a voluntary instrument that applies to all States and entities and to all fishers ... The objective of the IPOA is to prevent, deter and eliminate IUU fishing by providing all States with comprehensive, effective and transparent measures by which to act, including through appropriate regional fisheries management organisations established in accordance with international law' and also 'States should develop and implement, as soon as possible but not later than three years after the adoption of the IPOA, national plans of action to further achieve the objectives of the IPOA and give full effect to its provisions as an integral part of their fisheries management programmes and budgets.' Finally, it is to be noted that the IPOA contains some very specific suggestions: 'States should ensure that fishing vessels entitled to fly their flag do not engage in or support IUU fishing. A flag State should ensure, before it registers a fishing vessel that it can exercise its responsibility to ensure that the vessel does not engage in IUU fishing. Flag States should avoid flagging vessels with a history of non-compliance.'

The IPOA-IUU also specifies actions that shall be taken by all states to ensure that fishing is controlled by a formal system of authorisations and permits, that port states shall take action necessary to control activities at landing points, that the required research shall be undertaken by states.

That is a far-reaching and ambitious agenda and it will surprise nobody, I think, that it remains very far from being fully implemented even today, five years after the deadline for full implementation of national plans: each of us will individually judge the probability of this happening at any time in the foreseeable future. In 2004, when implementation should have been at least nearing completion, the FAO reported that of the states that responded concerning their levels of compliance, more than half had insufficient funds to proceed with planning their response, and one-third complained of a lack of manpower and skills to undertake planning.

To illustrate the consequences of these failures to comply with the IPOA-IUU, consider the situation in the area of just one of the FAO Regional Fisheries Bodies, the Fishery Committee for the Eastern Central Atlantic (CECAF); that is to say, the coastal regions from Cape Spartel, at the Straits of Gibraltar, to the mouth of the River Congo and including the entire eastern half of the central Atlantic Ocean.[40] This organisation, based in Accra, has a membership of 21 regional states and 13 with interests in the CECAF area: absent, of course, are the flag-of-convenience states in which many of the distant-water vessels are registered and also some states in which the commercial interests of such vessels are based. The resources in both the northern and tropical regions are reported by CECAF as being largely over-exploited or fully exploited and the fishery for at least one species should be halted entirely: as CECAF notes, this is a disastrous situation in a region where artisanal fishing is an essential element in the nutrition of a rapidly-increasing human population. Yet the Surveillance Operations Coordinating Unit, based in The Gambia, reported that about 34% of a total of 441 fishing vessels that had been observed at sea in 2002–3 were improperly marked and could not be identified.

A French parliamentarian[41] recorded her inspection in Las Palmas, on behalf of the EU Fishery Committee, of a reefer ship that was registered in Panama (although owned and operated by a state-owned Chinese company) which sported several names: *Lion Run* painted over *Sierra Grana*, as well as an IMO number that identified her as *Timanfaya*; she was unloading fish caught by 15 trawlers, some that had previously been identified fishing illegally off Guinea. The Spanish authorities, who suggested that she was really *Lion Run 21* (although unable to produce identifying documentation), pointed out that there were no regulations that required identification of the origins of fish landed in Las Palmas until they reach their point of first sale.

These operations off Guinea were, at least in 2005 and 2006, a masterpiece of complicity by national authorities in IUU fishing.[42] The port of Las Palmas acts as a port-of-convenience for reflagged vessels of all stripes, including for the many freezer trawlers (mostly Chinese-owned, mostly rustbuckets and mostly flying flags of convenience) that were then fishing on the wide Guinean shelf that lies between Sierra Leone and The Gambia: a public service organisation, the Environmental Justice Foundation (EJF) of London, reported locating 55 of these ships trawling on the Guinea shelf, of which 9 concealed their identity while at sea, in direct contravention of international law. They were working on the deeper parts of the shelf for what has been termed the subthermocline snapper fauna, their preferred targets being *Dentex gibbosus* and *D. dentex* – both characteristic of the fauna that occurs the length of the tropical West African coast, usually at depths from 70 to 100 m. These fish are frozen at sea, boxed and landed at Las Palmas, which, because it has the status of a free economic zone, has become a notorious port of convenience.

Subsequently, in the Billingsgate Fish Market in London, investigators from EJF located boxes of frozen *Dentex* with wrapping tape that identified them as having been frozen and packed aboard some of the trawlers that had previously been observed by EJF off Guinea. Of the ships so identified, fewer than half had been licensed to fish in the 200-mile zone of that country. Guinea has established a 12-mile 'no trawling' inshore zone to protect the sciaenid fauna that inhabits the tropical surface water above the thermocline; policing this zone to exclude foreign trawlers is difficult, although an interesting experiment in participatory surveillance was funded for some years by a British overseas cooperation agency: some inshore fishermen were provided with GPS equipment and radios to report incursions to patrol vessels. Although these vessels were unable to operate at night, when most incursions occur, the system does have some deterrent effect. I have been unable to discover whether the system has been continued and expanded to other coasts, as was planned.

Observations in the port of Las Palmas by EJF showed how this serves 'as a gateway through which illegally-caught fish can enter the huge European market. Illegal fish is mixed with legal, and fish caught by EU vessels mixed with that from other countries. It seems that once fish has been off loaded in Las Palmas, it can be transported anywhere within the EU without further inspection of its origin or legality.' The situation is said now to have attracted the attention of EU authorities who pointed out that five Spanish fisheries inspectors were too few to track the 360,000 tons of fish that transited the port annually with any confidence. Because much of this is trans-shipped without landing, it is impossible to control how its region of origin is reported, or even if it is reported to the FAO at all. Transfers of boxed fish products between ships have been observed at sea by EJF on the Guinea shelf, between ships fishing there both illegally and legally with Guinean licences.

Compliant port management at Las Palmas also permitted what has been called a dirty deal between the EU and Ireland to operate successfully[43]; this deal concerned the largest seiner/trawler anywhere in 2000, the *Atlantic Dawn* (144 m LOA, 7000 tonnes freezer capacity), built in Norway with Norwegian subsidies for an Irish businessman having political connections. The capacity of the Irish fleet at that time exceeded the EU allowance by 30%, so the vessel received a licence from Ireland to fish only in international waters, against the will of the EU Commission that launched two actions to prevent it. But a deal was arranged that increased the Irish fleet quota by just enough to accommodate *Atlantic Dawn* and to allow her to fish in EU waters for 3 months annually – thanks to the transfer of another trawler (owned by the same person) to Panamanian registry! The managers of *Atlantic Dawn* obtained a private licence to fish in Mauritanian waters for 9 months annually, in circumstances under which an EU licence would have been impossible to obtain.

However, the bonanza did not last for more than a few years. In 2006, the ship was boarded by Mauritanian authorities several times, detained and accused of fishing in an exclusion zone; she was subsequently withdrawn from West Africa to await the result of talks between EU negotiators and Mauritania concerning the inclusion of *Atlantic Dawn* in the EU fleet, which is permitted to take 205,000 tons of fish annually against a licence fee of €305 million. The EU fleet had been unable to take its full allocation in previous years, so quotas had been revised downwards in this agreement: one might be excused for wondering how *Atlantic Dawn* will fit into these arrangements. Then, despite reports that octopus were overfished by nearly a third, in 2006 government officials in Mauritania sold six more years' access to this fishery to 43 EU vessels for a sum that is said to be the equivalent of almost one-fifth of the national budget!

The movement of vessels flagged in one EU nation to fish in the waters of another by 'quota-hopping' has become rather common and widely condemned; this practice involves legally establishing a base in a port of another EU nation, and so obtaining access to the fishing quotas allocated to that nation. Although this practice is common, and at least serves the function of exploiting species less desirable in the markets of the host country, it is also a source of friction: recently, the then British fishery minister Tony Baldry tried to block what he called 'compulsory and substantial' cuts to the effective capacity of the British fleet until the EU took action to prevent Spanish and Dutch interests from quota-hopping into British waters – a system he stigmatised as 'crazy'.

SLASH-AND-BURN FISHING: THE EXTRAORDINARY LRFF (LIVE REEF FOOD FISH) TRADE

Far from the centres of fishery science, an extraordinary fishery feeds large fish into a complex marketing system of luxury food items destined for demanding consumers in China: one consequence is the rapid degradation of the reef ecosystems of almost the entire SE Pacific region and the disruption of traditional fishing societies.[44]

A demand has always existed for large demersal fish, mainly groupers, to be supplied alive to restaurants and high-quality fish markets in China, and this demand was satisfied largely from local coastal seas until towards the end of the last century, when coral reef species were introduced into the trade and commanded high prices, because they were said to be of superior flavour and texture. Large fish, sufficient to provide a meal for four people, are especially prized and are presented to diners alive in tanks, where they are chosen and sent to the kitchen for cooking: would it be reasonable to suppose that the bright colours and strong patterns of some of the most expensive species may have something to do with their popularity as dining-out fare?

This expansion in the LRFF trade from local to coral reef species of grouper[45] led to a very rapid expansion of reef fishing from just the South China Sea in the 1970s, to encompass the Indonesian and Philippine islands in the 1980s, and then down to the entire northern Australian coastline and the eastern Indian Ocean in the 1990s. Although 16 nations are involved in the export of live fish into this trade, the highest-priced fish come from Australia, the Philippines, Malaysia and Indonesia. It has been very difficult to estimate total LRFF flux, not least because locally registered fishing boats in Hong Kong did not (at least at one time) have to declare their landings in their home port! Traders make voluntary declarations of the volume of traffic to the Agriculture and Fisheries Department; the annual value of the LRFF trade was estimated at US$150 million at the turn of the century.

A very complex catching, buying and transportation system has evolved to deliver live fish to the end buyer, and has spun-off an equally complex aquaculture system that acts to hold large fish temporarily, awaiting transportation. Fish are taken by artisanal fishermen with hand-lines, traps, or, very frequently, by stunning the large groupers (and killing corals and the adjacent invertebrate fauna) with cyanide. Because groupers frequently form spawning aggregations, at locations and in seasons well known to local canoe fishermen, their populations are very vulnerable to a concerted attack induced by the presence of an itinerant mechanised vessel (often no more than a planked boat with an outboard), willing to buy their catches and participate to some extent in the fishing; the beach prices paid for live fish are 2–4 times higher than for fresh, dead fish. Attracted by the presence of an itinerant vessel willing to buy their catch, fishermen from nearby regions are quickly attracted. However, these arrangements are temporary, for the desirable species are very quickly fished out, and the itinerant buyer moves on, leaving the local fishing community with a broken ecosystem almost totally lacking large predatory fish.

The living groupers, taken during an episode of fishing, are held in floating traps until they can be transported in small vessels with circulating seawater tanks either directly to market, or to ports having air connections to Hong Kong or mainland China. Especially valuable live fish are then air freighted in plastic bags with an O_2 supply to the next buyer up the LRFF supply chain, until wholesalers in the fish markets display them in large glass aquaria to retailers and restaurateurs. The 15–40% of the fish bought by the iterant buyers that are too small for the luxury market are taken instead to aquaculture facilities for growing-up, later to enter the LRFF trade route in the same way as their larger cousins who went before: although taste testing has found no difference between these and wild-caught fish, the latter command much higher prices!

It goes without saying that such a pattern of fishing has something of the same consequences as slash-and-burn agriculture, because groupers have high longevity, mature late and often present little evidence of significant recovery after depletion. Almost worse than that is the problem – well-known in the Caribbean because of sports fishing for large, trophy fish – that some of the species taken are the dominant herbivores, such as parrotfish: it has been widely assumed that, in the absence of herbivores (both fish and invertebrates) at their natural abundances, any coral reef will become rather rapidly overgrown with macroalgae, to the detriment of the coral organisms themselves. This simple assumption may be faulty, as discussed in Chapter 9, but – all the same – a clear connection has been established between coral degradation and cyanide fishing in the Indo-Pacific region as with more destructive fishing with explosives and *muro ami*, or drag-lines that force fish to emerge from cover.[46] What will be left to the local fishing community in this vast region will, in the not-too-distant future, be only the fauna able to survive on algal reefs rather that the rich fauna native to the hermatypic coral reefs that their fathers fished.

Whether this elusive trade, as the Asian Development Bank put it, can be controlled and managed in any meaningful way is very doubtful, because the environmental consequences are so widespread and so difficult to quantify in economic terms. I confidently predict that the political will to take steps to control the taking of living fish does not, and will not exist, in any foreseeable future; in any event, most of the exporting countries lack the infrastructure to monitor the fishery and control it. The Bank takes the same view and suggests that an extensive management regime is not possible, noting that although the trading base of the LRFF trade may appear to be stable in a country for many years, the fishing grounds of this 'aggressive and intensive' enterprise are constantly shifting as each grouper stock is sequentially fished out, like other tropical species such as trochus or sea cucumbers.

INDIVIDUAL ENTERPRISE IN IGNORING REGULATIONS AND IN APPLYING POLITICAL PRESSURE

I would not go so far as Menachim Ben-Yami[47] as to suggest that few fisherman are innocent of poaching, but it has to be admitted that cutting corners is a characteristic of those who fish at sea, far from the eye of any regulator, and this has to be factored into any assessment of the probability of successful management. Some studies of the extent of cheating by fishermen at sea suggest that those who habitually cheat and those who never do form two small minorities, perhaps less than 10% each; the large majority may cheat occasionally, but only if a very favourable occasion presents itself.[48] I prefer a more complex model in which the regional social climate determines levels of compliance and am not surprised at observations that only 10% of New England lobster and clam fishermen are chronic violators while the rest habitually comply; in some other regions and some other fisheries, the percentage would certainly be reversed, and this unfortunate fact has been one of the principal obstacles to rational resource management. Even well-meaning and voluntary initiatives to protect resources by one group may be ignored by another; long ago, in 1883, a decision of some

Yarmouth trawlermen to avoid certain grounds when juvenile fish would be abundant was ignored by other trawlermen.

In many fisheries, regulations and quotas are routinely ignored or evaded by fishermen and boat-owners, and where this occurs outside the territorial jurisdiction of a coastal state there is very little that can be done to prevent it. Where it could be controlled, it is frequently tolerated because of the political repercussions of coming down too heavily on hard-working fishermen.

Regional patterns of relative compliance certainly exist, although violations occur in all oceans and off all coasts; Peru, the coasts of Somalia and Malaysia, and the East China Sea appear to be where the greatest numbers of cases of illegal fishing have been reported, while on the east and west coasts of North America fishermen are relatively compliant. Poaching in the Southern Ocean has also been recorded rather frequently.

In general, and at global scale, the level of IUU fishing in waters under national jurisdiction is probably inversely proportional to the degree of political stability in each nation. But there are exceptions, and one such occurs on the south coast of South Africa where a lethal triangle of drugs, unemployed coastal people and a very valuable mollusc fishery is proving impossible for the well-organised authorities to control: the product is subtidal *Haliotis midae*, abundant in the beds of giant kelp, representing the last undepleted abalone population left on any coast – the Californian abalones having been exterminated commercially more than a century ago, and the Peruvian during the past century.

These shellfish are prized as an aphrodisiac in China, where they command very high prices, and in recent decades have become the object of an illicit trade that both poachers and government scientists in South Africa estimate will eliminate the population within a very few years. The annual quota for licensed fishermen is totally insufficient to satisfy those who would participate, and the only investment required for participation as a poacher is a set of flippers and a pair of goggles. There are eager buyers around every corner offering cash or mandrake, a recreational drug[49] imported from the Far East, and the trade is controlled by local street gangs and Chinese smugglers. Although the South African fishery police are well-organised, well-equipped and supported by the military both ashore and at sea, the poachers easily slip through their fingers: bags of abalone are easily dumped from a diving-boat on the approach of a patrol vessel, and the arrival of police at a beach where poaching is occurring causes a general flight into the adjacent bush, so convictions are very difficult to obtain. Such are the realities of the interaction between organised crime and a highly valuable natural resource that are lost in global studies of the state of fisheries resources: it needs individual investigation to discover that imports of frozen and dried abalone into Hong Kong *alone* in 18 months in 2002–3 amounted to the equivalent of >1000 tons of fresh molluscs, although the legal TAC for the entire South African coastline for that period was only 350 tons; these imports originated in 5 African nations, in only one of which are abalones found. Almost all are judged to have had a South African origin, because organised smuggling by air into neighbouring countries is very easy from a myriad of bush airstrips.[50]

In the offshore fisheries it is, of course, skippers and fleet owners who make the decision to ignore quotas and gear regulations, or to poach in closed waters, rather than everybody who goes to sea to fish, and the decision to cheat (if not based on simple greed) is taken by balancing risks against financial rewards in each individual case; in some fisheries, where the catch is valuable, and the risks of being fined or losing the ship are low, illegal fishing may be the rule rather than the exception.[51] When the catch is exceptionally valuable, as in the Southern Ocean toothfish fisheries, the current levels of fines might have to be increased by an order of magnitude to become an effective deterrent.

Those who stay ashore also bend the rules and, once again, although I would not go so far as to suggest that all fisheries ministers are eager to modify management decisions to suit their own political agendas, it cannot be denied that many of them do just that to avoid the political discomfort imposed by fishermen or their constituents.

It is not difficult to find recent examples of this response: facing a meeting of turbulent trawlermen at Boulogne in early 2008, President Sarkozy reminded them of France's forthcoming tenure of the EU presidency, when he hoped to be able to satisfy their ambitions by having the whole question of quotas re-opened. Perhaps fortunately for the stocks, he found himself otherwise occupied with the war in Georgia, the banking crisis and other matters. But the pressure on politicians has been very heavy from the fishing industry in recent years, at least in France; trawlers have blocked ports and access of tankers to refineries, fishery offices have been trashed and filled with dead fish, and trucks containing fish imports from Spain have been stopped on the highways and their contents distributed to a grateful public, and so on. Fuel prices and restrictive EU quotas have been the main targets of their ire: one skipper summed up the sentiments of his colleagues: 'Un pêcheur sans poissons, c'est difficile à concevoir.'[52]

Such attitudes are inevitable in overcapitalised fisheries with a diminishing resource base, and with the bank knocking at the door: the skipper-owner of a splendid new French tuna seiner, one of a group built with subsidies, was recently caught on hidden camera by a Greenpeace crew which had pulled up alongside in the Mediterranean: 'Quotas', he said, 'are for the others, not for me. You don't seem to understand that I have the ship to pay off, and my crew to pay …' The owner of a similar seiner opened his books to journalists to show them that to cover his expenses he needed a quota of 200 tons, but had received only 140 tons for the season: in other words, he must cheat or go broke.

There are many ways to evade regulations at sea, or to ignore commonsense dictates to preserve the population, ranging from the simple and classical fine-mesh cod-end liner to the discarding of low-quality species to make room in the hold for higher-quality fish – a practice known in the trade as high-grading. Inspection at sea is usually ineffective and easily deceived by such tricks as duplicate logbooks – one for the owners, one for the inspectors.

Rather sophisticated ways of evading regulation have developed with the globalisation of fishing. Because it is topical and very complex, the way in which the bluefin tuna industry of the Mediterranean evades regulation is instructive. The social climate of the fisheries in this sea is very different from that of the regime of relative compliance off New England. This is a complicated fishery, technically very sophisticated, and with two quite different marketing objectives: large fish for freezing at sea, destined for the *Sashimi* market in Japan, where the value of the product is extremely (some would say indecently) high, and smaller tuna for sale to aquaculture enterprises which aim to grow the fish up for eventual sale into the fresh-fish markets of Europe. There is also a highly complex and, to the outsider, totally opaque commercial network involving enterprises and boats of many nations, from Spain to Turkey, on both the African and European coasts: in recent years, this fishery has become notorious for exceeding national quotas by very generous margins.

The evasion of regulation starts right at the top, as it were, for the number of ships actually working at sea significantly exceeds the numbers reported to comprise the national fleets: in 2007, Italy reported a fleet of 185 vessels, although 283 Italian ships were counted by the WWF at sea, actually or probably fishing.[53] A group of three Italian purse seiners observed fishing by Greenpeace included one, *Luca Maria*, not included in the ICCAT list of those licensed to fish.

The use of spotter planes to locate tuna concentrations has been prohibited by ICCAT in recent years as part of a new recovery plan for bluefin, yet about 10–15 of these remained based on the Italian island of Lampedusa in 2008 and were observed flying at sea over groups of seiners. The French frigate *Germinal*, on fishery patrol in June 2008, observed aircraft spotters on 27 occasions and photographed 4 for subsequent action; during the same patrol, 2 Turkish seiners were boarded while transferring 1350 live fish, of value about a million euros, to a floating container that was probably owned by Mitsubishi: neither ship carried any documentation of the provenance of these fish, although the crews did not resist inspection. The inspectors finally elucidated that the fish had originally been captured by the Italian seiner *Luigi-Perdre*, had been towed in their cage by a Turkish tug to the holding site, but exactly what was the role of the two Turkish seiners tied alongside the cage was never discovered, and nor was the final destination of the fish. Even more Byzantine was the affair a few days later of an old Croatian trawler found by *Germinal* to be towing a cage full of living bluefin; the skipper of the trawler resisted inspection and was unable to explain the provenance of his fish, but agreed to call someone 'who would know' on his VHF. After a conversation in Sicilian, he announced that two Italian and two Libyan seiners had each contributed to the 2000 bluefin now swimming in the 50 m diameter cage that he was towing, probably to the Croatian coast.[54]

Although in recent years the reported or directly estimated bluefin catches have been almost twice the total annual quotas, similar to those in the southern bluefin fishery discussed in the previous chapter, the real captures may have been very much higher: in 2008, the total capacity of the licensed fleet was about 50,000 tons although the total quota was only 28,500 tons; to the seiners registered in ICCAT nations must be added those, registered who knows where, that come through the Suez Canal, fish, transfer their catch to a reefer ship or a tuna cage, and depart whence they came. It has been estimated by independent observers that the real Mediterranean catch may well have been as much as 60,000 tons in recent years, despite the efforts of 50 patrol boats of all ICCAT nations and of 16 military aircraft, some of which share the same landing field on Lampedusa as the illegal spotter planes!

I have used anecdotal evidence here to emphasise the complexity of this fishery and the obvious impossibility of any rational control of fishing operations at sea or, indeed, of obtaining any real estimates of total catches in each season. The problem is obviously understood by

ICCAT, for their 2009 Revised Recovery Plan includes the following remedial measures, among others: (i) an ICCAT regional observer project will be established to ensure 100% observer coverage for all purse seiners over 24 m, all purse seiners involved in joint fishing operations and during all transfers to and harvesting from cages and (ii) video records of fishing and farming operations made by operators must be made available to observers and inspectors. One may be forgiven for scepticism concerning the probability that these and other new requirements will be observed.

However, even if poachers or cheats are apprehended, there is another obstacle to proper measures being taken that is not much talked about: clever defence lawyers, and the failure of judicial systems to convict poachers. I have seen no examples from the bluefin fishery, but the Uruguayan trawler *Viarsa* was chased for three weeks by Australian, British and South African ships after refusing orders to heave-to in the Antarctic fishing zone of Australia[55]; she was eventually boarded in the South Atlantic, 3000 km away, headed for home with about 100 tons of illegal Patagonian toothfish in the freezer. Two years later, on the grounds that *Viarsa* did not have gear in the water when intercepted, an Australian jury delivered a verdict of not guilty. However, a US$1.6 million fine was imposed by Uruguay against a trawler carrying 200 tons of toothfish valued at US$2,122,000 and Mauritius fined another ship US$2.4 million against a toothfish load valued at only US$440,000: in this case at least, the game was not worth the candle. Such fines have in no way deterred the illegal fishing that has brought this very valuable resource to endangered species status. The fundamental problem is that in many IUU situations the rustbucket fishing vessels are relatively less valuable than their catches so that an owner or operator will simply consider the fine as part of his operating expenses, replace the ship if it has been confiscated, and return to the fishery.[56]

Discarding is clearly not an issue in the bluefin or toothfish fisheries, although this is a major source of unreported fishing mortality, especially in multispecies fisheries. Discarding has customarily been thought to occur mainly when a skipper is high-grading, as was the case in the NW Atlantic cod-rush of the 1980s during which Canadian Coast Guard pilots, monitoring the Grand Banks fishery, were shocked by the long, white wakes of dead fish that they saw streaming out behind trawlers that were processing their catch. Such discards are difficult to prevent and incentives are created by two factors: the limited holding capacity of each ship, and the quota limitation system under which it is operating.

Obviously, each of these limitations creates an incentive to maximise the value of what is retained from the catch; machines to sort fish by size at sea may be used to automate the high-grading process and although banned by many, are still legal for the fleets of some nations. Each skipper has a number of difficult decisions to make to get the optimum balance between spending more days at sea to maximise the value of what is in the hold, or filling up rapidly and getting home to discharge a less-than-optimal catch. For obvious reasons, it is very difficult to quantify high-grading.

When large fish are scarce, trawler skippers have an incentive to evade the mesh-size regulations that are intended to enable smaller fish to escape alive from a trawl. These regulations depend on one of the fundamental activities of fishery science – that of measuring the escapement of legally under-sized fish through cod-end meshes of various dimensions, in various twines and knotted in various ways. Unfortunately, it is only too easy to constrain the free opening of meshes in a cod-end with a circular strap or by using a small mesh liner bag within the cod-end. The 1995 boarding of the Spanish trawler *Estai* in the NW Atlantic halibut fishery, and the dumping of the trawl by her crew, was discussed in Chapter 6. When recovered later, the net was found to have 114 mm mesh in the cod-end instead of the legal 130 mm, and also had been fitted with an 80 mm liner[57]. The very small halibut in the *Estai*'s hold were well below legal limit and even below first-spawning size, but were saleable and perhaps even preferred by buyers in Spain. Clear evidence for such a preference was also found when the temporarily stateless *Cristina Logos* was examined in the same region: she had several layers of small mesh liners in the cod-end, and in her hold were boxes of labelled either X, XX or XXX smalls, the latter clearly destined for a market preferring very small turbot.[58]

It is now realised that simple discarding (as opposed to high-grading) is an almost inevitable consequence of management by fixed single-species quotas in a multispecies fishery, such as much of the demersal trawling in the North Sea; this issue has become yet one more source of anger among European trawlermen who must frequently (or, they claim, habitually) dump a large proportion of their catch because it consists of species for which the fishery is closed, or for which they have no vessel quota. A notorious recent case is that of *Prolific*, a pair-trawler based in Shetland, which was filmed in 2008 by Norwegian coastguard aircraft while she dumped an estimated 5 tons of cod and other fish in UK waters. *Prolific* had previously been inspected and declared legal off Norway, in whose zone no fish may be legally discarded; it is not clear if the skipper was high-grading, or if he had no quota to land the cod and other species that he dumped; what I personally found very shocking was the nonchalance of the deck-hands, who continued to empty fish over the rail from stacks of boxes in clear view of the Norwegian helicopter that kept station alongside: dumping is, quite clearly, no more than a routine task for that particular crew.

This case is not unusual, for discarding rates from trawlers in the North Sea are very high, especially in fisheries specialised in species such as sole or prawns: it is not unusual to have to dump prime whiting, haddock and cod in such fisheries, and in the mixed fishery areas of the North Sea, as much as 40–60% must be discarded. The consequences of conflicts between regulations presently in force can be quite bizarre, as in the case of *Prolific*, which was forced to leave Norwegian waters carrying fish that would be illegal to have aboard in EU waters!

The drafting of regulations to reduce by-catch and discard rates is difficult, and several different formulations exist; one might think that an outright 'no discard' policy would be simple to draft, if hard to enforce, but there is no consensus about the best way to proceed: Norway prohibits discarding with the phrase 'it is prohibited to catch …', while the EU uses 'it is prohibited to have on board …'.[59] The EU stricture is easier to comply with than the ideal but slightly unrealistic Norwegian text, but each requires special legislative measures to be put in place, as shall be discussed in Chapter 10. But the reasons for discarding fish that could otherwise be utilised are very diverse, and discarding in order to conform to regulations probably accounts for only a small fraction of the global total, as suggested by Table 8.1.

The different processes reviewed in this chapter all contribute in their own ways to the decline and collapse of fisheries, and all may be involved in the various models of collapse identified by Mullon and his co-authors, but it would be difficult to suggest which of them contributes most importantly to observed population collapses and declines. Some are more amenable to solution than others, but I suggest that, taken as a whole, they comprise a problem – usually termed overfishing – which has no unique solution.

In the final chapter of this book, I will discuss some fisheries which, for quite different reasons, can broadly be considered to be sustainable – but I shall have to express pessimism that the models that they provide can be widely extrapolated to other regions or other fisheries.

Table 8.1. *Summary of the causes for the discarding fish at sea after their capture according to an analysis by FAO staff*

Biological
High species diversity of catches
Large juvenile year-class present
Area chosen for fishing has a high proportion of small fish
Reduction of older year-classes due to fishing
Target is not whole fish (e.g. roe, fins)
Species is poisonous or dangerous
Fishing operations
Hold space is too small for designated catch
Fish are damaged by gear
Non-selective gear in multispecies fishery
Discards higher at start of trip
Gear improperly or illegally rigged (e.g. mesh sizes too small)
High-grading to make room for better quality fish
Sorting non-target catch too time-consuming to be worthwhile
Species are uneconomical to freeze and process
Fish dumped if there is lack of demand at landing place
Regulatory
Fishing licence may prohibit retention of species (area, season)
By-catch quotas not established by regulatory authority
Laxity of regulation, inspection and policing

ENDNOTES

1. Koeller, P. (2003) The lighter side of reference points. *Fish. Res.* **62**, 1–6.
2. Ainsworth, C.H. and U.R. Sumaila. (2004) Intergenerational valuation of fisheries resources can justify long-term conservation: a case study in Atlantic cod (*Gadus morhua*). *Can. J. Fish. Aquat. Sci.* **62**, 1104–10.
3. Bowering, W.R., *et al.* (1997) Changes in the population of American plaice off Labrador and Newfoundland: a collapsing stock with low exploitation. *Fish. Res.* **30**, 199–216.
4. Worm, B., *et al.* (2006) Impacts of biodiversity loss on ocean ecosystem services. *Science* **314**, 787–90.
5. Mullon, C., *et al.* (2005) The dynamics of collapse in world fisheries. *Fish Fish.* **6**, 111–20.
6. Hilborn, R. and C.J. Walters. (1992) *Quantitative Fisheries Stock Assessment.* New York: Chapman and Hall.
7. Prince, J.D., *et al.* (2008) Contraction of the banana prawn (*Penaeus merguiensis*) fishery of Albatross Bay in the Gulf of Carpentaria, Australia. *CSIRO Mar. Freshw. Res.* **59**, 383–90.
8. See, for example, Pauly, D. (1995) Anecdotes and the shifting baseline syndrome. *Trends Ecol. Evol.* **10**, 430.

9. Schrank, W. E., *et al.* (2003) *The Costs of Fisheries Management*. Farnham: Ashgate Press.
10. Koeller, P. (2008) Ecosystem-based psychology, or how I stopped worrying and learned to love the data. *Fish Res.* **90**, 1–5.
11. Finlayson, A. C. (1994) *Fishing for Truth – A Sociological Analysis of the Northern Cod Stock, from 1977 to 1990*. Newfoundland: Memorial University Press.
12. Rijnsdorp, A. D., *et al.* (2006) Sustainable use of flatfish resources: solving the credibility crisis in mixed fishery management. 6th Flatfish Ecology Symposium, Maizura, Japan.
13. Sinclair, M. (1987) *Marine Populations: An Essay on Population Regulation and Speciation*. Seattle, WA: University of Washington Press.
14. Reeves, S. A. (2003) A simulation study of the implications of age-reading erors for stock assessment and management advice. *ICES J. Mar. Sci.* **60**, 314–28.
15. Reeves, op. cit.
16. ICES (2000) Report on the Study Group on Baltic cod ageing. ICES CM 2000/H:01.
17. Beamish, R. J. and G. A. McFarlane. (1983) The forgotten requirement for age validation in fisheries biology. *Trans. Am. Fish. Soc.* **112**, 735–43.
18. Dickey-Collas, M., *et al.* (2004) Evaluation of Stock Assessment models. RIVO Report CO23/04 (84 pp).
19. Rijnsdorp, A. D., *et al.* (2006) Sustainable use of flatfish resources: solving the credibility crisis in mixed fisheries management. Presented at 6th Flatfish Ecology Symposium, Maizuru, Japan.
20. Pastoors, M. (2005) Evaluating fisheries management advice for some North Sea stocks: is bias inversely related to stock size? ICES CM 2005/V20.Rin.
21. However, see the next chapter for an account of the curious evolution of the decision-making process in Canada in the period now known there as 'after the collapse'.
22. I am indebted for these observations to a DFO scientist who may prefer not to be identified.
23. Rosenberg, A. A. (2007) Fishing for certainty. *Nature* **449**, 989.
24. Francis, R. I. C. C. and R. Shotton. (1997) 'Risk' in fisheries management: a review. *Can. J. Fish. Aquat. Sci.* **54**, 1699–715.
25. Butterworth, D. S. and M. O. Bergh. (1993) The development of a management procedure for the South African anchovy resource. *Spec. Publ. Fish. Aquat. Sci.* **120**, 83–99.
26. At last count!
27. Finlayson, A. C. (1994), see above.
28. Hilborn, R., *et al.* (1993) Current trends in including risk and uncertainty in stock assessment and harvest decisions. *Can. J. Fish. Aquat. Sci.* **50**, 874–880.
29. *The Independent*, London, 29 November 2008.
30. 'I confirm that I'm not convinced …'.
31. 'Its just like the bolsheviks – worse that the 1940s!'
32. *Le Monde*, 21 June 2008.
33. Hønneland, G. (2004) *Russian Fisheries Management; The Precautionary Approach in Theory and Practice.* Leiden: Brill Publishers.
34. For example, see Tabira, N. (1996) Common fishery right and coastal fishery management in Japan. *Mem. Fac. Fish. Kagoshima Univ.* **45**, 37–42.
35. Yatsu, A. (2008) Fisheries management and regime shifts; lessons learned from the Kuroshio/Oyashio Current System. (Open files, Jap. Fish. Res. Agency.)
36. This section draws very heavily on one of the few recent analyses of this problem: Alder, J., *et al.* (2001) Compliance with international fisheries

instruments in the North Atlantic. In *Fisheries Impacts on North Atlantic Ecosystems: Evaluations and Policy Exploration*, Ed. T. Pitcher. University of British Columbia.
37. 0 = no compliance, 1 = some, 2 = full compliance.
38. Pitcher, T., *et al.* (2009) Not honouring the code. *Nature* **457**, 658–9.
39. See: International Plan of Action to prevent, deter and eliminate illegal, unreported and unregulated fishing. FAO, 2001.
40. For much of the information in this section, I am indebted to the 2004 submission to FAO 'IUU fishing in West Africa' prepared for CFFA and GreenPeace by Hélène Bours.
41. Aubert, M.-H. (2007) Report on the implementation of the EU Plan of Action against IUU (2006/2225(INI) EU Parliamentary Committee on Fisheries, Final Report 29.1.2007.
42. This section is based on the report 'Pirate Fish on your Plate' published in 2007 by the Environmental Justice Foundation, London (ISBN No 1-904 523-12-9); available at www.ejfoundation.org.
43. See 'Atlantic Dawn', *The Ecologist* (01/04/03).
44. This section is based on the report to the Asian Development Bank in 2003 'While stocks last: the live reef food fish trade'.
45. The most desirable species are *Epinephalus lanceolatus*, *Cromileptes altivelis*, *Cheilinus undulatus*, *Plectropomus leopardus* and *P. maculatus*.
46. Aronson, R.B. and W.F. Precht. (2006) Conservation, precaution and Caribbean reefs. *Coral Reefs* **25**, 441–50.
47. Ben-Yami, M. (2006) The lawless oceans. *World Fishing*, March 2006.
48. Unpublished theses, University of Rhode Island.
49. Mandrake, a methamphetamine, is widely consumed in South Africa.
50. http://www.iss.co.za/pubs/papers/105/Paper105.htm.
51. Sumaila, U.R., *et al.* (2006). Global scope and economics of illegal fishing. *Mar. Policy* **30**, 696–703.
52. 'Its hard to imagine a fisherman without fish!'
53. Associated Press, 7 October 2008.
54. *Le Monde*, 21 July 2008.
55. BBC News on 28 August 2003 and 5 November 2005.
56. Sumaila, U.R., *et al.* (2006), op. cit.
57. Harris, M. (1998) *Lament for an Ocean*. Toronto: McLelland and Stewart.
58. Novaczek, I. (1995) The turbot war. *Samudra*, April issue, 27–30.
59. Kelleher, K. (2004) Discards in the worlds marine fisheries: an update. *FAO Fish. Tech. Paper*, **470**, 1–131.

9

Why don't some fish populations recover after depletion?

'I know the human being and fish can coexist peacefully.'

George W. Bush, Saginaw, Mich., 29.9.00

Simple population models incorporating density-dependence suggest that a collapsed population should begin to recover when the fishing fleet turns to other, more rewarding areas, so that mortality on the depleted population is relaxed. Unfortunately, this is not a sure outcome. The uncertainty of recovery, and the reasons why some collapsed populations remain stubbornly collapsed, are the topics approached in this chapter.

As early as the end of the nineteenth century, Garstang had reported that populations of plaice in the southern North Sea were significantly reduced in abundance compared with earlier years, and larger fish were relatively less numerous[1]; consequently, overfishing became both a catchword, and a worry. But the inadvertent great fishing experiments of 1914–18 and 1939–45 both produced clear evidence that a reduction in effort can indeed result in almost complete recovery of fished populations; post-war investigations in 1919, led by Graham,[2] and in 1946 by a larger team, showed conclusively that 4–5 years of rest from fishing had permitted the plaice populations to recover their early abundance, as recorded in data on catch rates.[3] The stock biomass increased by a factor of five during the first conflict and by a factor of three during the second, and the modal lengths of adult fish also increased by about 20 cm during each of these wars. These recoveries of plaice populations, and evidence from elsewhere in the North Sea, were discussed at the International Overfishing Conference of early 1946, very soon after the end of hostilities. These recoveries, after just a few years' relief from fishing, might encourage us today except for the fact that in neither of the pre-war periods was the plaice population in the southern North Sea as heavily depleted as many stocks are now. Today, of course, it would be difficult to take remedial measures comparable to the removal of all fishing vessels from an entire sea for a significant number of years.

Although this chapter will be principally concerned with special processes that may prevent the recovery of depleted populations, and the potential for action to be taken to encourage recovery, it should be emphasised at the outset that by far the most frequent cause of failure to recover is no more than a continuation of fishing mortality, which may occur even after action has been taken by a management agency to protect the population. This is certainly the case in many depressed fisheries, especially in warm seas lacking any effective management regime.

But the same can also occur where – by global standards – management should be effective, but where continued fishing mortality on a depleted population represents either the effect of unrecorded by-catches in a fishery directed at a different species, or of a deliberate ignoring of fishing regulations. Both of these effects have probably helped to slow the recovery of depleted Atlantic cod populations in regions where, in principle at least, we should expect management to be as effective as anywhere; a study of ICES, DFO and NMFS stock assessment data for the 1990s and later shows that although the nine largest populations in the NE Atlantic have had the potential to increase at rates of 40–60% a year during this period, their growth has been restrained by continued fishing pressure.[4] The situation in the NW Atlantic is more complex, because five cod populations located in the Gulf of St. Lawrence and on the Grand Banks have shown little net increase even in the absence of fishing since around 1990, and some of them exhibited a rapid drop in potential productivity during the 1980s; four other populations have had exhibited the same potential as the European cod and have performed similarly under continued fishing pressure.

As elsewhere, the continued depletion of cod populations that have the potential to recover is caused both by accidental capture in fisheries targeted at other species and also by deliberate poaching. Only very recently, a Russian trawler was arrested on the outer Grand Banks with 50 tons of prime cod in her hold; she was taken into St. John's where the skipper was charged with, among other things, the release of used engine oil at sea. Because she was intercepted outside the Canadian EEZ, the cod could not be seized, and the trawler departed whence she came.

The depletions discussed in the previous chapters are, in many cases, so profound that many analysts have begun to discuss the probability of extinctions, either of individual populations or entire species. The evidence is not good, but it is generally supposed that extinction is relatively unusual in the ocean compared with on land. Unfortunately, discussion of the probability and definition of extinction risks degenerating into what seems to be rather sterile formality of the categorisation of the dangers of extinction so dear to organisations like the IUCN or AFS. It is not very useful to suggest that long-lived, deep and valuable species, like the Patagonian toothfish (that are, incidentally, called 'white gold' by some in the industry) 'will be extinct within five years'.[5] The habitat of this species is the deep shelf of the stormy Southern Ocean, where its distribution is more patchy than continuous, and where adequate assessment data to determine the real status of the stocks are neither available nor likely to be available in the future, so the state of the toothfish stocks remains unquantified in any formal sense, and any predictions concerning their future can be little more than guesses: the only real evidence of stock depletion that we have is the observation that the trawlers are continually constrained to shift grounds.

In fact, we have rather few careful studies that document the behaviour – and probability of extinction – of species that have been reduced by fishing to very small population size, and the studies that we do have are not uniformly distributed in all seas. One useful case, however, is that of the barndoor skate (*Raja laevis*) of the NW Atlantic. This very large species was never abundant but, until the 1970s, was a regular by-catch in trawl fisheries on Georges Bank and the Scotia Shelf. It was claimed that these skate were very close to extinction at the end of the twentieth century,[6] because no specimens had been taken in research surveys of the St. Pierre Bank for the entire period 1978–98; this much-quoted study suggested that a population of around half a million barndoor skate had dwindled to less than 500 individuals during the 1970s. This species became a paradigm for a commercially extinct fish and 'the barndoor skate', wrote the authors, 'could become the first well-documented case of extinction in a marine fish species'.

They wrote too soon, however, because some recovery of the population was observed in the early 1990s in the data of the US NMFS assessment surveys of Georges Bank,[7] and by 2005 the population had already increased to about a half of its biomass in 1965 as suggested by reports of by-catch of skate; a parallel repopulation of the adjacent portions of the Scotian Shelf was indicated in the results of the

Canadian DFO assessment surveys.[8] This case history suggests that the evocative term 'extinction' (which has the precise meaning of the death of the last individual of a species or population) has been used too freely by those advocating stricter controls on fishing: exaggeration does not improve their already excellent case. The history of recorded by-catches of this species strongly suggests that the observed pattern of changing abundance represented a natural change, either of population numbers or in the preferred location of the population.

Apart from extinctions and continued fishing pressure, the questions before us in this chapter are: (i) to what extent is the process of population collapse reversible, (ii) how frequently does recovery occur – and after what delays, and (iii) what are the controls on recovery? Responses to the first question are not unanimous: an estimate of the rate of recovery of depressed fish populations has been made by Hutchings on 232 populations for which spawning stock biomass and fishing mortality can be obtained from the Myers data archives[9]: his results were not very encouraging. The mean historic decline of biomass of these populations combined was 83%, a figure that was close to the decline in individual groups of species and a result that matches what we might expect from analysis of landings data, as discussed in Chapter 7.

Among the 232 populations, there were data for 90 stocks of fish that extended over a 15-year (say, three generations) period after collapse, where this is defined as the steepest period of decline in the data set; these stocks were examined for evidence of recovery 5, 10 and 15 years post-collapse. Only 8% had fully recovered after 15 years, while about 40% were still in decline, these being preferentially those that had previously suffered the steepest rates of decline. Further, those populations for which fishing mortality decreased after collapse could not be distinguished in this analysis from those which continued to be fished, suggesting that a moratorium on fishing may not be a sufficient response to population collapse; the general conclusion of this study was that little or no recovery is likely to occur for at least three generations after a marine fish population has experienced declines of >60% of its earlier spawning biomass.

Formal remedial measures have been taken in response to population collapse in a limited number of instances; Caddy isolated 33 examples of collapsed fish and invertebrate populations for which remedial measures such as closed areas, by-catch prohibitions and closure of fisheries had been taken: of these, 13 of 21 stocks of fish and 9 of 12 populations of invertebrates subsequently showed a significant trend towards recovery.[10] Remedial measures are therefore not to be ignored by management, as shall be discussed in the next chapter.

Depleted fish populations cannot be considered in isolation from the regional ecosystem of which they are a part, and in which fishing continues. Consequently, those species that are liable to be taken as by-catch are less likely to recover than those that are usually the target of a specialised fishery: perhaps for this reason, clupeids are generally found to be more likely to recover than gadoids. Hutchings' study did not admit of the possibility of natural cycling of population size, as occurs in species such as the collapsed capelin populations of the Barents Sea in which, unusually, a strong recruitment event (as occurred in 1974, at a time when the population was depleted) normally leads to a rapid regrowth of the stock biomass (as indeed occurred in the following year), to levels higher than the pre-depletion biomass. So it is often difficult to evaluate the consequences of remedial measures that may be taken after a population collapse.

However, in some instances, we can be sufficiently sure of the course of events to isolate the effects of climate and of remedial measures: after the collapse of the NE Pacific halibut populations in the 1970s, a recovery occurred only when environmental conditions became exceptionally favourable for reproduction to occur, even though vigorous measures had already been taken to protect the population and encourage its recovery by (i) lowering harvest levels to 75% of computed surplus production, (ii) permitting only by a restricted number of specially licensed vessels to exploit the species and (iii) prohibiting by-catch by all other fishing vessels operating in the region. Something similar appears to have occurred in the northern shrimp fishery off Newfoundland and Labrador, for which a traffic light system for monitoring was introduced in response to reported low abundances; yet it seems to have been the relative absence of cod predation and a favourable environmental regime that boosted stock recovery in this case rather than the remedial measures.

In other cases, a recovery does appear to have been induced by remedial measures, even where environmental conditions remained relatively inhospitable throughout; such a case may be that of the 4X5Y haddock stock that had been reduced to low biomass by a combination of poor environmental conditions (inducing low growth rate and condition factor) with excessive fishing mortality. The progressive reduction of fishing mortality over a 10-year period appears to have been associated with unusually strong recruitment that quickly induced a recovery of the population and an increase in the area that it occupied. The Icelandic cod decline of the mid 1970s appears to have been reversed largely by adoption of strong and novel remedial measures: individual transferable quotas (ITQ) for the target species to reduce capitalisation and to induce social surveillance to reduce high-grading, and by the allocation of cod-equivalent quotas for by-catch species in the cod fishery.

Generally, depleted populations of small pelagic species appear to respond better to remediation than those of demersal species, and not only because the by-catch problem is not very important in these fisheries (see above); Caddy suggests that shoaling pelagic species appear to maintain good compensation of growth rates even at small population sizes, citing King, Gulf and Atlantic mackerel, Icelandic and North Sea herring and Barents Sea capelin. I shall return to this suggestion below.

DEPENSATION, PREDATION AND OTHER OBSTACLES TO POPULATION RECOVERY

It is axiomatic that a population of animals that is reduced to very small numbers may not reproduce as effectively as when they are present in normal abundance. Depensation, or the Allee effect,[11] is most simply expressed as the consequence of reduced survival and reproductive rates of individuals when their density is anomalously low, usually when a population is reduced by predation or disease: general ecology emphasises the consequences of density of individuals in their habitat. There is, in the ideal case, a critical level of density below which individual reproductive performance is progressively and irreversibly degraded until the population becomes extinct. Above this critical level, which represents an unstable point of equilibrium, recovery of the population is possible to the maximum density permitted by the habitat, and this represents a second unstable point. There remains great uncertainty, even when depensation can be detected, concerning the process that produces the characteristic rates of survival and reproduction, and it is becoming evident that – in marine organisms at least – these are very diverse.

Depensation may be identified by empirical observations, or else by numerical analysis using a modified stock–recruitment model.[12] The analytical study by Myers and others has been very influential in subsequent discussions of depensation in fish, because the results suggested that depensation is relatively unusual[13]: that it could be detected in only 3 of 129 stocks is a result that has been much quoted. This study was based on a test that compared the normal Beverton–Holt spawner–recruitment function to a similar model modified to include depensatory recruitment so that one parameter controls depensation in the recruitment curve: depensation is indicated when this parameter takes values >1. The problem with this study is that of the 129 spawner–recruitment data sets available, only 26 were satisfactory for the test: of these, only 3 suggested depensation: Icelandic herring and two Pacific salmon populations. For the rest, the authors noted that 'reductions in fishing mortality rates by managers should enable currently remaining stocks to rebuild, unless environmental changes occur ... that alter the underlying dynamics of the stocks'. It is worth noting that it had been independently noted that the stock–recruitment relationship in Icelandic herring is linear or even depensatory at low population sizes.[14]

More usefully for management, Liermann and Hilborn[15] suggested how to enter similar analytical tests into Bayesian analysis: spawner–recruitment data from many populations of a single taxon enables the estimation of the variability of depensation within taxa and suggests that because the tails in the distribution of the depensatory parameter extend into the range suggesting depensation, managers should allow for this possibility.

The Myers data archives were revisited some years later with rather different results.[16] Simple plots of individual stock–recruitment sets ($N = 303$) were submitted in graphical format to three fishery scientists for allocation to three models for a declining population: (i) insufficient evidence ($N = 152$–202), (ii) no depensation ($N = 38$–93) or (iii) possible depensation ($N = 44$–112). These results suggested that depensation may occur much more widely than thought by Myers and his colleagues, although about half of the cases attributed to the possible depensation category were somewhat equivocal and might have represented a situation in which environmental effects were significant.

Ecological studies of the performance of organisms at unusually low individual densities demonstrate empirically the existence of a wide variety of mechanisms by which rates of survival and reproduction are decreased at small population size: the most frequently cited mechanisms include inefficient external fertilisation of ova in both invertebrates and fish, the inefficiency of shoal formation in gregarious species at low individual densities, and higher relative mortality due to predators, in cases where predator numbers are not controlled by changing abundance of the depleted prey species.

Mating systems in both fish and invertebrates usually require that individuals should be present at greater than a threshhold density, although cases are known in which a species survives extraordinarily low densities. A threshold was observed for giant conchs (*Strombus gigas*) in shallow water in the Bahamas[17]; scuba investigations of their reproductive activity at 54 sites revealed unambiguously that mating does not occur when density of adult conchs is $<$56 ind./ha, and that spawning does not occur at $<$48 ind./ha. As individual density increases, reproductive activity increases to an asymptote at around 200 ind./ha, which is the ideal density for this heavily fished species in this habitat. Conch populations, therefore, do not recover from depletion rapidly.

More complex is the mechanism of depensation in the mating system in Atlantic cod (*Gadus morhua*). In this species, although fertilisation is external, eggs and sperm are not broadcast but are released during the course of a complex reproductive behaviour pattern involving both male and female fish.[18] In spawning aggregations, males and females pair, ventral surface to ventral surface with some clasping by the pelvic fins of the male, the two fish matching their swimming speeds as eggs and sperm are released; the mating pair are often accompanied by satellite male fish, also releasing sperm of which some are successful in fertilising ova. Lower overall population size is accompanied by lower density of fish in spawning aggregations and lower success in the induction of mated pairs of fish, since pairs must be of approximately similar size: cod are batch spawners in which delays in the release of eggs after ovulation incurs over-ripening and heavy loss of their viability. Experimental evidence from captive spawning aggregations indicates that the percentage of ova that are fertilised is a function of the number of males (from 1 to 4 individuals) that participate in the event. Depensation in wild populations is indicated by observations of the density of fish within spawning aggregations of the northern cod population in 1992, when it was very depleted: density varied by 2 orders of magnitude, but all densities observed at sea were lower than the average density of individuals in the tank experiments. The authors of this important study conclude that the fertilisation rate in Atlantic cod does, indeed, decline with abundance even as variance in fertilisation rate increases.

Somewhat similar spawning aggregations are characteristic of many large demersal species and, in tropical seas, aggregating and spawning groupers exhibit even more complex behaviour than Atlantic cod. At Little Cayman in the Caribbean, Nassau grouper (*Epinephalus striatus*) form very large spawning aggregations at winter full moons in which male fish dominate, at least in spawning aggregations that have not been heavily fished.[19] The behaviour of individual fish in the aggregation is very complex, and involves changing between four colour phases at different stages in the evolution of the aggregation: barred (normal, territorial), bicolour (submissive), white belly and dark. At the start of the 4-day period after full moon, the aggregation of dark individuals begins at sunset, but on day 5 after the full moon, herding of dark fish by bicolour individuals is seen; these follow and surround the dark fish, nudging them up into the water column. On spawning nights, the shoal forms a cone at about 20–30 m depth, in which individual males perform a variety of courtship swimming patterns. Spawning bursts of 1–7 bicolour fish chasing a single dark fish involve the group in episodes of rapid vertical swimming, during which gametes are released. Such episodes, once initiated, are frequent and may come to involve larger numbers of fish of both sexes. At spawning sites where fishing has been heavy, the numbers of males accompanying each spawning female is much reduced; this appears not to be a consequence of the protogynous hermaphroditism of this species. In this instance, the most important Allee effect in a depleted population may result from interference with the instantaneous sex ratio of the population; it is predictable that such complex behaviour patterns will not persist in very small aggregations, comprising an abnormal sex ratio: a depensation mechanism is thus identified.

The formation of schools at dawn and their dispersion when the ambient light is too low to maintain them is characteristic of the behaviour of many small species of pelagic fish; observations at night of both anchovies and capelin suggest that inter-fish distances are at least $\times 20$ greater at night than by day; at night, the fish have been observed to be dispersed horizontally at a preferred depth, so forming layers of individuals.[20] Dispersal at night is a requisite for efficient feeding on dispersed planktonic organisms, and aggregation during daylight hours is an effective device to minimise predation rates by visual confusion of larger predatory fish. Daytime schools, such as those of anchovies, are tightly organised and closely packed, so that information transfer concerning external dangers may pass rapidly across the school as waves of agitation; schooling is a highly effective general mechanism for reducing the rate of capture of individual schooling fish by individual predatory fish.[21]

The effectiveness of a fish school as a predator refuge increases with school size, which varies significantly in each species, and is more effective at high population densities than low. In this way, schools differ from physical refuges which are less effective at high population density, because accommodation within them is limited: both relatively and absolutely, losses will be progressively greater at small school size, and this will create depensation in the stock–recruitment relationship.[22] Reduced to absurdity, this argument suggests that a predator can easily catch a single anchovy, swimming alone, while to identify visually and catch an individual in a tight school is much more difficult.

When the population density of a schooling fish is reduced, not only are individual schools of smaller size (and are thus less protective of the individuals within them), but their assembly at dawn after a night of individual feeding may also be less efficiently and rapidly completed than when the population is larger and the density of scattered individuals higher. This has been well observed in planktivorous populations of young sockeye salmon during their lacustrine phase. Sockeye exhibit cyclic population density of almost 2 orders of magnitude in these juvenile fish, while their predators (rainbow and lake trout, burbot) are relatively long-lived and their population sizes are much more stable. Thus, there are alternating periods of low and high sockeye abundance, relative to their predators. During periods of relatively low density, young sockeye have difficulty reforming schools at dawn: at worst, they may have a low probability of encountering and joining a school before meeting a predator: an encounter model demonstrates that, as population density increases, the probability of prey–prey encounters rises faster than that of prey–predator encounters.[23] This is a recipe for a depensatory effect at low prey density.

The general case of relatively stable predator populations feeding on prey populations that are, at the same time, being depleted by constant fishing mortality has major implications for depensation, because of the necessary relative increase in the rate the pattern of predation. Many suggestions for the dynamic interaction between target species, their predators and fishing are on the table, but are very difficult to evaluate. In the trophic triangle scenario, the adults of the target species feed to a significant extent on a small forage fish which is itself in competition with juveniles of the target species; reduction in the adult population of the target species will permit an unwanted population increase in the forage species, increased food searching time of juveniles of the target species, and hence their reduced survival through this lagged depensation mechanism: this scenario is the basis for the recently minted cultivation/depensation hypothesis to explain recruitment depression at low population size.[24] Some support for the hypothesis is afforded because this mechanism would solve the classical ecological question of how juvenile individuals of large species of fish survive predation by smaller species as they grow: their parent generation maintains such species at sufficiently low population levels for the survival of their own juveniles. This technique has long been familiar in connection with strategies for stocking lakes, in which the objective is to achieve a balanced prey–predator community of species.

I suggest that there is an abundance of evidence for the existence of a wide range of ecological mechanisms and processes that may be evoked in explanation of what appear to be depensation effects in depleted populations.

ON RE-INVENTING NATURE AND MAKING NEW SPECIES

Fisheries science gives approval to two methods of increasing the potential yield of fish and invertebrates to be obtained from the sea: both are thought to improve the performance in this respect of a natural or pristine ecosystem. The first, of course, is the reduction of the population biomass of each exploited species to some fraction of the pristine level computed by a logistic model to obtain a maximum yield. Some version of this mechanism is central to conventional fishery science, although the then-Soviet scientists took a more expansive view of it than was done in the West. It was thought in Russia that rational exploitation of a population required management measures that were designed to raise its productivity by ensuring the maximum output of fish biomass for a minimal use of its food resources: this could be achieved only, they thought, by a 'proper control of the age composition of the population', as Nikolsky put it.[25] The Russian scientists of the mid-twentieth century also placed much emphasis on ensuring that potential food was not wasted, quoting Danish investigations that showed how much of the production of the benthos passed into blind alleys, such as echinoderms, that are not consumed by commercial fish; they noted that much potential fish food was wasted in boreal seas by the consumption of a large fraction of the production of plankton by ctenophores and medusae, that are themselves little eaten by fish.

Such observations led them to propose a second method for increasing the output from marine resources that was little discussed in the West – that of the introduction and acclimation of exotic species into the ecosystem to better use 'wasted' food resources; it was thought that it should be possible to introduce species that would prove to be valuable additions to the fisheries but would not consume the same food as valuable species already present and already utilised. Although the Soviet scientists believed that marine ecosystems could be managed in rather simple ways, they seem to have had little apprehension that natural balances might go awry in ways that could not be predicted, and that might be disastrous both for fisheries and for the natural ecosystem.

Acclimation of novel species to enrich natural ecosystems was perceived principally as being useful in fresh waters where the natural fauna of lakes and rivers reflected not necessarily their potential to accommodate species, but rather the history of their ephemeral connections with other water bodies in post-glacial times. But some marine acclimations were performed, and some of the consequences are with us still, notably the burgeoning population of very large Pacific red king crabs (*Paralithodes kamschaticus*) presently munching their way southwards along the deep shelf west of Norway. During the 1960s, a massive Soviet acclimation programme released 1.5 million zoea-stage larvae, 10,000 1–3-year-old juveniles and 2600 adults into the Kolafjord on the Russian Barents Sea coast; these rapidly created a new and valuable fishery for the Russians, and within 30 years the crabs had rounded North Cape and invaded the western coast of Norway.

Paralithodes is indigenous along the shelf edge everywhere from Kamchatka to Alaska and performs seasonal reproductive feeding migrations between 300–400 m and the shallow shelf, feeding massively on benthic invertebrates: molluscs, asteroids and polychaetes. Although their presence on the Norwegian coast has created a new fishery there, it has also brought practical and social problems to the existing fisheries: loss of bait from long-lines set for bottom fish, and serious by-catch problems in gill-net and trawl fisheries. The long-term effect on the benthic fauna of such a major new predator is unpredictable, but it is safe to say that major changes in the ecosystem of the shelf and slope down to 300 m can be anticipated. What the knock-on effects on commercial species of fish will be is even more unpredictable. The benthic component of the NE Atlantic boreal ecosystem is undergoing a major re-organisation.

There may be cases where the total transformation of a benthic ecosystem by the removal of the dominant fish populations has resulted, paradoxically, in an apparently stable and profitable fishery for invertebrates that previously supported the fish; this appears to be the case in some shallow-water muddy-bottom habitats in tropical seas, especially in the SE Asian region, where a situation resembling a monoculture of penaeid shrimps has been induced by the removal of fish by trawling. The massive and continual withdrawal of shrimp biomass for the elaboration of a variety of sun-dried and processed high-protein food products appears – at least superficially – to be a sustainable activity. Because it is comprised of multiple small-scale units, the catching sector may perhaps be sufficiently sensitive to changes in stock size – and hence fishing success – as to have some capability of self-adjustment of effort.

Serious as the potential problems associated with re-inventing nature in this way may be, they pale alongside those associated with making new species.[26] By this, I mean the transformation by fishing of the natural life-history characteristics of exploited fish species, so that they come to differ from those that were evolved in response to the exigencies of the environment. Some aspects of this were discussed in Chapter 3. Because fishing mortality is not uniformly imposed across all year-classes, but falls preferentially on larger, older fish, a fished population comprises fewer year-classes, each growing more rapidly than in a pristine population. By this process, coldwater fish come to have life-history characteristics more appropriate to warm seas: Stergiou described this process as the 'tropicalisation of fish stocks'.[27]

It is axiomatic that the natural longevity of each population of each species should be one of the wider set of characteristics that have evolved to enable it survive in the variable environment of a particular region of the ocean. This axiom is consistent with the observations, discussed above, that different age structures are characteristic of different parts of each species habitat; this structure is observed to change under changing external stresses, such as when a species is invading new habitat. In fact, we observe that the age structure of each population takes only a limited range, and will return to the undisturbed state when unusual environmental stress is removed, as Lotka suggested in 1925: 'Now, age distribution is indeed variable, but only within restricted limits. Certain age distributions will practically never occur ... There is, in fact, a certain stable age distribution about which the actual age distribution varies, and towards which it returns if through any agency disturbed therefrom.'

Generally, after several decades of intensive trawling, none of the larger species in a fishery retains an age distribution that is appropriate to the exigencies of the natural environment in which it lives.[28] Truncation of natural age structure must be a strong violation of ecosystem balance, and what Walsh and his associates termed a maladaptive change of life history characteristics, yet the extent to which some populations have been truncated by fishing (right under the noses of apparently competent fish stock managers) is quite remarkable. Common sense would suggest that a species that had evolved a longevity of 25–30 years to survive in its natural habitat would be at a disadvantage if it was reduced to only about 10 year-classes, yet this is just what happened to the cod populations of the NW Atlantic at the end of the twentieth century, despite the application of what was considered at the time (at least by some) to be state-of-the-art fisheries management.

This event has become the classical demonstration of the effects of industrial fishing on the populations of a long-lived fish, the cod (*Gadus morhua*) of the SE Labrador shelf and the Grand Banks east of Newfoundland, together comprising NAFO Divisions 2J, 3K and 3L; these stocks suffered a 99% reduction in abundance of mature fish in the closing decades of the last century. It was much the same story elsewhere in the region: from the southern Grand Bank to the eastern Scotia shelf the stocks were reduced by 92–98%, and those of other regions from 23% (southern Gulf of St. Lawrence) to 78% (western Scotia Shelf).

In 1960, the 2J3KL stocks comprised about 6.0×10^6 tons biomass (or about 2.5×10^9 individuals supported by about 1.25×10^6 recruits annually), but was reduced to only about 0.5×10^6 tons by the early 1990s.[29] Cod landings peaked at about 1.4×10^6 tons in 1969, but then rapidly collapsed, only to recover somewhat after the unilateral Canadian extension of national jurisdiction in 1976 and the entry of the strong 1973–5 year-classes. Unfortunately, subsequent poor recruitment and higher fishing mortality quickly ended population renewal. Strong year-classes again occurred in 1986–87, but these failed to stem the decline. A total fishing moratorium was finally imposed in 1992 but, as will be discussed in a later chapter, these cod populations – along with others in the NW Atlantic – have not recovered as rapidly as anticipated, for reasons rather poorly understood.

The simple decline in biomass of the northern cod does not reveal the truly dangerous aspect of the collapse, which was the progressive and almost complete loss of older fish (Figure 9.1), as well as a decline in age at first maturity from about 7 years in 1980 to about 5 years in 1992. At the end of this period, the observed proportion of females that were mature at age 5 progressively increased from $<1.0\%$ in 1982–83 to about 3–5% in 1992–2000 and in 1992, for the first time, some 4-year-old females were found to be mature.[30] This was probably only the end of a much longer tale, for the median age at maturity of NE Atlantic cod was reduced by fishing from 10–11 years of age in 1920–40 to about 7.5 years in 1970–80,[31] and it seems very reasonable to extrapolate the data for NW Atlantic cod back along the same trajectory to suggest that a reduction of more than 50% in age-at-maturity had been induced by half-a-century of industrial fisheries on the Grand Banks.

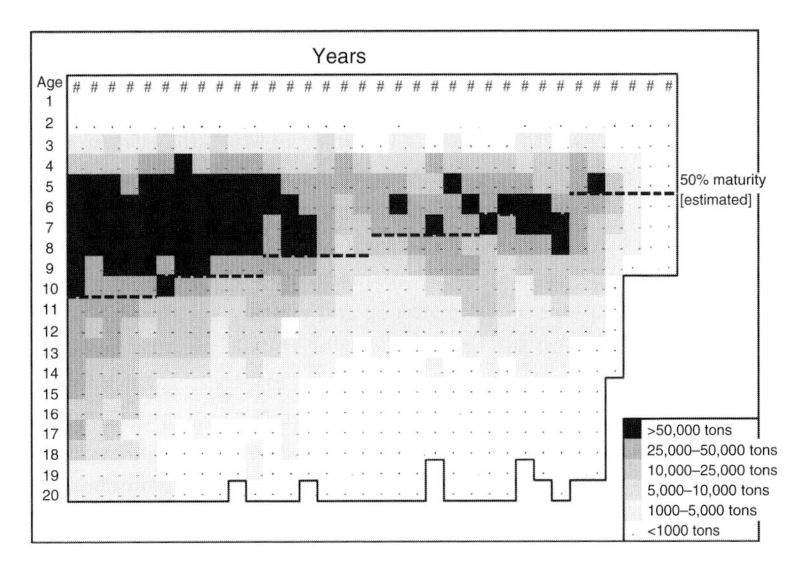

Figure 9.1 Diagram illustrating the progressive loss of older year-classes and the earlier maturation of NW Atlantic cod during the period of heavy fishing and consequent stock collapse on the Grand Banks; data from NAFO Division 2J3KL.

Finally, during the period of most rapid population collapse of Grand Banks cod, the weight-at-age of 5-year-olds sampled in autumn research surveys declined progressively from about 1.5 kg to about 1.0 kg, but subsequently increased again to near the 1982 weights. This has been interpreted as representing a response to mean regional water temperatures, although the Canadian investigators noted 'that historic trends in condition indices are complex and poorly understood'. Rapid response of characteristics such as growth rate and size-at-age are to be expected to be induced by fishing because of the plasticity of the phenotype of fish to environmental changes, and such changes in life-history traits are widely reported to have been induced by fishing. Law discusses several examples of phenotypic changes in growth rate, weight-at-age and age-at-maturation from the NE Atlantic fisheries, such as increases in *Solea* associated with increase in food availability and both increases and decreases in *Platessa* due to density-dependent factors.[32]

In some cases, there appears to be a relatively simple relationship between growth rate and population abundance, as in the case of NW Atlantic mackerel (*Scomber scombrus*) where mean weights-at-age of the 1978–82 and 1984 cohorts were found by Overholtz to be significantly and negatively correlated with population density[33]; the 1982 cohort was both exceptionally large and also exceptionally slow-growing. In this case, it is assumed that growth rates of young individuals conform to the relative sizes of the incoming year-class to which they belong, but their growth rate is mediated by total population size once the juvenile fish enter the offshore adult habitat.

However, in other populations (especially those under high fishing mortality) the situation may be more complex, as in almost all North Sea populations of cod, haddock and whiting. For these, an ICES Working Group found the anticipated, albeit weak, negative relationship between population size and individual growth rates which might have been due (as they put it) to feeding competition, but they suggested caution because changes in fishing effort in certain regions could also produce the same effect. They also suggested that enhanced survival of small individuals in good year-classes might depress the mean size-at-age of the individuals and thus 'mimicking the effects of density-dependent growth'. Nevertheless, population abundance appears to be a simple predictor of sexual maturation at ages 2 and 3 for silver hake in the NW Atlantic, and growth at ages 1–2 of the same species is negatively correlated with population size.[34] Density-dependent growth has also been unequivocally demonstrated in the early ocean life of sockeye salmon in the North Pacific where both total sockeye abundance in the Gulf of Alaska and also within-stock abundance are equally important in setting asymptotic body size of individual fish. These differences can be very significant, for Canadian sockeye may show as much as a 20% decrease in body weight from long-term mean weights at high abundances of conspecific fish in the Gulf of Alaska.

The phenotypic consequences of the collapse of the cod populations in the NW Atlantic would be worrying enough, because the life-history traits of the remnant populations diverged so strongly from those that had evolved naturally to enable the northern cod to master its environment, but worse is to come: to what extent are the observed changes really only phenotypic? Genetic change can be induced rapidly in experimental fish populations so as to produce maladaptive changes in multiple traits, and it reasonable to assume that similar effects are induced by fishing, which is a special case of the induction of selective pressure by human harvesting or hunting of wild populations. Unfortunately for us and for the fish, such pressure induces genetic changes far more rapidly than those that are induced by natural environmental change. The breeder's equation or $R = h^2S$ (where h^2 is the relative heritability of a trait, and S is a measure of selection pressure) has been tested by observation, using data describing the traits of 40 systems of exploited prey, both terrestrial and marine, and including fish. The results are simple, but critical for our expectations concerning the potential recovery of depleted fish populations: they are (i) that hunting generates mortality rates far greater than natural rates, and (ii) that hunting generates selection pressures that act in the opposite direction from those that are imposed by natural mortality.[35]

Indeed, as the authors of this study comment: 'Phenotypic traits ... are constantly moulded by changes in the environment and by numerous agents of natural selection ... however, modern humans have emerged as a dominant evolutionary force': the rate of genetic change in these 40 systems, due to hunting or harvesting, is as much as 300% faster than change due to natural agents. Other human-induced evolutionary changes, such as those caused by pollution, are up to 50% faster than natural change.

The keys to the striking selection differential between harvesting-induced genetic changes and those induced by natural changes are, once again, very simple: (i) recorded harvesting mortality may be as much as 400% greater than natural mortality, and (ii) harvesting mortality falls principally on reproductive adults while the pattern of natural mortality emphasises predation on young pre-reproductive individuals in both terrestrial and marine vertebrates, thus creating selection pressure towards early maturation and reproduction – exactly the juvenescence or tropicalisation that we observe in so many fish populations. It is noteworthy that, among the 40 systems available for these studies, it was the demersal fishery of the NW Atlantic that recorded the highest harvesting mortality rates on adult individuals.

Even in the ideal case in which regrowth of a population at rates that are suggested by the logistic equation and the phenotypic traits of the population, we cannot expect a return of the population to anything like the condition prior to the fishery-induced collapse. The selection-differential imposed by the ambient natural environment in which the regenerated population now finds itself will be very slight compared with the selection pressure previously induced by the fishery. So the best that may be expected in the case, for instance, of the NW Atlantic cod would be a population of small, fast-growing and early-reproducing fish quite unlike those that our ancestors fished on the Grand Banks and elsewhere. In such a case, we can expect that the ecological changes in the natural environment induced by fishing, such as benthic disturbance, will return to the natural pre-fishery condition significantly faster than the genetic change induced in the fish population.

Fish are known to possess sufficient genetic variation to permit the evolutionary selection by fishing of traits such as age-at-maturation although it may be difficult to distinguish selection of early-maturing genotypes among survivors from the consequences of fishery-induced genetic evolution.[36] Nevertheless, it has been suggested that the collapse of the NW Atlantic cod populations was a major 'selective episode' that significantly modified the genotype: recourse to analysis of probabilistic reaction norms (which define life-history traits such as the probability that individuals will mature at a specified age) suggests that because these shifted towards early maturation during the 1980s, but without concomitant increased growth rates, a genetic change had been induced in the populations. The potential of fishery-induced evolution of this kind has the potential of reducing the productivity of the population and reducing the stability of its reaction both to environmental change and to fishing mortality.

So much for theory and inference from observation: the critical experiments that demonstrate how rapidly genetic change may be induced by fishing have been performed on Atlantic silversides (*M. menidia*), a small, commercially exploited species[37]; in only five generations, selective harvesting of large or of small individuals had produced genetic changes in several life-history traits. Removal of large individuals induced slower growth rate and smaller adult size, together with substantial declines in egg size, larval growth rate and viability, and lower food conversion and growth efficiencies. Behavioural and structural changes were also observed. The fitness correlates of the population from which small individuals were removed changed in the opposite sense, with a faster-than-expected growth rate to a larger adult size, and so on. These changes reflected the differences that are observed in unfished populations from high and low latitudes. As the authors of the study remark, fishing wild populations will not change their genetic structure so rapidly as in the reported experiments because they are typically longer-lived than *Menidia*, have overlapping generations, and are harvested in a less precise manner than in the experiments. But that is cold comfort when faced with the much longer period over which wild populations have been subjected to a large fish removal pattern of mortality. These experiments recall those on populations of small tropical fish (*Poecilia reticulata*) living in high- and low-mortality habitats on Trinidad; here, the rates and direction of genetic changes were similar to those induced by commercial fisheries.[38] This is probably a general case in fish, as shall be discussed below, being associated with the high temporal variability of the marine environment, and the consequent diversity of genotypes in fish populations. Inevitably, if and when fishing mortality is removed, the surviving population must have life-history parameters ill-matched to the exigencies of the habitat and a period of evolutionary adjustment must intervene before population recovery can occur; the notion of selecting for a specified set of life-history traits in order to give 'the maximum total yield following evolution ... the harvest pattern called the evolutionary stable optimal harvesting strategy (ESOHS)'[39] seems to be no more than yet another fisheries cloud-cuckoo-land.

So, I think we may conclude that although there is at least one reason why the phenotypic plasticity of fish, associated with their indeterminate growth pattern, does give some comfort to the notion of fisheries sustainability, there are rather more contrary reasons to believe that their indeterminate growth renders fish populations even more vulnerable to fishing mortality than would be the case if their growth pattern, like that of terrestrial mammals, was determinate.

BROKEN ECOSYSTEMS

The marine environment is nothing if not variable, and big changes in conditions occur much more swiftly at sea than on land, as was discussed in Chapter 5; pristine marine ecosystems appear to exhibit a relatively high degree of resilience, recovering structure and function rather readily after perturbation. Thus, when observed changes in an ecosystem are found to be the result of a periodic shift in global wind patterns, we may think of the change as benign because it is naturally reversible: when the change is caused by the effects of fishing, pollution or engineering works, we may think of it as malignant, because it may be irreversible within a period of interest to the present generation.

It is axiomatic that recovery of a depleted population of fish or invertebrates requires that the natural habitat, for which its form and behaviour have progressively evolved, should still exist; to the extent that this habitat has been strongly modified, then it is less probable that population recovery will occur. It will be no news to the reader that the effects of fishing on the marine habitat have been profound and damaging, although the extent of these effects has been fully understood only in recent years: malignant change in the marine environment is now, unfortunately, rather common in all seas.

Development of an understanding of the mechanisms and consequences of fishery-induced ecosystem changes takes us back to the sources of ecological theory and the perturbation of ecosystem structure by the removal of keystone species, whose activities are critical to the balanced structure of the ecosystem. There are also some outstanding difficulties in reaching anything like satisfactory conclusions. For one thing, there are remarkably few studies of the mechanism of disturbance in the marine ecosystem on which we can rely and, for another, there are remarkably few time-series observations of ecosystems detailed enough and sufficiently long that we can be sure that what we are observing is truly novel. Certainly, we can rely on the theory of the keystone species *sui generis* as a taxon whose ecological effects are disproportionate to its relative abundance on the basis of some classical studies of near-shore ecosystems, such as the time-worn example of interaction between sea otters, urchins and kelp in the NE Pacific, and the experimental removal of a predatory starfish (*Pisaster ochraceus*) in a bay on the Washington coast: subsequently, its most important prey, the mussel *Mytilus*, overgrew, almost as a monoculture, the previously diverse benthic fauna.[40]

However, we cannot rely on recent statements concerning the novelty of apparently extraordinary conditions, such as the proliferation around Japan in recent years (2007–8) of the very large rhizostome medusa *Stomolophus nomurae*, usually reported as a novel occurrence of jellyfish caused by overfishing or climate change. In fact, mass occurrences of this species are not novel and were reported regionally – but not in the international media like the present outburst – as having formed mass swarms in the Japan Sea in 1959 and also 20 years earlier than that.[41] As at the present time, fishing was interfered with during the 1959 outburst; remedial measures currently being undertaken, such as seining and cutting the masses of captured individuals with long knives, seems a futile exercise that serves perhaps only to release gametes in less-than-fully-desirable locations.

Although they have captured public imagination in recent years, perhaps we should not be surprised at the occurrence of swarms of medusae in the sea. Swarming in summer, at least in mid-latitudes, is an essential part of their life cycle of alternating asexual and bisexual generations. The benthic polyps of the asexual generation, attached to solid substrates such as pier pilings, synchronously produce many ephyrae larvae by strobilation in spring to initiate a sexual medusoid generation, the individuals of which later initiate the next asexual generation during mass spawning events in summer: growth from 5 mm *Aurelia* larvae in April produces 20 cm medusae by July. Reports of mass occurrences of jellyfish, having nuisance value, have been commonplace ever since such events were noticed and recorded; nevertheless, there is some indication of their having increased in relative importance in recent decades from many regions. Regular fluctuations of numbers are recorded for *Pelagia noctiluca* in the Mediterranean; swarming episodes having a periodicity of about 12 years from 1785 to 1985 have been recorded for this species. Public concern for these events at the present time certainly contains a component of Pauly's shifting baseline syndrome; journalists and the public have very short memories, and changes in environmental conditions of all sorts are today too readily judged to be entirely novel and extraordinary. I can personally verify that children paddling on the British south coast in the 1930s were very careful not to be stung by *Aurelia* swarms!

Some of the cases of inhabitual swarming of medusae that have been reported are due to invasive species, such as the very large (80 cm diameter) Lessepsian immigrant *Rhopilema nomadica* in the eastern Mediterranean: not known to swarm in its original habitat, this species is now a nuisance to bathers and fishermen from Egypt to Greece.[42] It is probable that transport of *Aurelia aurita* from Europe was widespread even by early steamships, so that swarms of this jellyfish are now cosmopolitan and create difficulties for ports and power stations. The proliferation of *Aurelia* and *Mnemiopsis* into the Black Sea is a classical horror story[43]; this basin lacks predators of medusae, so that these now dominate the pelagic ecosystem to the detriment of native clupeids.

It is not yet clear exactly what is the role of fishing in the occurrence of swarms, despite the many statements in the media blaming either fisheries or climate change; one candidate is the heavy mortality of sea turtles whose populations were greatly reduced during the twentieth century but which were, at one time, a major source of predation on large medusae and siphonophores. Nor is it clear to what extent the apparent proliferation of swarms of medusae is effective in depressing fish populations by predation on their larval and post-larval stages in the plankton, despite some reports of significant effects – as on North Sea herring recruitment.[44] Nor, finally, is it clear what the food-web consequences are for a pelagic ecosystem dominated by a trophic dead-end, in the case of large organisms having no predators.

It has been suggested that very heavy fishing mortality might reduce planktivorous fish biomass sufficiently as to induce a regime shift to dominance of medusae that would be only weakly reversible. Cury and co-authors discuss the progression during almost 30 years of the northern Benguela ecosystem towards such an undesirable result[45]; the original dominants, sardine and anchovy, were heavily fished during the 1970s and their biomass was largely replaced by zooplanktivorous mackerel and mesopelagic species of fish, together with inhabitual abundance of large (12–27 cm) medusae of *Aequorea* and *Chrysaorea*, also zooplanktivorous. Hypoxic conditions, perhaps induced by an excess of phytoplankton over consumption in the modified ecosystem, later maintained this species mix and hindered clupeid recruitment and so regrowth of their original populations. Resource surveys in 2003 showed that the northern Benguela had, indeed, passed the transition point from a fish- to a medusa-dominated ecosystem.[46] The biomass of the original sardine and anchovy population is estimated as having been about 17 million tons, but was found to be only about 0.8 million tons during these surveys, together with other pelagic species for a total of 3.6 million tons: the biomass of jellyfish, almost entirely of *Aequorea forskalia*, was an astounding 12.2 million tons. Earlier but less comprehensive surveys had suggested even larger biomasses of jellyfish, of about 40 million tons for the two genera *Aequorea* and *Chrysaorea* combined. One might suppose from their physical appearance that the organic content of fish and jellyfish was greatly different, and hence that the budgetary impact of jellyfish in a carbon-based ecosystem would be rather low, but the two life forms are more similar than you might think: teleosts are around 65–75% moisture content, and jellyfish around 95–98%. So the biomass of fish and jellyfish recorded in the 2003 assessment surveys was rather similar, at around 1.1–1.2 million tons each.

The consensus of those who have studied this situation is that a regime shift from a fish- to jelly-dominated pelagic ecosystem has occurred that may not be reversed rapidly, and that it was most probably a consequence of the extraordinarily heavy extraction of biomass from the original planktivorous sardine and anchovy populations during the 1970s so leaving unconsumed zooplankton biomass available to nourish medusae. This is an excellent example of a malignant regime shift even if, as the authors of the 2003 survey point out, the data represent only a snapshot of a highly complex system.[47]

More complex regime shifts than this relatively simple example may occur whenever the structure of the pristine regional ecosystem is perturbed by the removal of what we may call a keystone group. Fishing does not normally modify the abundance of oceanic herbivores directly, since these are dominated by millimetre-scale zooplankton much too small to interest us as potential food, and so we have comfortably supposed in the past that trophic cascades of the kind observed on land or in lakes after the removal of top predators do not often occur in the open sea, so that the base of the trophic pyramid that supports the fish in which we are interested is relatively unmodified by fishing.[48] But the collapsed cod-dominated ecosystem of the NW Atlantic has taught us a lesson: trophic cascades can occur in marine ecosystems if we push them hard enough.

In this case, four trophic levels and (just perhaps) nutrients were involved, and the cascade was driven by removal of top predators on the eastern Nova Scotia shelf 'thereby meeting the requirements of top-down control and indirect effects with multiple links' in trophic cascades.[49] A transition occurred in the mid 1980s associated with a major reduction, both absolute and relative, of the biomass of the ecosystem-structuring demersal fish community: fishing pressure had been so heavy that populations of cod, haddock, hake, pollock, cusk, flounder, and skate were collectively reduced by two orders of magnitude to a small fraction of their biomass that was characteristic of the 1970s, when large fish had been dominant in their populations. By 1990 there were 95% fewer large (>8 kg) and 65% fewer small (<8 kg) demersal fish in the ecosystem, and their areas of relatively high abundance had contracted. The extent of the collapse cannot be sufficiently emphasised: reductions in cod biomass, relative to historical highs, at the time of the closure of the commercial fishery in 1992–93 were massive here as elsewhere in the NW Atlantic – 91–99% on the Grand Banks, 95% in the Gulf of St. Lawrence and 50–84% from Nova Scotia to southern Newfoundland. Of all the extraordinary events in this story, that of the re-opening of the commercial fishery (quota 9000 tons) by DFO in 1999 is the most unlikely; this little-known, limited and politically charged fishery was a failure because, even though the season was extended, the quota could not be reached: there were not enough fish in the water. It has been suggested that the exploitation rate on cod during this brief episode reached 20% on 3-year-old fish and 35% on fish older than 5 years.

As the cod populations collapsed on the eastern Nova Scotia shelf, the biomass of small pelagic fish increased (<375% in individual abundance) with herring, capelin and sand lance dominating in different regions, while benthic invertebrates (shrimp, snow crab), once the prey of demersal fish, also increased strongly. There is also evidence of a statistical correlation between these changes and zooplankton (negative) and phytoplankton (positive) biomass in response to the increase in pelagic fish and larvae of benthic invertebrates. In the same years, bottom water temperature was anomalously low in relation to the long-term record of this region; stratification was progressively enhanced from the mid 1980s onwards.

In addition to the striking collapse of large fish biomass, there was a simultaneous reduction in the physiological condition of the survivors, which became lean; all species among a list of about 50 species showed this tendency, but about a dozen were strongly affected, among them cod; individuals of this species exhibited deformation and injury to the mouth region, as if from having been forced to subsist on benthic infauna rather than their habitual diet. Such observations led to the formulation of a remarkable thesis[50] that might be called the 'starving ocean' model.

Limnologists are comfortable with the possibility that a significant part of the phosphorous content of a lake may be bound up in the biomass of fish and large invertebrates, and that phytoplankton growth depends on its recycling, but the idea has rarely been discussed in relation to the ocean because of the enormous reservoir of dissolved inorganic nutrients, lying below the nutricline and constantly regenerated there. Removal of fish biomass may seriously deplete the available stock of phosphorous and reduce the potential for seasonal growth of the lake plankton.[51] In the semi-enclosed, shallow basin of the Baltic Sea a similar balance between living biomass and total available nutrients has been observed, in which fish may compete with phytoplankton for phosphorous.[52]

However, on the Nova Scotia shelf, and perhaps in similar situations, it is not inorganic nutrients that are at issue, but the quantity of organic material that must be recovered from regional phytoplankton production by the benthic ecosystem in order to balance regional losses caused by removal of demersal fish. During the period of the fishery from 1960 to 1990, a total of 9×10^9 kg of organic material had been withdrawn from the demersal fish and invertebrate component of the eastern Nova Scotia shelf ecosystem; by the early 1990s, demersal biomass was 2 orders of magnitude lower than in the 1980s. This removal occurred at a rate that was approximately equivalent to the regional rate of primary phytoplankton production, so that the overall removal of biomass could only have been sustained if the entire regional phytoplankton production had passed directly to the surviving demersal fish and invertebrates, and been assimilated by them through a three-step food chain: this, of course, did not and could not happen, so the accumulated biomass of the entire benthic ecosystem must have been progressively eroded.[53]

It is unclear to what extent this admittedly theoretical process may have contributed to the observed regime shift to an ecosystem in which the structuring effects of demersal fish became very constrained, nor is it clear if the devolved state of the ecosystem is maintained by the general depletion of biologically useful benthic biomass or by some other constraint, but the continuing poor physiological state of the remnant demersal fish suggests that this may be the case.[54]

Although we do not have the same level of ecosystem analysis for the Grand Banks as we have for the Nova Scotia shelf, this region is perhaps equally depleted of benthic/demersal biomass, and has also perhaps suffered a major regime shift; however, here the indications are a little different.[55] A period of unusually cold water was reversed only after the turn of the century and coincided with major changes in abundance and habits of capelin, a major food item of 2J3KL cod: adult capelin were small and beach spawning was reduced and late, and it is only in very recent years that these trends have been reversed. The DFO stock assessment surveys report that the cod biomass in areas 2J/3KL in 2008 remained as low as 4–5% of the 1980 biomass, that overall mortality was very high until 2003 when it began to improve slightly, and that the very slight increase observed in recent years is due almost entirely to growth of the 2002 year-class.

It is not at all clear under what conditions regime shifts such as these are reversible, especially given the role that the anomalous oceanographic regimes may have played in restraining the return of habitual benthic/demersal dominance, but it is likely that a return to the original regime will not occur until there is a very unusual recruitment event in one or more of the large demersal fish populations.

If we choose to extend the bounds of the fishery ecosystem to include all aspects of the original habitat, then the notion of broken ecosystems may be applied very much more broadly, for – especially in coastal regions – the effects of our activities, and not only of fishing, has been very damaging – and has been so for many centuries, as discussed in Chapter 6. There are today very few inshore or estuarine regions where the fish and invertebrate populations exist in anything resembling pristine conditions, and the rate of the degradation of inshore regions is very fast: who would have thought only a few years ago that wide sandy beaches from Galicia to the Solent would now be covered in late summer with a deep green sludge, a green tide of decaying, H_2S-producing[56] *Ulva armorica*? One might think oneself in the Venice lagoon ... farmers, like fishermen, have a lot to answer for, because this is but one consequence of the degradation of littoral ecosystems by the run-off of excess fertilisers from croplands and run-off from feed lots.

The general result of all this must be to restrain the probability of the recovery of a population of fish or invertebrates reduced to small population size by fishing. Generally, we may assume that much of the degradation of inshore and estuarine habitat is caused by organic and chemical pollution, by the consequences of engineering works or the reduction in river flow; it is difficult to measure how much is due directly to fishing operations. What has been called 'microbialisation' of coastal habitats, due directly to an increasing run-off of both inorganic nutrients and organic material from terrestrial sources, has damaged habitats like Chesapeake Bay and the Baltic Sea that 'are now bacterially-dominated ecosystems, totally different from those of a century ago'.[57] In such places, we cannot hope that a collapsed population will recover as the logistic model would suggest; oyster- and mussel-rearing or collecting in such regions has become, in many places, a difficult operation beset with episodes of anoxia and toxic dinoflagellate blooms and may be sustainable only with great difficulty – and some luck.

In cleaner seas, both shallow and deep coral reef habitat is being heavily and progressively degraded, in some regions principally as a direct result of fishing; deep coral habitat along the shelf edge at all latitudes and shallow-water coral reefs in tropical seas distant from great river mouths are currently being degraded very rapidly. In the case of deep reefs, this is due almost entirely to the effect of trawling for cold-water demersal fish discussed in Chapter 7; the gear[58] used at depths of >200 m beyond the shelf edge or on the summits of isolated seamounts is necessarily very heavy and destructive of the long-lived, complex and delicate epibenthos of this habitat, rich in corals such as *Lophophelia*. That the fish populations themselves are rapidly depleted is now well-known, but we have very little information on the relationship between young recruits of these species and their benthic habitat: will the almost complete destruction of branching corals and associated epifauna restrain the recruitment and regrowth of orange roughy populations after depletion? I think we have no way of knowing, but it would be unwise to think that their original habitat might be irrelevant to that process.

It is a matter of public knowledge and concern that the hermatypic coral reefs of tropical seas are being degraded rather rapidly, and while this is not of concern to any of the great fisheries of the world – apart from the infamous fishery for large groupers to be sent alive to oriental dining tables, discussed in Chapter 8 – it is of concern to many artisan-scale fishermen throughout the tropics, although their activities are part of the problem. The possibility of arresting the further degradation of coral reefs by protecting them from further fishing shall be discussed in Chapter 10.

The reef ecotope of cool seas is dominated by oysters such as *Crassostrea virginica* of the western North Atlantic coasts or *Ostrea edulis* of European seas; massive oyster reefs dominated Chesapeake Bay (they are said to have been capable of filtering the entire water column every 3 days!) and were, of course, dredged for the market. After the introduction of mechanical dredging in the 1870s, the reefs were progressively destroyed so that the fishery collapsed during the 1920s; the loss of their filtering capacity then induced a massive eutrophication of the bay, recorded in the progressive dominance of benthic over planktonic diatoms in sediment cores.[59] Similar histories have been recounted for the Baltic and the Dutch estuaries; perhaps the oyster fishery should be viewed as a general model for the fisheries of the future – from exploitation through collapse to farming captive populations? However that may be, there is clearly no realistic way back to the previous state in the collapsed cold-water oyster fisheries.

On the open continental shelf, benthic habitat destruction by trawling gear has been a concern for a very long time, but we have almost no measure of the effects on fish ecology of the changes undoubtedly wrought in the benthic invertebrate ecosystem. These effects have been most extensively studied on the level sea bottom of the North Sea: in the central regions, aerial surveillance data showed that beam trawls passed over the entire area between 0.2 and 6.5 times each year, there being a significant relationship between the degree of disturbance and the rate of decrease of benthic invertebrate biomass. As would be predicted, species whose individuals that might be broken by impact with trawling gear, such as bivalves and spatangids, were more severely depleted than those, such as annelids, less likely to be suffer individual damage.[60] Surprisingly, in this region at least, a very significant overall reduction in benthic biomass due to trawl damage was not associated with a change in the trophic level of the community, nor in the relationships between the organisms at different trophic levels within the community, because – the authors of this study suggest – there is a replacement of less-robust to more-robust species within each trophic level, and a shift from a spatangid- to a polychaete-dominated community. This shift from larger to smaller species will also involve an increase in secondary productivity compared with the rate of primary production in the superjacent water column.

A very similar result was obtained on the Newfoundland banks by a study of the benthic fauna dominated by clams and polychaetes in sandy deposits (3 years, 200 grab samples, 246 taxa) taken before and after trawling: species numbers and biomass both declined, but there was little evidence of long-term effects; trawling and natural disturbance had the same signatures.[61] This result was confirmed by later findings of studies of physical disturbance of North Sea benthos, which showed that sandy deposits recover most rapidly from disturbance, while soft, muddy deposits take longer to recover both physically and biologically: in-filling by soft oozes occurs widely on disturbed areas of softer sediments and contributes to their relatively long recovery rates.[62]

Since the degree of change induced in a benthic ecosystem by trawling depends not only on relative trawling intensity but also on the type of deposit, and hence benthic community that is impacted, it is no surprise to find that the effect of trawling on benthic invertebrate communities differs significantly between regions. Between the 1920s and the 1990s, long-term data on benthic communities show that on two grounds (Dogger and Inner Shoal) benthic composition remained essentially unchanged, while on three others (Dowsing Shoal, Great Silver Pit and the Fisher Bank) significant changes in the relative taxonomic abundance occurred. Perhaps surprisingly, in light of the often expressed concern for the effects of otter boards and other dragged engines on the sea floor, there was no large-scale loss of sensitive organisms: the time scale of benthic re-organisation after disturbance appears – in this region at least – to be shorter than the scale of disturbance.[63] We should remember, also, in the context of trawl damage that in very high latitudes (where benthic communities have similar taxonomic structure on similar deposits as in mid-latitudes) the bottom deposits at shelf depths are roiled almost continuously by the dragging of the feet of icebergs or smaller fragments of ice. Around Antarctica, such ice scours are ubiquitous: as one study puts it[64]: 'In summer, millions of icebergs from sizes smaller than cars to larger than counties ground out and gouge the sea floor and crush the benthic communities ... [which] have had to recolonize local scourings and continental shelves repeatedly, yet a decade of studies have demonstrated that they have (compared with lower latitudes) slow tempos of reproduction, colonisation and growth.' Although the effect of trawling on the sea bed has been described in alarmist terms that compares its effect to clear-cutting a forest,[65] perhaps we shouldn't spend too much more time worrying about the effect of trawls dragged on the bottom – an environmental problem that was first raised by the Commons of England in the fourteenth century! It is only on deep-sea hard-bottom habitats, such as the seamounts and along the shelf break, in the cold-water coral reefs habitat, that the effects of trawling on the habitat should probably cause us much concern. However, despite the evidence,

I am sure that research funds will continue to be directed towards this problem for which there is probably no technical solution: nobody is going to propose that trawling should be banned outright, and (as shall be discussed in Chapter 11) in those few seas where rational management has a chance of succeeding, no-go zones for trawlers, to protect unusually sensitive habitat, are easily identified and are accepted by the industry.

Given the wide variety of problems that could prevent the simple rebuilding of a population once fishing mortality is relieved, it is perhaps more surprising that some populations do, in fact, rebuild successfully than that some others should fail to do so. Potential remedial measures shall be discussed in the following chapter as part of the proper response of fisheries management to the present situation in the fisheries.

ENDNOTES

1. Garstang, W. (1900) The impoverishment of the sea – a critical summary of the experimental and statistical evidence bearing upon the alleged depletion of the trawling grounds. *J. Mar. Biol. Assoc. UK* **6**, 1–69.
2. Borley, J. O. (1923) The plaice fishery and the War. *Fish. Invest. Ser. II*, **5**(3), 54pp.
3. Margetts, A. R. and S. J. Holt. (1948) The effect of the 1938–45 war on the English North Sea trawl fisheries. *Cons. int. explor. Mer. Rapp. Proc.-Verb.* **122**, 26–46.
4. Hilborn, R. and E. Litzinger. (2009) Causes of decline and potential for recovery of Atlantic cod populations. *Open Fish Sci. J.* **2**, 32–8.
5. Animal Welfare Institute, *Ann. Quart.* **51** (3).
6. Casey, J. M. and R. A. Myers. (1998) Near extinction of a large, widely-distributed fish. *Science* **281**, 690–2.
7. NEFSC/NOAA (2006) Status of Fishery Resources off the northeastern US: flounders (http://www.nefsc.noaa.gov/sos/).
8. Simon, J. E., *et al.* (2002) *DFO/CSAS Research Document* **2002/070**, pp. 67.
9. Hutchings, J. A. (2000) Collapse and recovery of marine fishes. *Nature* **406**, 882–5.
10. For this suggestion and much of what follows in this section, I am indebted to Caddy, J. F. and D. J. Agnew. (2004) An overview of recent global experience with recovery plans for depleted marine resources and suggested guidelines for recovery planning. *Rev. Fish. Biol. Fish.* **14**, 43–112.
11. Allee, W. C. (1931) *Animal Aggregations, A Study in General Sociology*. Chicago, IL: University of Chicago Press.
12. Curiously, some authors distinguish between an Allee effect obtained by observational evidence and depensation observed in analytical models: see Frank, K. T. and D. Brickman. (2000) Allee effects and compensatory population dynamics within a stock complex. *Can. J. Fish. Aquat. Sci.* **57**, 513–7.
13. Myers, R. A., *et al.* (1995) Population dynamics of exploited fish stocks at low population levels. *Science* **269**, 1106–08.
14. Jakobsson, J. (1980) Exploitation of the Icelandic spring- and summer-spawning herring in relation to fisheries management. *Rapp. P.-v. Réun. Cons. int. Explor Mer.* **1977**, 23–42.

15. Liermann, M. and R. Hilborn. (1997) Depensation in fish stocks: a hierarchic Bayesian meta-analysis. *Can. J. Fish. Aquat. Sci.* **54**, 1976–84.
16. Walters, C. and J.K. Kitchell. (2001) Cultivation/depensation effects on juvenile survival and recruitment: implications for the theory of fishing. *Can. J. Fish. Aquat. Sci.* **58**, 39–50.
17. Stoner, A.W. and M. Ray-Culp. (2000) Evidence for Allee effects in an over-harvested marine gastropod: density-dependent mating and egg production. *Mar. Ecol. Progr. Ser.* **202**, 297–302.
18. Rowe, S., *et al.* (2004) Depensation, probability of fertilization and the mating system of Atlantic cod (*G. morhua*). *ICES J. Mar. Sci.* **61**, 1144–50.
19. Whalen, C., *et al.* (2004) Observations of a Nassau grouper (*Epinephalus striatus*) spawning aggregation site in Little Cayman, including multi-species spawning information. *Env. Biol. Fish.* **70**, 305–13.
20. Aoki, I. and T. Inagaki. (1988) Photographic observations on the behaviour of Japanese anchovy *Engraulis japonicus* at night in the sea. *Mar. Ecol. Progr. Ser.* **43**, 213–21.
21. Gerlotto, F., *et al.* (2006) Waves of agitation inside anchovy schools observed with multibeam sonar: a way to transmit information in response to predation, *ICES J. Mar. Sci.* **63**, 1405–17.
22. Clarke, C.W. (1974) Possible effects of schooling on the dynamics of exploited fish populaions. *J. Conseil* **36**, 7–14.
23. Parkinson, E.A. (1990) Impaired school formation at low density: a mechanism for depensatory mortality in sockeye salmon. *Ministry of Environment, British Columbia, Fisheries Management Report.* **99**, 1–14.
24. Walters and Kitchell (2001), op. cit.
25. Nikolsky, G.V. (1969) *Theory of Fish Population Dynamics.* Edinburgh: Oliver and Boyd, p. 323.
26. Longhurst, A. (1998) Cod: perhaps if we all stood back a bit? *Fish. Res.* **38**, 101–08.
27. Stergiou, K.J. (2002) Overfishing, tropicalisation of fish stocks, uncertainty and ecosystem management: Ockhams razor. *Fish. Res.* **55**, 1–9.
28. Stearns, S.C. (1976) Life history tactics: a review of the ideas. *Am. Nat.* **108**, 783–90.
29. See the COWESIC 2003 assessment and status report on Atlantic cod available at www.sararegistry.gc.ca.
30. DFO Canada Stock Status Report **A2–01** (2000).
31. Law, R. (2000) Fishing, selection and phenotypic evolution. *ICES J. Mar. Sci.* **57**, 659–68.
32. Law, op. cit.
33. Overholtz, W.J., *et al.* (1990) Impact of compensatory responses on assessment advice for the Northwest Atlantic mackerel stock. *Fish. Bull.* **89**, 117–28.
34. Helser, T.E. and T.P. Almeida. (1997) Density-dependent growth and sexual maturity of silver hake in the north-west Atlantic. *J. Fish. Biol.* **51**, 607–23.
35. Stenseth, N.C. and E.S. Dunlop. (2009) Unnatural selection. *Nature* **457**, 803–04; and see also Darimont, C.T., *et al.* (2009) Human predators outpace other agents of trait change in the wild. *Proc. Natl Acad. Sci.* **106**, 952–4.
36. Olsen, E.B., *et al.* (2004) Maturation trends indicative of rapid evolution preceded the collapse of the Northern cod. *Nature* **428**, 932–5, and see also Hutchings, J.A. (1999) The influence of growth and survival costs of reprodicion on Atlantic cod. *Can. J. Fish. Aquat. Sci.* **56**, 1612–23.
37. Walsh, M.R., *et al.* (2006) Maladaptive changes in multiple traits caused by fishing: impediments to recovery. *Ecol. Lett.* **9**, 142–8.
38. Reznick, D.N. and C.K. Ghalambor. (2006) Can commercial fishing cause evolution? Answers from guppies. *Can. J. Fish. Aquat. Sci.* **62**, 791–801.

39. Mayhew, P.J. (2006) *Discovering Evolutionary Ecology: Bringing Together Ecology and Evolution*. Oxford: Oxford University Press.
40. Paine, R.T. (1966) Food web complexity and species diversity. *Am. Nat.* **10**, 63–5.
41. This section relies very heavily on the following work: Mills, C.E. (2001) Jellyfish blooms: are populations increasing globally in response to changing ocean conditions? *Hydrobiology* **451**, 55–68.
42. It is worth noting that a congeneris species *R. esculenta* from the Pacific is eaten in China: cut into slices, sun-dried, it is served finely shredded with cucumber as a salad.
43. Longhurst, A. (2006) *Ecological Geography of the Sea*, 2nd edn. San Diego, CA: Academic Press.
44. Lynam, C.P., *et al.* (2005) Evidence for impacts by jellyfish on North Sea herring recruitment. *Mar. Ecol. Progr. Ser.* **298**, 157–67.
45. Cury, P. (2004) Regime shifts in upwelling ecocystems: observed changes and possible mechanisms in the northern and southern Benguela. *Progr. Oceanogr.* **60**, 223–43.
46. Lynam, C.P. (2006) Jellyfish overtake fish in a heavily fished ecosystem. *Curr. Biol.* **16**, R492.
47. Lynam, Chris, pers. comm.
48. Cury, P., L. Shannon and Y.-J. Shin. (2001) The functioning of marine ecosystems. *FAO Reykjavik Conf. on Resp. Fish. Mar. Ecosyst. Doc. 13*, 1–22.
49. Frank, K., *et al.* (2005) Trophic cacades in a formerly cod-dominated ecosystem. *Science* **308**, 1621–3, and also see Choi, J.S. (2004) Transition to an alternate state in a continental shelf ecosystem. *Can. J. Fish. Aquat. Sci.* **61**, 505–10.
50. This was one of the suggestions of Debbie Mackenzie, a concerned amateur naturalist who took on the DFO establishment and who is the unacknowledged author of suggestions very close to the thesis of biomass depletion published later by Choi *et al.* (op. cit., note 49); the website on which she argued her case is unfortunately no longer available.
51. Bartell, S.M. and J.F. Kitchell. (1978) Seasonal impact of planktivory on phosphorous release by Lake Wingra zooplankton. *Verh. Int. Ver. Limnol.* **20**, 466–74; Schindler, D.E. (1992) Nutrient regeneration by sockeye salmon fry and subsequent effects on zooplankton. *Can. J. Fish. Aquat. Sci.* **49**, 2498–506.
52. Hjerne, O. and S. Hansson. (2002) The role of fish and fisheries in Baltic Sea nutrient dynamics. *Limn. Oceanogr.* **47**, 1023–32.
53. Choi, J.S., *et al.* (2004) Transition to an alternate state in a continental shelf ecosystem. *Can. J. Fish. Aquat. Sci.* **61**, 505–10.
54. Choi, J.S., *et al.* (2005) Integrated assessment of a large ecosystem: a case study of the devolution of the eastern Scotian shelf, Canada. *Oceanogr. Mar. Biol. Ann. Rev.* **43**, 47–67.
55. DFO Newfoundland and Labrador Region: 2008 Assessment of northern (2J3KL) cod. *Science Advisory Rept. 2008/034.*
56. The death of a horse and the collapse of its rider on the strand in Brittany in 2009, has been attributed to this noxious gas.
57. Jackson, J.B.C., *et al.* (2001) Historical overfishing and the recent collapse of coastal ecosystems. *Science* **293**, 629–38.
58. Trawls and otter-boards developed for this fishery are variously described as rock-hoppers or canyon-busters.
59. Newell, R.J.E., *et al.* (1988) in 'Understanding the estuary: advances in Chesapeake Bay Research', Chesapeake Bay Research Consortium.
60. Jennings, S., *et al.* (2001) Impacts of trawling disturbance on the trophic structure of benthic invertebrate communities. *Mar. Ecol. Progr. Ser.* **213**, 127–42.

61. Kenchington, E. L. R., *et al.* (2001) Effects of experimental otter trawling of the macrofauna of a sandy bottom ecosystem on the Grand Banks of Newfoundland. *Can. J. Fish. Aquat. Sci.* **58**, 1043–57.
62. Dernie, K. M., M. J. Kaiser and R. M. Warwick. (2003) Recovery rates of benthic communities following physical disturbance. *J. Anim. Ecol.* **72**, 1043–56.
63. Frid, C. L. J., *et al.* (2000) Long-term changes in the benthic communities on North Sea fishing grounds. *ICES J. Mar. Sci.* **57**, 1303–09.
64. Barnes, D. K. A. and K. E. Conlan. (2007) Disturbance, colonization and development of Antarctic benthic communities. *Phil. Trans. R. Soc. B* **362**, 11–68.
65. Watling, I. and E. A. Norse. (1999) Disturbance of the seabed by mobile fishing gear: a comparison to forest clear-cutting. *Conserv. Biol.* **12**, 1185–97.

10

Is the response of the fishery science community appropriate?

'I have spent most of my life opposing the hubris of managing the ocean and its contents: a losing battle, I'm afraid ...'

Sidney J. Holt[1]

Around the turn of the century, the delicate state of fisheries resources and the uncertain future of the industry had captured the attention of scientists, economists and the general public in a way that it had never done previously, perhaps because it was part of the wider public concern for the natural environment that was then developing. Two different but related threads emerged in the discussions: (i) evaluations of the probability that adequate supplies of sea food could be sustained if wild populations remained heavily depleted, and (ii) proposals for novel approaches to stock management that might perhaps avoid what seemed to many to be an inevitable global collapse of fishery resources.

This is not the place to try to add substantively to the speculation concerning the trajectory that the supply of seafood is likely to follow in the future and what balance will be achieved between landings of wild-caught fish and the production of fish in sea-ranching facilities. A very comprehensive look into that particular crystal ball, published as recently as 2005,[2] suggested that the future of fisheries is already with us: the authors predicted that collapses of global financial markets and oil price shocks would each modify rather strongly the future trajectory of fisheries. As I write, the world is already struggling with collapsing financial markets, and we are also wondering what happened to the brief oil price shock that caused French fishermen to blockade their ports only last summer to demand fuel subsidies! Now, their problem is quite different because of declining consumer demand; wholesalers are currently dumping unsold inventory, and ex-boat prices are collapsing – so the fact that fuel prices have also collapsed is probably cold comfort. Declining demand has also hit Nova Scotia lobstermen: on one occasion, after grumbling about the situation on the VHF while at sea, a group on the south shore decided spontaneously to quit fishing and return to port to protest the low prices offered by wholesalers.

However, what we are concerned with here is not the future of the fishing industry, but rather the response of the fisheries management science community to the evident management failures of the late twentieth century. Concern about population failures in the recent past has been putting pressure on fisheries science to propose alternative management models, on the not-necessarily-correct assumption that faulty management science was the central factor in the depletion of global fish populations. So, in recent years, we have seen an unprecedented flow of new proposals for rational management intended to address some of the perceived failures of management. Although the problems are complex, just a few principal themes have emerged in this ongoing debate: (i) acceptance that one should err on the side of caution when making management decisions, (ii) recognition that fish populations are an integral part of complex and poorly understood marine ecosystems and (iii) an understanding of the consequences of excess capitalisation in many fisheries. These are the issues addressed in this chapter.

OPTIMISM IS THE RULE – BUT WHO CALLS THE TUNE?

Almost without exception, contributions to the flood of analytical studies of fisheries problems have had one common characteristic – that of optimism: the idea that fisheries might not be sustainable and that fishing might not be manageable in the real world has been admitted only on the fringes of the debate. Optimism is perhaps predictable, given the nature of the community engaged in the discussions: administrators of regional organisations charged with fishery management, scientists employed by them and charged with devising and deploying scientific methods for stock management, researchers and teachers in university departments of fisheries science and, finally, economists and business analysts working with the fishing industry. For a wide variety of reasons, we cannot expect such a community to respond other than with an Obamian '*Yes, we can!*' And so it has been.

Some have suggested that the organisation of fishery science has in the past permitted too great a freedom to scientists to determine their own research priorities, and that in a competitive, market-driven future, fishery science should abandon its classical model (publicly funded, investigative) in favour of a different model (problem-solving, partially industry-funded). At the same time, a lack of clear separation between science and advocacy increasingly threatens the independence of science and its potential to develop a proper understanding of the novel trajectories now being taken by fisheries.

Suggestions along these lines were made quite recently in a major study of the potential future states of fishery science,[3] although this did not take full account of the extent to which a major and progressive shift had already occurred in the performance of fisheries science, and how it is used by at least some major RFMOs: the consequences of this evolution are not yet clear, but the omens are not good.

In at least one nation, the management of research conforms to this prediction very closely so that, once again, the future is already with us. In post-collapse Canada, the decision-making process in DFO was quickly re-aligned to integrate the opinions and priorities of industry; this was done by the replacement of the internal scientific advisory committee by a council of academics and industrialists; DFO scientists sat on this council only *ex officio* and only because their presence was necessary to interpret the data in the scientific stock assessments for which they were responsible. The in-house DFO scientific advisory committee that was replaced had been the principal source of management advice since extension of jurisdiction in 1977. The new council guided the DFO minister in his negotiations with NAFO, so advice received by him was based not only on scientific stock assessments but also on the opinions and priorities of fishermen and the processing industry.

After 2003, regular advice was no longer formally sought from this council, as stock management devolved even further from a science-based procedure towards an ad-hoc, case-by-case, pragmatic activity which responded to political, social and industrial priorities as well as to science-based advice on stock status. At the same time, because the Auditor-General had criticised the DFO for not setting clear management objectives, Objective-Based Fisheries Management (OBFM) was formally embraced. This was intended to enable clear and rational objectives to be set that would be sensitive to the Precautionary Approach (PA)[4] and to biological and economic reality. The FAO Code of Conduct for Responsible Fisheries that was promulgated in 1995 had already incorporated the PA as one of its requirements; this Code formally requires that 'States and users of living aquatic resources should conserve aquatic ecosystems. The right to fish carries with it the obligation to do so in a responsible manner so as to ensure effective conservation and management of the living aquatic resources.' This should have been a defining time in fishery science because the lack of adequate scientific information, or failure properly to define reference points, could no longer be used as a reason for not taking management action.[5]

During this same period, NAFO meetings had evolved from the science-based model of ICNAF, which it replaced in 1977, to one in which 'diplomatic negotiations and trade-offs to solve management problems of straddling stocks came to assume much greater importance ... in some instances supplanting scientific analysis and review'.[6] NAFO followed the pattern of the Canadian DFO in only reluctantly implementing principles to which it had been formally committed, such as the PA or EBFM; both organisations have a shared emphasis on developing greater flexibility in their response to social, economic and political imperatives as these evolve.

Associated with this trend has been a general loss of credibility in the utility of scientific knowledge, by both the interested public and fisheries managers, that stems back to the years of the collapse. A very clear case in which scientific advice was ignored was the re-opening of the fishery for several cod populations after the moratorium, a decision that responded not to favourable scientific assessments, but rather to the fact that the period of income support for Atlantic fishermen had terminated and small coastal communities faced great difficulties. In this, and in similar cases of decisions that were made contrary to scientific opinion, the administrative strategy appears to have been either to denigrate science or to make it invisible.[7]

Research management tends now to be increasingly bureaucratised with the intention of facilitating the provision of specific answers to specific questions, rather than increasing our knowledge of complex natural systems and deepening our understanding of them. This process threatens the autonomy of science and has been generalised by Christophorou,[8] who comments that the 'detachment of many managers of scientific institutions from science and from practicing scientists and their constant and almost total preoccupation with the customers' leaves them no time (or, I would add, any inclination) to ponder about the state of understanding and factual knowledge of the problems facing them.

Although I have emphasised here the evolution of the management of the scientific endeavour in Canada, because it is here that I am most familiar with the process, the same pattern of bureaucratisation of science in management organisations is almost general and is associated with a progressive unwillingness to admit individual and expert responsibility in determining directions and expenditures in science, and the progressive evolution of formal accountability and programme planning.[9]

Although the tools of management have continually evolved over the last half century, as discussed in Chapter 1, the novelty of the debates in the past 15–20 years has been the variety of novelties placed on the table for evaluation which has, of course, strayed far beyond formal discussion in the refereed literature or in meetings arranged by management agencies or scientific societies; it now animates a world of websites devoted to supporting various initiatives, which have become illustrative of the level of social acceptance of, for instance, 'ecosystem-based fishery management'. One of the earliest roots of this movement was in the Management Strategy Evaluation (MSE) approach of the International Whaling Commission that sought to place the mechanics of stock assessment within the wider framework of rational management of a 'fishery' by the assessing the consequences of all management options and considering trade-offs; it will involve the use of performance indicators of all players in the game, including the environment that supports the resource.

Of course, the successive development of novel fashions in fisheries management is but a reflection of our times: 'Over the last two decades', wrote Pierre Lemieux,[10] 'business executives have successively embraced management by objectives, employee empowerment, business process engineering, core competencies, and six sigma – not to mention the Japanese model and business ethics'. All of these are no more than codified common sense, of course, and at least the first one listed by Lemieux has leaked across into the management of fisheries!

THE NEW ECOLOGICAL IMPERATIVE: PARADIGM SHIFT OR FEEDING FRENZY?

Ecosystem-based fishery management (EBFM) is a strange phenomenon for several reasons[11]: it has been widely endorsed by international organisations, governments and government agencies as a rationale for a new approach to management, and also by large numbers of university scientists, yet 30 years after its initial publication, debate continues about exactly what it is, and how exactly it should be implemented. Despite this, the general principle has been widely accepted, to the extent that the World Summit on Sustainable

Development (Rio+10) held at Johannesburg in 2002 required that signatory nations should develop an ecosystem approach to fishery management by 2012.

The recent discussions concerning an ecological approach to management appear to exaggerate the extent to management was restricted to consideration of the population dynamics of one target species; in fact, the very fruitful science of fishery oceanography was built on the proposition that environment mattered in stock management. In the early 1960s, the new Bureau of Commercial Fisheries facility on the cliffs above Scripps was named the 'Fishery-Oceanography Center'; research on oceanography, and on the ecology of tuna and clupeid populations of the eastern North Pacific, was greatly expanded in the new laboratories and the new research vessel that went with them. However, most of the knowledge of environmental effects on fish populations that resulted from fisheries oceanography, as a correspondent has suggested to me, was knowledge that was not acted upon, because assessments have been habitually run with assumptions of constant fishing mortality.

Even so, formal proposals were being made by the 1970s that single-species management, however performed, was not a satisfactory solution and that we required a multispecies approach that was sensitive to environmental conditions.[12] A principal concern of the early studies was the effect of fishing at one trophic level on the biomass and productivity of other levels, especially because fishing preferentially targets those organisms near the top of the pyramid that are normally subject to relatively low natural mortality.[13] But, even with this evidence of wider thinking, the formalism of an MSY was retained in many North American fisheries.

The earliest document that I can locate that formally proposed a widening of this vision and a move towards an EBFM format was the NOAA-NMFS Program Development Plan 'Ecosystems Monitoring and Fisheries Management' published in April 1987.[14] Although debate about management options for fisheries was active in the fisheries community at this time, formal EBFM was ignored by almost everybody except some government agencies for almost a full decade: curiously, it was only in 1995 that the first scholarly paper on this subject appeared,[15] perhaps responding to the progressive development of suitable software. Subsequently, the number of such contributions increased exponentially to reach a plateau of around 200 papers a year in the new century (Figure 10.1). After a similarly short gestation, many national agencies and RFMOs adopted the principle in their statements of intent concerning management techniques.

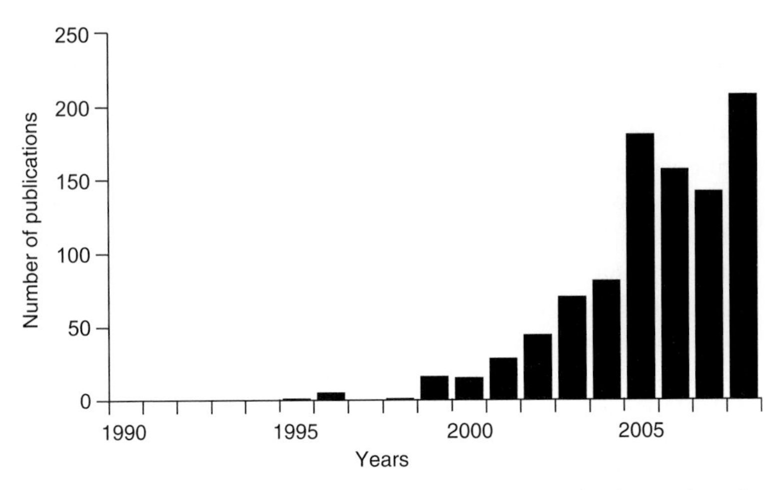

Figure 10.1 The evolution of an informational cascade: the number of scholarly papers that discussed the concept of 'Ecosystem-based Fishery Management', or EBFM, per annum after the first such paper in 1995.

The NOAA-NMFS text of 1987 was drafted by Gary Sharp and bears re-reading today, because very little that has been written since adds much to his basic suggestions: 'The purpose of this initiative', he wrote, 'is the orientation of the NOAA fisheries program to encompass a multi-species/ecosystem approach and provide the basis for developing a forecasting capability ... the central concern underlying the shift of the focus of NOAA fisheries programs towards one which supports ecosystem monitoring as a basis for fisheries ... management ... is the need to develop a capability to forecast changes in these ecosystems.

'All living organisms are part of biologically and environmentally-linked species groups ... loosely termed a marine ecosystem from which they extract their needs and to which they in turn contribute support for other components. Fluctuations in the biomasses of these organisms are strongly influenced by both internal and external factors ... over periods of decades without a commercial fishery.

'Just as most species are not harvested in isolation, they cannot be managed in isolation ... manager must understand inter-species relations if he is to manage inter-related fisheries. A holistic approach to research, monitoring and management of living marine resources is now feasible because of advances in technology and systems modeling methods ... timely for a shift in Federal program efforts towards a multi-species/ecosystem approach that will support a forecasting capability.

'Seven regional marine ecosystems (RMEs) with U.S. jurisdiction ... provide the framework for development of the NOAA program for the ecosystem monitoring and fisheries management (N Atlantic shelf, S Atlantic and Gulf of Mexico shelf, Atlantic oceanic, California Current, Pacific oceanic, Gulf of Alaska and Bering Sea, Antarctic).'

This Program Development Plan responded not only to the social spirit of the times, but also to movement within the marine science community, for these were also the years when the concept of formal Large Marine Ecosystems (LMEs) was introduced by Ken Sherman of NMFS, with the intention that these should become the focus of integrated studies and models. The earliest public manifestations of the LME movement appear to have been special sessions held at the AAAS meetings of 1984 and 1987.[16] User-friendly modelling software (notably EwE[17]) to perform simple simulations of regional ecosystems was becoming freely available at this time, as were cheap desktop computers on which to perform the simulations. This convergence led to assumptions that EBFM would necessarily involve holistic ecosystem modelling and, as Caddy remarked, 'wherever EBFM is discussed you will see trophic models offering pre-digested solutions to your modeling needs'.[18] There were, and still are, those who are wary of the uncritical use of such software, yet EwE is very seductive and is widely used in the EBFM context. Unfortunately, simulations are in trophic currency and this can simulate only one aspect of the biota: excluded are simulations based on biological size spectra, or involving economic and political factors, yet these may be essential aspects of any regional problem.

Rather strangely, when the notion of EBFM was formally reviewed by NOAA in a Report to Congress in 1998 (as mandated by the Magnuson-Stevens Act of 1996), no recommendation for quantitative ecosystem simulation was made. Instead, it was suggested that each Regional Fishery Council should delineate the ecosystems for which it was responsible and, for each, to develop a Fisheries Ecosystem Plan, that would serve as a metric against which individual stock management plans could subsequently be measured. To do this would require obtaining knowledge to characterise the 'biological, chemical and physical dynamics' of each ecosystem; conceptual models of the food web were then to be developed together with metrics describing the 'habitat needs of different life stages of all plants and animals that represent the significant food web' although no suggestions concerning simulation models were made. Councils were also enjoined to 'assess how uncertainty is to be defined', to 'develop indices of ecosystem health as targets for management' and to 'describe long-term monitoring data and how they are to be used'.

Considering the breadth of these tasks, it is not surprising to read of the extent of the progress that was made in the next five years. A NOAA document of 1998 entitled 'Technical Guidance for Implementing an Ecosystem-based Approach to Fisheries Management' noted that seven of the regional Councils had taken up the recommendations and only one was concentrating, instead, on improving its habitual single-species approach to management. The authors of this document noted that as the concept of ecosystem-based management was explored, it became progressively more complex, principally because of the realisation of the full regional ecological complexity that exists within each LME (or its equivalent) and also of the necessary interaction with social and economic considerations that are, in effect, an integral part of each ecosystem.

Other agencies used a more relaxed version of EBFM in their policy statements. The Canadian Oceans Act of 1997 specified only the definition of Ocean Management Areas 'which are to be managed for the aggregate of ocean uses, where ecosystem features are to be considered and a precautionary approach is to be applied to stock management'. The FAO, in reviewing EBFM for member states in 2002, noted that ecosystems are complex and provide 'goods and services beyond those of benefit to fisheries'; it was suggested, therefore, that it would be better to discuss an 'ecosystem approach to fisheries' or EAF, a term that had been coined at the FAO Technical Consultation on EBFM in Reykjavik in 2002 to satisfy those member states which objected to the ecosystem becoming the new objective of fisheries management, rather than the fish. The Reykjavik report discusses institutional requirements, principles, objectives and implementation issues of EAF, but makes mention neither of regional ecosystems, nor of their modelling.

The EU Commission took a different stance again in their Fisheries Ecosystem Plan, which leaves formal EBFM somewhat on the back burner. Fisheries Council Regulation 2371 states that the objective of the Common Fisheries Policy is to ensure sustainable exploitation of fishery resources by application of the precautionary principle and will aim at progressive implementation of an ecosystem-based approach: critics have pointed out that this text does not specify how far progressive implementation is to be an extension of present practices and how much is intended to be novel.[19] One of the Work Packages associated with the Fisheries Ecosystem Plan seeks to identify key species and key habitats that characterise each region; included are the seabirds, marine mammals, fish, benthos and zooplankton thought to be the major drivers of ecosystem functioning in each region. Once identified, metrics or ecosystem indicators are to be identified for each significant species: these are to represent the state of the system and are philosophically similar to the reference points and indicators discussed in Chapter 1.

These proposals from major fisheries agencies are very different from the approach supported by many individual marine scientists, especially those people who are committed to regional ecosystem modelling with EwE software. At a 2004 symposium, a group of these scientists presented model analyses of ecological structure and performance for 10 regional ecosystems, or types of ecosystems, representing all seas and all latitudes. The convenors suggested that this symposium 'may very well be mentioned favourably when the history of the incorporation of ecosystem considerations into fisheries management is written some day'.[20] Such an approach to the problem of incorporating ecological relationships in fisheries management is central to the many volumes now devoted to studies of the LMEs. Most of the contributions to the study and synthesis of regional ecosystems published in this series were written with EBFM firmly in the authors mind: in some cases, this as explicit, as in Pauly's analysis of the consilience of mass-balance modelling in oceanography and fisheries management, a coming together of disciplines from which new suggestions for coastal area management may be expected to evolve.[21]

Considering the wide range of activities that have been induced by the exploration of the twin concepts of precaution and ecosystem management, it is perhaps not surprising that it has become progressively more difficult to specify exactly what is, and what is not, to be part of a resource management system sensitive to integrated natural processes, and as late as 2002 we were reading papers with titles that asked, rather plaintively, 'What does Ecosystem-based Management mean?' and 'EBFM: what is it and how can we do it?'.[22] The answer, of course, differs depending on who is asking these question: at least for fishery scientists, the definition of Peter Larkin probably comes very close to their thinking, being centred on multispecies interactions in the context of a variable physical and chemical environment.[23] The confusion over terminology is such that in a study of the implementation of EBFM in Australia, the heads of the federal and state resource agencies felt compelled to draft a comparative analysis of the terminology already in use globally before they felt able to discuss EBFM; in fact, they decided to retain the term Ecologically Sustainable Development (whatever that might mean), already used in their departments.

Several NGOs have become involved in the debate, which has tended – especially in their hands – to become somewhat theological. I sometimes find myself wondering how many acronyms can dance on the head of a pin? Such a case is the debate organised recently by the NGO Marine Ecosystems and Management, an information service based on the University of Washington; the issue to be resolved is whether EBM or EBFM has priority? One contributor suggests that in most cases, an ecosystem approach to fisheries is a necessary but often not sufficient part of the sustainable use of aquatic ecosystems: 'managing the whole ecosystem is best but managing individual parts can still be useful even when the whole cannot be achieved'. Another remarks that 'most commonly there will be a number of sectors impacting on an ecosystem and real progress will only be made if they can be addressed simultaneously within a broad EBM'. Yet another wrote that the logical extension of attempting to balance diverse social objectives is that fisheries management must be embedded within a broader ecosystem approach.

The scientist-turned-administrator is indeed more likely to emphasise the integration into fisheries management not only of ecological considerations but also new objectives designed to maximise sustainable yields and economic returns from the fishery, together with a sensitivity to employment and other social concerns.[24] A specimen of this language that I especially cherish is to be found in the 2007 EurOcean report of the Aberdeen Declaration – A New Deal for Marine and Maritime Science: 'An all-embracing maritime policy aimed at developing a dynamic maritime economy in harmony with the marine environment, supported by sound marine science and technology, which allows human beings to continue to reap the rich harvest from the oceans in a sustainable manner.' Having admired the meticulous drafting, one might well ask 'what is there left to say?'

Fortunately, there is a lot left to say: consider, for example, the apparently simple problem of metrics. Initiatives like the EU Fisheries Ecosystem Plan require the development of what Steele calls ecometrics to characterise the state of individual species in a regional ecosystem, of its physical state and of all associated social and economic activities. These, somewhat to extend Steele's argument concerning the diversity, productivity and resilience of marine ecosystems,[25] are incompatible because we cannot assign values in a common currency to, say, the seabird population of the North Sea or to the by-catch in a directed fishery, that enables them to weighed against metrics describing unemployment in Lowestoft, or the economic viability of a company that markets frozen fillets. In this case, integrated systems modelling is incompetent to render advice, and we appear to be condemned to continue either with the management methods of the present day in which response occurs preferentially to the process that is creating the greatest political pain, or else with an EBFM system in which management responds to information concerning the fishery, the industry and the environment but lacks quantitative integration.

Even in single-species management, there has always been a level of uncertainty in the relatively simple data that are used in setting catch limits. However, the accumulated uncertainty about ecological states and economic outcomes associated with any EBFM procedure increases as the number and complexity of indices required by the procedure themselves increase. The associated costs of research and monitoring that is required to quantify the indices must also increase accordingly, and the extent to which such costs will be realistic in tomorrow's world is anybody's guess: fisheries science and management require heavy capital investment both ashore and at sea, beyond the realistic capabilities of many maritime nations. In Canada, the significant run-down of the facilities and personnel of oceanographic and fisheries facilities during the 1990s was part of across-the-board economies by the federal government that allowed it, finally, to balance the national budget after decades in deficit.

In this context, Paul Dengbol has memorably described the complex problem associated with the introduction of EBFM as 'the noble art of addressing complexity and uncertainty with all onboard and on a budget'.[26] More bluntly, in many regions it is probable that expenditures on research and management have passed the limits of cost-efficiency based on the value to society of fisheries outputs; in any foreseeable world, many of the Utopian plans for rational fisheries management that have been discussed in the last 20 years will fail to find financial support unless a political value can also be adduced.

Meanwhile, the debate continues, for analysis of the potential and content of EBFM still has great attraction for many fisheries researchers, perhaps especially among those who do not go to sea to try to do it: not only have fisheries agencies assembled study groups to discuss the potential of EBFM, but so also academies of science and learned societies have individually published theme issues of their proceedings devoted to analysis of an ecosystem approach to fisheries. Although there is great divergence among the individual opinions that have been expressed, there is no general rejection of EBFM – as I suggested above, we have to expect optimism for the future, both from individuals and agencies, after the relatively inglorious past of fisheries management. A Policy Forum published in *Science* in 2004, presents and discusses EBFM as if it were a recent proposal, and agrees (rather prissily) that it 'should move forward now despite current uncertainties about ecosystems and their response to human actions because the potential benefits of implementation are as large as or greater than the potential risks of inaction'.[27]

Even today, new analyses are being funded: a group based at UC Santa Barbara announced very recently its intention to develop a unifying terminology and a common analytical framework for assessing marine fisheries and ecosystem change, to be presented to 'managers and government agencies to help them understand the progress that has been made'.[28] I find it very easy to be cynical about such initiatives, for these are certainly not the issues that are responsible for the general malaise in fisheries management. It would be easy to dismiss the general enthusiasm for EBFM and the general acceptance that it is a practicable way of managing resources as yet another example of conformity. Cascade theory suggests that group cohesion is maintained by adherence to a relatively small amount of original information, by imitation and by an incentive to disregard new information.[29] At any rate, it is quite difficult to find contrary discussion concerning EBFM and it would certainly be far more difficult to get critical reviews accepted for publication than supportive studies: for one thing, it would be hard today for an editor to find peer reviewers with an entirely open mind, but that's another story ...[30]

But, in the end, what is most remarkable about the flood of studies related to EBFM that have appeared in the last 15 years or so is the relatively tiny proportion of authors who have discussed any concrete results obtained by the application of the new management theory; one could easily conclude that EBFM, or EAF, or whatever other acronym took your fancy, was an entirely theoretical proposition. Even for some cases that have been thought to be notable successes of EBFM, one is left wondering what has changed from earlier days. The notion that classical single-species management was an example of extreme reductionism and was done in total isolation from information concerning other organisms than the target species, or of the changing state of the environment, is not correct, although often remarked; at least in the European school of management by analytical models, the implications of changing external factors likely to modify growth, reproduction and mortality were constantly examined[31] – although, as I have suggested earlier, with means that might appear derisory today.

Concern was expressed during the 1970s (and well before EBFM had come on stage) that commercial catches of krill around the Antarctic continent could have consequences for species that preyed on them – seals, seabirds, whales and fish – and, consequently, the CCAMLR (which was established in 1982) is currently attempting to monitor the effects of krill fishing on these organisms within small-scale management units (SSMUs). This is not, however, any more than a natural evolution from what was already understood 35 years ago and is scarcely a triumph of EBFM, the more so since both Japan and South Korea oppose the presence of observers aboard krill fishing vessels.

The region of the Benguela Current off Angola, Namibia and South Africa is said to be a good example of EBFM in practice. The LME project established in this region in 2004 was specifically directed at forecasting the performance of the ecosystem of the Benguela Current in relation to the effects of fishing and changing external forcing.[32] The results of this investigation, supported by EwE simulations of changes in the ecosystem forced by observed changes in the circulation, ensured that our understanding of this region was brought in line with what we knew about, say, the responses of the California Current to fishing, changing wind stress and the mutual dynamics of sardine and anchovy populations 30–40 years ago.

The Benguela LME project also stimulated the establishment of a three-nation Benguela Current Commission that is intended to address transboundary issues concerning population depletion, habitat restoration and coastal pollution. Once again, one is left wondering where exactly is the novel content in this undoubtedly useful progress; international fishery commissions are hardly a new idea, and many others had already occupied themselves with fish habitat and ocean conditions long before being required to do so by the advent of formal EBFM. Reluctantly, one has to conclude that much of the discussion concerning this theory of management represents no more than old wine in a new bottle, appropriately labelled.

This impression is strengthened when one attempts to locate serious attempts to implement an ecosystems approach to management; almost universally, accounts of implementation turn out on examination to be no more than yet another analysis of what EBFM really means and how the agency concerned might move forward to its implementation; on the NW Atlantic shelf, American and Canadian plans exist to implement a 'pilot project to test implementation approaches for Integrated Management (IM)'. This is to be done with projects on the eastern Scotian Shelf by DFO and in the Gulf of Maine by

NMFS: it is reported in internal DFO documents[33] that the Scotian Shelf project has explored both governance frameworks and the development of conceptual and operational ecosystem objectives; unfortunately, this project evolved before national policy guidelines were available, so it was necessary to retrace its steps to be compliant with these. And so on and so on ... one wonders how serious people can have sufficient patience to handle such nonsense.

However, this region has given us a possible confirmation of the utility of one approach to management that offers a way of integrating a highly diverse field of information into an informal model, thus bypassing the wholly impractical suggestions of holistic and numerical simulation of ecosystems and their socio-economic relations. This is a system of indicators of ecosystem status, based on Caddy's proposals for a 'traffic light' management approach.[34] Using numerical indicators derived from assessment surveys and from the industry, a wide range of variables could theoretically be integrated into an assessment of the risks associated with increasing or maintaining present effort. Each indicator would be a very simple three-level numeration of the variable represented in the system as a single indicator: the original authors insisted that it would be of use only in situations where management was committed to a set of decision control rules to determine what action was required.

Such a system was implemented by Peter Koeller, after an extensive set of simulation exercises on several shrimp populations on the Canadian east coast. Subsequently, it has been used by him in producing assessment advice for the shrimp populations of the Scotia Shelf: he suggests that this process enabled him 'to stop worrying and love the data' and, by this, I take it that he meant just getting on with the job.[35] This system of traffic lights integrates annual values of 23 indicators, each of which describes the state of the physical environment, of the fishery, and also relative shrimp productivity and absolute shrimp abundance; for each indicator, three states (good, indifferent and bad) are defined at the 0.33 and 0.66 percentiles of the value used. Data for about half of the indicators extend back to 1989, and from 1995 the full set is available. The data are incorporated into a matrix (Figure 10.2)[36] representing (i) the observation relevant to each of the (in this case) 23 indicators, (ii) the interpretation of the observation and (iii) the three-level evaluation; in this way, numerical data of very different kinds can be integrated into the same decision matrix. The simulations suggest – as we might expect – that the approach would perform better with higher numbers of indicators, even when these are poorly measured.

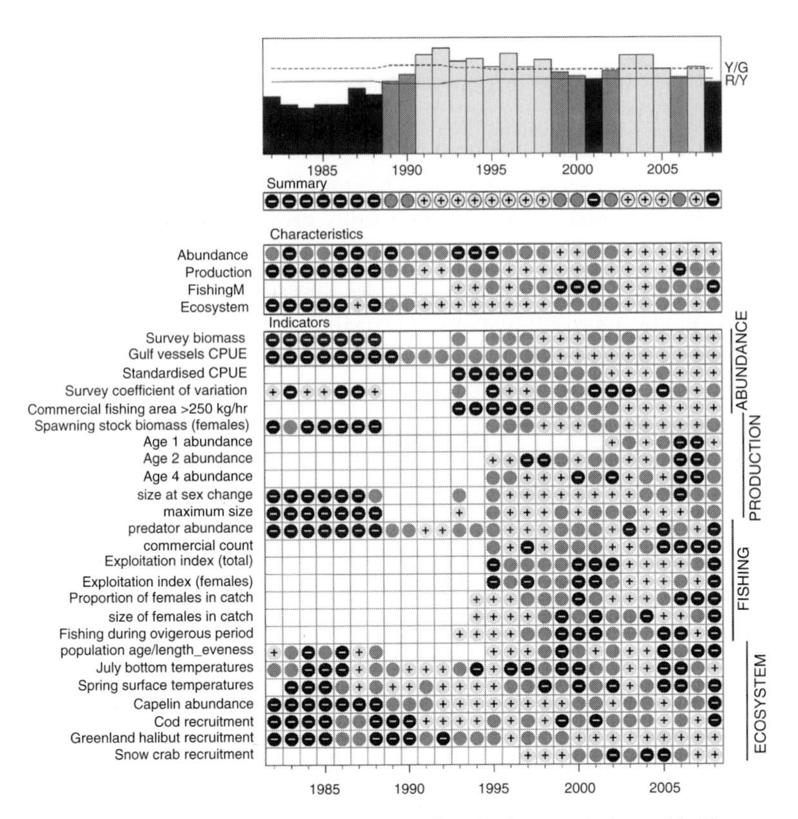

Figure 10.2 'Traffic light' diagram of stock characteristics and indicators obtained from northern shrimp (*Pandalus borealis*) stock assessments; in the original figure, the icons are red, green and amber (courtesy of Peter Koeller).

Koeller remarks that although his intended purpose in using this approach was only to find a way of communicating complicated and diverse information in a transparent manner, what emerged was unexpected: the computed TAC for shrimps exactly tracked his mental integration of the indicators and his judgement of how the population would respond. This appears to me to be a much saner method of integrating complex information into management than the various holistic modelling approaches that have been suggested, and is as near to an ideal EBFM approach as we can get in the real world. This is particularly evident if it is associated, as Caddy suggested, with simple harvest control rules that are consistent with the dynamics of the managed population – of whatever species – and with the political and social requirements of the management agency. In the case of this shrimp population, the control rules are simple: the target is to set the TAC so as to keep the exploitation rate at about 15%, modifying it according to the indications from the traffic light matrix. In 2009, this suggested a significant reduction, so the TAC was reduced from 5.0 K to 3.4 K tons.

For only one fisheries region can I locate a fundamental commitment by a fisheries agency to management with an eye firmly fixed on the state of the ecosystem, and where minimal time has been wasted in formal analysis of potential approaches to the problem: this has resulted in a singularly successful fishery but, for reasons that will become clear, I reserve a discussion of this region until my final chapter.

THE END OF OPEN-ACCESS FISHING?

Property rights over natural resources are as old as history, but enhanced interest in them has developed among fisheries economists and management agencies during the last 30 years or so. Already during the 1950s, several economists had discussed the potential danger that the common pool status of sea fisheries might lead to over-capitalisation, to reduced rents and depleted resources and this is, of course, the opposite side of the same coin as Hardin's tragedy of the commons. They suggested that some form of restricted access or property rights would have to be accepted by governments but, because of the complex legal situation resulting from the common public right to fish, fisheries management agencies reacted only slowly to this suggestion, even when the excesses of industrialised fisheries confirmed the expectations of the economists.

Apart from political reluctance to open a very controversial issue, the conflict between restricted and open access to resources was (and is) not at all easy to resolve: recall that in the Commonwealth and in the United States, the public right to fish has been guaranteed constitutionally since Runnymede in 1215, and that in many other nations the same right is likewise enshrined constitutionally. Even today, the sale of public fishery resources to individuals raises the hackles of right-wing political opinion, especially in North America; this is rather strange considering the history of the allocation of federal lands to homesteaders, on which the agricultural landscapes of the USA and Canada were built. On the other hand, among economists, environmental groups and liberal political opinion, resource privatisation in the fisheries is not to be criticised.

When restricted access to fisheries was first discussed in modern times,[37] national agencies could have accorded private access to common-property resources only inside the narrow coastal zone, within cannon-shot of the shores, that is the territorial sea: beyond that, restricted access, but not property rights, could be established only in fisheries to which the national fleet had assured access under an international agreement. It was only with the implementation of UNCLOS 3 in 1976, and the establishment of 200-mile national Exclusive Economic Zones, that governments could seriously address the question of property rights or restricted access in the offshore areas. This still left, of course, by far the greater part of the ocean as the 'common heritage of mankind' where nations may apply restrictions on the actions of their nationals only where their access to regional fisheries is controlled by an RFMO.

Despite all these difficulties, it was inevitable that one response to the problem of over-capitalisation of fisheries and stock depletions would be the establishment of some form of limited entry, and many versions of that idea subsequently evolved.[38] Progressively, this evolution occurred along a continuum from (i) open access fisheries to (ii) regulated common-property fisheries with access by limited-entry licensing, and thence to (iii) privatised or exclusive fisheries, which leads naturally to the idea of individual transferable quotas (ITQs), each representing a share of the total allowable catch of a species.[39] These shares are allotted, or sold, to a person or to a boat by the government on behalf of the people (who are the owners of the resource), and are transferable by sale to another party, as if they were property. Superficially, this seems like a simple solution to the problem of common-property resource depletion, but it does raise highly complex issues (economic, social, political, financial) and some scepticism has been generated concerning its real efficacy. Some serious blunders were made by national agencies during the implementation of transferable quota systems. In New Zealand, the levels of the first ITQs were based on the landings made by the new owners in previous years, so that when the TAC had to be reduced to conserve stocks, quota buy-back cost the government some NZ$30 million in 1986 alone, a cost that was obviously unsustainable. It took some time for the system to adjust to the reality that TACs, and therefore ITQs, must be a variable quantity.

Two of the most sensitive political problems to be resolved in establishing property rights are who shall benefit by the sale of public goods, and how shall the individuals who shall be offered the same goods be selected among the many who might benefit from them and want them? This is not a trivial problem, for the monetary values of ownership of resource licences or quotas may be very high indeed: an accountant's valuation of the individual worth of the 81 mid-shore licence holders of the New Brunswick snow crab fleet is around US$6 million – a windfall, indeed, since these limited-access licences were issued free. Such was the effect of the federal DFO's selection of this group as a model partner, to the exclusion of 4000 other regional boat owners who exploit multi-species and lobster fisheries, but are excluded from all except 10% of the snow crab resource: this policy 'made millionaires of relatively few people, and the battles that have ensued have been dramatic and emotional'.[40] These people have picked up some of the costs of the relevant science and management to the benefit of the DFO budget, but some might think that the nation or nearby local communities should rather have benefited from this windfall.

There is a great deal of diversity in the legal status of ITQs: do they comprise property or just entitlement? At one extreme, in Canada, they are carefully termed fishing privileges by the DFO, and their permanence depends on continued political support in Ottawa. At the other extreme, in Australia, the quota right is regarded as property by the courts. In Iceland, the indefinite nature of the tenure of ITQs has led to a market in quota trades that operate through brokerages. In New Zealand, each ITQ is a constitutionally protected property right of the holder in perpetuity. There are significant consequences of these differences for those in the fishery: the Canadian Senate Committee on Fisheries, discussing ITQs, was told that under the New Zealand system, banks regarded ITQs as collateral against loans, while in Canada fishermen still have to borrow money from fish merchants.

Important social consequences arise from the trading of ITQs, a form of management that was initially welcomed very widely in Iceland, although it was only a privileged few who obtained the free allocations. Capitalisation of these led to their accumulation in the hands of corporations and investors and concentration of fishing operations, a result that had not been foreseen by the public. In Alaska, the rationalisation of the crab fishery by ITQs created much anger in coastal communities left out in the cold, as they claimed, in favour of a few individuals and corporations. New entry to ITQ fisheries may become prohibitively expensive, because the necessary quota rights add very greatly to the cost of a vessel: consequently, once a deckhand, always a deckhand, as people in the business of fishing are starting to complain. This is a significant change in the traditional way of life at sea in the fisheries, where intelligence and hard work could bring individual rewards in the past. Fishing communities themselves are threatened economically – and may disappear – if their resources are alienated to interests in the city, as has happened significantly in Iceland.[41] Elsewhere, as in Canada, this problem may be even more important in the case of aboriginal coastal communities on the Pacific and Arctic coasts, and the Canadian Senate has enjoined the DFO not to extend the regime of ITQs until the needs and rights of such communities are assured.

But here, we are concerned principally with the problems in the fisheries caused by over-capitalisation and depleted stocks, not social problems: to what extent have ITQs increased the effectiveness of management? One anticipated (and planned) consequence of ITQs was the reduction of capacity in those fisheries to which it has been applied, and to reduce the intensity of the 'race to fish' in order to maximise catches before the total quota is filled; this has indeed occurred and is certainly a major advance in management.

However, the overall effectiveness of ITQ management in improving stock status is very difficult to assess. A recent study[42] that appears to demonstrate conclusively the effectiveness of ITQs in slowing population collapse is unfortunately very simplistic, although it has attracted some media attention.[43] It was based on data describing the reported landings from 11,135 individual fisheries of which only 121 were determined to be managed by full ITQs. The authors used a simple correlation between the rate of population collapse in two groups of fisheries of such different sizes that it is rather unconvincing as to causation, especially in the light of the selective nature of the fishing nations included. I have not been able to obtain the list that these authors used, but a slightly different enumeration gives the following pattern of 148 ITQ fisheries: Iceland – 23, New Zealand – 68, Australia – 23, eastern Canada – 5, Atlantic coast USA – 3, Gulf of Mexico – 1, Peru/Chile – 6, California – 1, Oregon/British Columbia – 14, Alaska – 4. Weak ITQ systems, in the form of limited entry licences, are also in place in Namibia, the Netherlands and probably elsewhere. These are all nations where, for other reasons, one would expect to find a management performance superior to the dismal global average!

It has long been anticipated that although the introduction of ITQs must reduce fleet size, it might do nothing to reduce the rate of discarding low-grade fish so as to enhance the value of what is landed against each individual quota – in fact, it was feared that this practice might be enhanced since the total landings of each boat or enterprise were now limited. It had been hoped that this tendency might be offset by a changing attitude at sea, on the grounds that ownership of the resource might enhance the perceived need for individual stewardship. Examination of what has actually happened in each nation that uses ITQs suggests mixed results, although they are largely positive[44]: evidence for an increase of discard rate at sea is equivocal, but discarding does continue in almost all these fisheries. In the New Zealand fisheries, it has been noted that larger proportions of non-target species are landed by boats carrying observers than by those without, and discards are still described as being substantial in the Australian fisheries. Off British Columbia, discard rates are also reduced in the presence of observers, and also reduced when ITQs were originally introduced. It should also be remembered that in the NW Atlantic during the cod crisis, some of the scandalously heavy discards were from Canadian boats already operating under limited entry licences, now described as a soft ITQ system. In fact, incentives for discarding must always be present whenever limits are imposed on landings but not on catches: video surveillance at sea or on-board observers are the only watertight means of ensuring that the practice ceases. It is supposed that the tighter the national ITQ system – in the sense of property rights – the greater the incentive to individual stewardship, including the disincentive to discard.

It is in the area of enforcement that the greatest advances have been made by the imposition of ITQs. In several countries, enforcement, monitoring and even management may be done now by specialised companies that have been approved but not funded by the relevant government agency. In New Zealand, the Challenger Scallop Enhancement Company manages the southern scallop fishery on behalf of the holders of ITQs and, in Canada, companies that are certified by the DFO are responsible for monitoring landings in ITQ fisheries. On the west coast, the ITQ holders in the sablefish fishery help to pay for stock assessment surveys while, in Namibia, the industry pays for onboard observers.

In none of the nations that have established ITQ systems in their fisheries does there appear to be have been anything other than a positive effect on stock status so that, purely from a biological point of view, this appears to be a positive move in fishery management. That it is economically sensible is also undeniable, although governments may find themselves responsible for social problems – and resultant heavy costs – in coastal communities: it is very easy to dismiss this as only what is to be expected when any major industry is re-organised, but the social difficulties may be considerable for those who have lived for generations in small outports with nothing to live on but the sea, and some rough summer grazing for a cow.

MARINE PROTECTED AREAS: NEITHER PANACEA NOR RED HERRING

The setting aside of areas of the shelf seasonally or permanently for the protection of fish of other valued organisms is also as old as history, but has been taken up again in the debate concerning future arrangements at sea. However, much of the discussion has been more banal than progressive, and some of the estimations of the potential of marine reserves are at best naïve. For some authors, the use of protected areas is part and parcel of the general popular movement in favour of ecosystem or habitat protection, and hence related to EBFM, while for others it is very much associated with the management of exploited fishery resources in the sense of total sanctuaries. Nor, unfortunately, do the discussions manage to avoid semantics, although I shall not bother the reader with such interesting problems as whether marine protected areas (MPAs) are a special class of marine reserves, or vice versa, nor with how properly to define a network of marine reserves.

Many readers will recognise the origin of the title of this section in one of the most level-headed analyses of the potential of MPAs to assist in solving today's problems in marine resource management: like its author, I conclude that their potential lies somewhere between the extremes suggested, and that MPAs will be one essential tool among many that should be used in the future.[45]

These protected areas are often discussed as if they were simply areas set aside as no fishing zones, or else for the protection of fragile habitat like a coral reef and most of those already established do, indeed, have this objective. However, the term has been used much more widely than that, as is indicated by the national inventory of MPAs in the regions under US jurisdiction. These include almost 1700 sites, ranging in size from the $<1.0\ km^2$ of the *USS Monitor* wreck in New York harbour to the $956{,}969\ km^2$ of the Aleutian Islands Habitat Protection Area. The degree of protection that these sites enjoy from fishing varies widely: a total ban exists in 162 sites, commercial fishing is banned in 143 and sports fishing in 18 sites (totalling some $21{,}200\ km^2$ or about one-hundredth of the total area of continental shelf of the United States), while in 756 others there is some level of restriction on fishing. By referring to the organisation of MPAs on the American coasts, I do not mean to suggest that this is necessarily the ideal situation, but rather to emphasise that protected areas can no longer to be discussed as if they belonged to a theoretical future; they are here, now, and it should be assumed that what is possible to arrange on North American coasts is possible everywhere – given only sufficient political will. But it is important to have rational expectations for what protected areas can deliver. Are our expectations simply the conservation of critical habitat of endangered species of fish or invertebrates, or are they rather the general maintenance of sustainable fish populations on the continental shelf in the case where classic stock management has failed?

Those who believe that fishing effort control alone will suffice to manage populations will, perhaps naturally, oppose any recourse to MPAs, while those who support the use an ecosystem-based approach to fish stock management will be proponents of MPAs, which are often perceived as one technique for incorporating ecosystem objectives into fisheries management.[46] These people may be joined by those (for example, fisherfolk) who fear that the effect of population re-growth within an MPA may not be felt very far from their boundaries, so that they will experience very little profit from losing the 20–40% of the total range of their target population, this being the percentage of a species habitat that is most often quoted as being essential for inclusion within a protected area.

The simple fact that restricted areas have been used in the social management of fisheries since antiquity and continue to be so used in less-developed fishing societies today is evidence that their effect is positive on at least small spatial scales. One of the more ingenious examples that I have encountered is that of the poaching lobstermen on the South Shore of Nova Scotia, a problem solved by their nineteenth-century parish priest who allocated use-areas to each man – together with a common area to be used only by those whose own patch failed to produce an equitable catch.[47] That community-protected areas can still be established today off subsistence-level fishing villages is illustrated by the example of Ucunivanua in Fiji, where villagers established an exclusion area for *Anadera* clams that produced dramatic increases in numbers and size of these organisms in just a few years within the area and, by spillover, increased catch rates outside it.

This example demonstrates what, in fact, we should expect everywhere that part of the marine ecosystem is totally protected: there is a very rapid and very general positive response of all previously depleted biota within the protected area, which should be observable at a wide range of spatial scales. It must be emphasised that the simple observation of increased individual size and population abundance in protected areas does not provide unequivocal evidence of the effectiveness of an MPA[48]; the experimental design of any comparison the between control site and experimental areas is fraught with difficulty, as are before versus after comparisons in which the possibility of environmentally forced changes are usually ignored. It should not be neglected that species whose abundance changes may be responding to an environmental change.

Nevertheless, examples are assembled from protected areas (or regional fisheries closures) off South Africa, Corsica, Spain, New Zealand, Kenya, Bahamas, Jamaica, US Virgin Islands, Florida, Newfoundland, California and the State of Washington; in all these places, more-or-less rapid growth of a depleted population was observed when compared with the performance of the same species outside the protected area. Observations that predatory species respond fastest, and detritivores most slowly, is perhaps no more than indicative of the initial level of depletion.

Typical is the response of populations of demersal fish in the no-go security area off Cape Canaveral, which are now more abundant by a factor of 13 than outside the area and are also individually much larger than before the closure. A spillover effect is significant along a 200 km stretch of coastline, where anomalously large specimens of fish (compared with the local population) now occur regularly; the length of the period between closure of the area and the first appearance of large fish nearby corresponds closely with the longevity and growth rates of the species concerned. The same effect has also been noted in very large closed areas, as after the 1994 closure of 17,000 km^2 on Georges Bank to all forms of fishing liable to damage demersal fish or their habitat: this closure quite rapidly turned around fishery-induced declines of haddock, yellowtail and witch flounders and especially of scallops, whose densities after five years had reached an order of magnitude greater than those in fished areas on Georges.

The increased interest in MPAs has led to increased interest in the spatial dimensions of the biology of individual species, or ocean neighbourhoods in the sense of Palumbi, following Sewell Wright: 'the area centered on a set of parents that is large enough to retain most of the offspring of those parents. If adults move widely, neighborhoods are large and diffuse. If adults are sessile and larvae are restricted in their dispersal, then a neighborhood might be small and distinct. Other species with sedentary adults and highly dispersed larvae may have large neighborhoods if long-distance dispersal is common or small neighborhoods if long-distance larval dispersal is rare.'

It goes without saying that the effective size of species-specific neighbourhoods is highly diverse, and also that MPAs must be more effective for those species having neighbourhoods whose dimensions match those of the protected area; the rule-of-thumb suggestions that have been made concerning the proper size of MPAs in relation to the area occupied by an entire species are thus seen to be derisory. Unfortunately, we are very ignorant concerning neighbourhood size, especially with regard to the dispersal areas of planktonic larvae, so that generally the best we can do is to examine the genetic structure of subpopulations, assuming that the distribution of these demonstrates the area covered by larval drift, juvenile settlement and adult migration. The ecology of dispersal and adult migration is very complex, yet must be one of the keys to the successful planning of a system of MPAs.

Besides any effect on fish populations, a general evolution of their benthic habitat towards the pristine state has also been noted in many MPAs; the increase of predatory fish populations in coastal closed areas off New Zealand, Tasmania and Chile led in each place to a reduction in the density of grazing echinoids. Consequently, the shallow-water urchin barrens characteristic of these areas prior to protection once more grew forests of large kelp that now begin to approach the pristine state.

The special case of the response of tropical coral reefs to protection perhaps merits special attention. These are complex and fragile ecotopes in which a balance between hermatypic corals, benthic algae, herbivores and predators must be maintained. A healthy hermatypic reef has a high coverage of coral organisms and an abundance of fish, both large predatory species and herbivores, together with urchins that graze algae and, in some species, living coral material. This balance is disturbed by the removal of large fish by sports or artisanal fishing and rapidly results in overgrowth of the coral organisms by algal turf, or their bioerosion by some species of urchins.

The rate of degradation of hermatypic reefs is today causing serious concern, as in the Caribbean where a progressive decline of large staghorn (*Acropora*) and elkhorn corals has been noted during the past 40–50 years. In the 1950s, the reefs were relatively pristine on the north coast of Jamaica and supported large populations of both herbivorous and predatory fish; by the 1970s, the fish were largely gone including – most importantly – the smaller, herbivorous surgeon- and parrotfish. Nevertheless, the large corals remained healthy, being kept clear of macroalgae by the last remaining abundant herbivore – the echinoid *Diadema antillarum* – whose population had exploded after the predatory fish were removed. In 1983, a then unknown pathogen removed the urchins and by the 1990s the corals had largely been overgrown: less than 3% of the area of the reefs still supported live staghorn coral.[49,50] This is the classic textbook description of coral reef dynamics and response to fishing, and should be reversible if protection can be applied – and, indeed, recovery in protected areas of at least the fish populations has been recorded, as in the US Virgin Islands.[51]

Although it is generally assumed that full recovery of corals will occur once a sufficient biomass of herbivorous fish has been restored: such is the herbivory scenario of Jackson that has been generally accepted as the mechanism involved in the destruction and recovery of reefs, at least on their deeper slopes. A recent study[52] rejects the herbivore hypothesis as the sole cause of this degradation, pointing out that the primary cause of the disappearance of *Acropora* corals in shallow water in the Caribbean was a pathogen, the white-band disease that opened up large areas of reef flat to algal growth that could not be controlled by herbivores; it is now widespread in the warm seas. Other mechanisms are also probably involved in the almost global decline of the health of reef ecosystems: a massive survey of the this process in the Indo-Pacific, based on 6000 individual surveys of 2600 reefs over the period 1968–2008 shows that reef degradation has been much greater than anticipated: reefs having >60% living coral cover, and thus resembling the pristine state, are now an anomaly representing less than 2% of all reefs and, progressively, coral coverage has been declining at about 2–3% annually in the Indo-Pacific since 1993.[53] Contrary to expectation, regions that are thought to be relatively well managed, such as the Great Barrier Reef, are now no longer exceptional in terms of active coral cover.

More generally, corals also react to seawater temperature increases of 1–2°C above ambient for a week or two, or to unusually low-salinity seawater, by expulsion of their zooxanthellae, the symbiotic photosynthetic organisms that that inhabit their tissues. This leads to bleaching of the entire coral organism, and (after some weeks) to death by starvation: this is a natural process, associated very strongly with ENSO events, as in 1997 and 2002 and, as I write, in 2008, although it may also be responding more generally to the current global increase of seawater temperature.

Parenthetically, given that fishing is at least one cause of degradation of reef ecology, and the reduction in coverage by live coral, it should be no surprise that the fringing reefs of the atolls of the Chagos archipelago are said to be the most pristine remaining: the human population of these atolls was evacuated to Mauritius and other places in the late 1960s on the orders of the British government of the day, so as to establish a NATO surveillance and transit base on Diego Garcia. The rest of the Chagos atolls have been uninhabited since then so that the fish populations of the reefs have been rested for almost 50 years, which is plenty of time for large herbivorous fish to establish ecologically correct population sizes again. It is something of a moral dilemma to know how to balance the current agitation in Britain for the return of the Chagos population to their atolls against the proposal to declare their home a Marine Protected Area (or Marine Monument in the current US terminology) where fishing would be prohibited. How should one balance the rights of the coral ecosystem to remain functional against the islanders' rights of return to take up fishing again?

Although each of these new approaches has its (literally) ardent supporters who claim that it represents the way of the future in fish stock management after a bad period of fumbling and stock depletions, it is premature either to support the enthusiasm of these people or to suggest – on the other hand – that these new approaches will come to be (at the worst) semantic exercises, or (at the best) techniques of very limited application. All have the potential for rendering stock management more effective than it has been in the past, and each may be able to assist in cases where classical stock management has failed, although none can claim to represent a universal cure.

It is unfortunate that at least part of the discussion concerning the development of new methods has taken a path towards scholasticism, especially in the case of EBFM, and also in the maze of political/economic theory in the case of privatisation of resources. Nevertheless, each does represent a new sensitivity to ecological and social reality that holds some promise for changing management for the better.

Unfortunately, discussions of remedies for the fishing crisis generally neglect the simple fact that over a very large part of the ocean there is no effective management; by far the greatest number of fishermen – be they industrial or artisanal – go to sea in situations where there is neither effective control over their methods and catches, nor over the reporting of their activities. Even the artisanal fisheries that in the past had been the sustenance of coastal villages from Africa to Labrador can no longer be depended on, often solely because of changing social conditions. The post-Colonial population explosion along tropical coastlines would alone have put enough pressure on local fish populations to empty the shallow seas of resources, even without the incursion of new tools: monofilament gill nets, outboard motors and the rest. There is always some patching to be done, as often by benevolent organisations as by the state, to encourage fishermen and their wives to form fishing and marketing cooperatives, but the basic problem of too many people and too few fish cannot be resolved.

The same is also true at a larger scale, because the demand by the swelling (I almost wrote swollen) population of developed countries cannot be supplied from their own seas. Europe is one of the largest producers of fish products from its own EEZ, yet must import a very large fraction of its own requirements, and the same is true for other large centres of population in developed or rapidly developing nations. The outlook for rational management of wild populations to satisfy this increasing demand cannot be said to be very bright and we can be confident of only one thing: that fish farming will play an increasingly important role in our nutrition – but that introduces another set of problems that are beyond my scope here.

As shall be briefly discussed in the final chapter, the percentage of the nations of the world that exercise any credible level of management is dismally small and it requires more idealism than I possess to assume that the new approaches to management discussed in this chapter, and so often hailed as the new way forward, have any chance of generating a change in stock management at anything approaching global scale. The reality at sea is otherwise.

ENDNOTES

1. Holt, S.J. (2006) The notion of sustainability. In: *Gaining Ground: In Pursuit of Ecological Sustainability*, Ed. D.M. Lavigne. Guelph, Ontario: International Fund for Animal Welfare, pp. 43–81.
2. Garcia, S.M. and R.J.R. Grainger. (2004) Gloom and doom? The future of marine capture fisheries. *Phil. Trans. R. Soc. B* **360**, 21–46.
3. Ibid.
4. Sidney Holt suggested to me that the notion of 'precaution' is old hat in fishery science, and was used by, for example, the IWC scientists in the 1970s – and thus long before it was formalised and popularised.
5. Richards L.J. and J.-J. Maguire. (1998) Recent international agreements and the precautionary approach; new directions for fisheries management science. *Can. J. Fish. Aquat. Sci.* **55**, 1545–52.
6. Shelton, P.A. (2007) The weakening role of science in the management of groundfish off the east coast of Canada. *ICES J. Mar. Sci.* **64**, 723–9.
7. David Schindler, Evidence to the Standing Senate Committee on Fisheries and Oceans, Ottawa, 11 May 2004.
8. Christophorou, L.G. (2001) *Place of Science in a World of Values and Facts.* Series: Innovations in Science Education and Technology, #10, New York: Springer.
9. Once again, I apologise to the reader for being parochial – but the discipline-oriented Fishery Oeanography Center at La Jolla, that I had the honour of

directing in the 1960s, evolved into the objective-oriented Southwest Fisheries Center during my tenure – and against my wishes and advice: this was a direct result of Robert McNamara, and his PPBS system of administration that spread like wildfire through federal agencies.

10. Lemieux, P. (2003) Information cascades: why everybody thinks alike. *Le quebécois libre*, 2.11.03.
11. Here, I use 'EBFM' as a shorthand for all the many manifestations of the phenomenon: at times, it has seemed that there are as many acronyms as there are authors writing about the application of ecology to stock management. The differences in definition offered have been fiercely defended but trivial.
12. Mercer, M.G. (1982) Multispecies approaches in fisheries management. *Can. Spec. Publ. Fish. Aquat. Sci.* **59**, 1–168.
13. May, R.M. (1979) Management of multispecies fisheries. *Science* **205**, 267–77.
14. This document had been preceded in 1986 by an FAO document, authored by John Caddy and Gary Sharp, which discussed an ecological framework for marine fisheries investigations.
15. Count based on keywords in title of articles listed by Google Scholar.
16. Sherman, K. and L.M. Alexander. (1986) *Variability and Management of Large Marine Ecosystems*. AAAS Selected Symposia 99. Boulder, CO: Westview Press.
17. Christensen, V. and D. Pauly. (1992) ECOPATH II – A software for balancing steady-state ecosystem models and calculating network characteristics. *Ecol. Modelling* **61**, 169–85.
18. Caddy, J. (2008) EBM opinion: EBM is about more than just managing fisheries. *Mar. Ecosyst. Mgt.* **2**, 7.
19. Document JNCC 03 D14 (December 2003) of the UK Joint Nature Conservancy Committee.
20. Christensen, V. and D. Pauly. (2004) Editorial: Placing fisheries in the ecosystem context, an introduction. *Ecol. Modelling* **172**, 103–07.
21. Pauly, D. (2002) Consilience: concept and some digressions. In *The Gulf of Guinea Large Marine Ecosystem*. Leiden: Elsevier, LME series, ed. J.M. McGlade.
22. See, for example, *Bull. Mar. Sci.* **70**, 589–611 and *Fisheries* **27**, 18–22.
23. Larkin, P.A. (1996) Concepts and issues in marine ecosystem management. *Rev. Fish. Biol. Fish.* **6**, 139–64.
24. See, for example, summaries of the revised EU Common Fisheries Policy.
25. Steele, J.H. (2006) Are there eco-metrics for fisheries? *Fish. Res.* **77**, 1–3.
26. Dengbol, P. (2002) The ecosystem approach and fisheries management institutions: the noble art of addressing complexity and uncertainty with all onboard and on a budget. *IIFET Paper* **171**, 1–11.
27. Pikitsch, E.K., *et al.* (2004) Policy Forum: ecosystem-based fishery management. *Science* **305**, 346–7.
28. See www.nceas.ucsb.edu/projects/12109 (B. Worm and R. Hilborn, NCEAS Project 12109, 'Finding common ground in marine conservation and management').
29. Lemieux, P., op. cit.
30. See a discussion of this problem in *Fish. Res.* **86**, 1–5.
31. Holt, S.J. (2009) A comment to MEAM concerning an article by D. Pauly. Marine Ecosystems and Management, a bi-monthly information service of University of Washington, available at http://depts.washington.edu/meam/
32. Shannon, V, *et al.* (eds) (2006) *Benguela: Predicting a Large Marine Ecosystem*, Amsterdam: Elsevier.
33. And also see Rutherford, R.J., *et al.* (2005) Integrated ocean management and the collaborative planning process: the Eastern Scotia Shelf Integrated Management (ESSIM) initiative. *Mar. Policy* **29**, 75–83.

34. Caddy, J.F. (1998) Deciding on precautionary management measures for a stock and appropriate limit reference points as a basis for a multi-LRP Harvest Law. *NAFO SCR Doc. 8, SN 2983*, 13 pp.
35. Koeller, P. (2008) Ecosystem-based psychology or how I learned to stop worrying and love the data. *Fish. Res.* **90**, 1–5.
36. A full description of this approach is given in DFO document CSAC/SAR 2008/85 from which this image was obtained, courtesy of Peter Koeller.
37. One of the earliest discussants was Robert May (1980) in *Nature* **287**, 675–6.
38. A very good introduction to the whole question of rights-based fishing is the NATO ASI held in Reykjavik in 1988, published as Neher, P.A., *et al.* (1989) *Rights-based Fishing*. Berlin: Springer-Verlag.
39. For a useful review of property rights in fisheries, see Leal, D.R. (2004) *Evolving Property Rights in Marine Fisheries*. Lanham, MD: Rowman and Littlefield.
40. Michael Belliveau, Exec, Sec., Maritime Fishermens Union, 1998, quoted in Standing Senate Committee on Fisheries (Canada) 3rd Report, 2006.
41. Copes, P. and G. Palsson. (2000) Challenging ITQs: Legal and political action in Iceland, Canada and Latin America – a preliminary overview. Oregon State University, Proceedings IIFET, pp. 1–5.
42. Castello, C., S.D. Gaines and J. Lynham. (2008) Can catch shares prevent fisheries collapse? *Science* **321**, 1678–81.
43. For example, *The Economist*, 20 September 2008.
44. Arnason, R. (2002) *A review of international experiences with ITQs*. University of Portsmouth, Future Options for UK Fish Quota Management – A report to the Ministry for the Environment, Food and Rural Affairs.
45. Kaiser, M.J. (2005) Are marine protected areas a red herring or a panacea? *Can. J. Fish. Aquat. Sci.* **62**, 1194–9.
46. Sainsbury, K. and U.R. Sumaila. (2003) Incorporating ecosystem objectives into management of sustainable marine fisheries, including best practice reference points and use of marine protected areas. In: *Responsible Fisheries in the Marine Ecosystem*, Eds. M. Sinclair and G. Valdimarsson. Rome: FAO pp. 343–61.
47. Recounted in Charles, A. (2000) *Sustainable Fishery Systems*. Oxford: Blackwell Science, p. 370.
48. This section owes much to reviews by Palumbi, S.R. (2004) Marine reserves and ocean neighbourhoods. *Ann. Rev. Envir. Resour.* **29**, 31–68, and Gell, F.R. and C.M. Roberts. (2003) Benefits beyond boundaries: the fishery effects of marine reserves. *Trends Ecol. Evol.* **18**, 448–55.
49. Jackson *et al.*, op. cit.
50. Zabel, R.W., *et al.* (2003) Ecologically sustainable yield. *Am. Scient.* **91**, 15–157.
51. Beets, J. and A. Friedlander. (1999) Evaluation of a conservation strategy: a spawning aggregation closure for red hind, *Epinephalus guttatus*. *Environ. Biol. Fish.* **55**, 91–8.
52. Aronson, R.B. and W.F. Precht. (2006) Conservation, precaution and Caribbean reefs. *Coral Reefs* **25**, 441–50.
53. Bruno, J.F. and E.R. Selig. (2007) Regional declines of coral cover in the Indo-Pacific: timing, extent and subregional comparisons. *PlusOne* **2** (8) e711.

11

Conclusion: sustainability can be achieved rarely and only under special conditions

'A traveller bought a little fish and gave it to his inn-keeper to prepare; asked if he had cheese and vinegar and olive oil, the traveller replied that if he had those things he would not have bought the fish . . .'

Plutarch on *The Eating of Flesh*

Does anything in the previous chapters provide what we are looking for – a mechanism that supports the assumption that marine ecosystems are able to 'produce a surplus that we can share, year for year', provided always that they are not excessively transformed by fishing or some other activity?

What most strongly supports this thesis is the characteristic biology of teleost fish, which are very remarkable organisms, because of the interaction between a reproductive mechanism that is unique among vertebrates and lifelong, indeterminate growth. Although the significance of their very high fecundity for sustaining fisheries may have been misinterpreted in the past, the biology of those teleost fish that adopt a 'fecundity', rather than a 'bestowal', reproductive strategy may be essential in maintaining populations of large predatory fish in cold seas with high primary production but a very low diversity of other fish species: as discussed in Chapter 2, this reproductive strategy of teleosts may be the essential ingredient in our great northern fisheries.

Because their relative fecundity increases progressively throughout life after maturity, a depleted adult population of teleosts may be able to increase more rapidly in numbers of individuals when external conditions change for the better than could, for instance, a depleted population of terrestrial mammals.

Teleosts are also capable of anomalously strong density-dependent increase in weight-at-age, since their pattern of indeterminate growth

(Chapter 3) enables them to respond throughout life to improved feeding conditions: organisms having a determinate growth pattern can respond in this way only during their youth. This capability of teleosts must enhance their ability, compared with mammalian populations, to recover from a fishery-induced population decline if fishing mortality is reduced or removed.

Then, the relative openness of the marine habitat compared with terrestrial habitats (Chapter 4) suggests an explanation for the simple observation that many teleost populations can indeed tolerate some level of fishing pressure. Perhaps marine organisms, including fish, have a greater tolerance than terrestrial vertebrates to mortality imposed by a novel predator that has invaded their pristine ecosystem by natural means. If this is correct, it may explain why fishing – representing an invasive and novel predator – appears, at least subjectively, to be better tolerated than hunting by invasive humans is by terrestrial mammals.

Although there can be no doubting the truth of the general assumption that the growth of fish populations can be manipulated to increase their productivity by imposing additional mortality on top of what they suffer naturally, whether we can compute appropriate levels of additional mortality is altogether another matter. So it is hard to know how much confidence we can have in statements like the following, taken from an EU Council regulation prepared by the Commission for the long-term management of European hake: 'Rather than targeting specific biomass levels it has proven necessary to provide for a situation where the stock may fluctuate unpredictably around the recovery target ... therefore proposed to fix the objective of exploiting the stock at a fishing mortality rate of 0.17 per year. This is in accordance with scientific advice that fishing at this rate will provide the highest yields from the stock.' But, more importantly, what confidence can we have that this fishing rate will not be exceeded when the fleets get to sea and, if it is, that this fact will be recorded so that the advice to be modified next time around? Indeed, the list of those whose immediate interest is to exceed any set level of mortality is legion, and the activities of ship owners, wholesalers, politicians and the rest are far more likely to be the cause of the mismanagement of fisheries than is erroneous scientific advice.

Despite such problems, the notion of sustainability remains the foundation of fishery science and is also inherent in the remarks taken as my text[1] concerning the properties of marine ecosystems and the surplus biomass that we may take by fishing, but before we really get to sustainable fishing, it may be useful to take a small diversion to consider what we really mean by sustainable – a word very easy to use without much thought. If you think you know what it means, I suggest a quick consultation of the dusty OED on your shelves, where you will find rather different and less politically charged meanings from the present-day 'harvesting a resource so that it is not depleted or permanently damaged' of online dictionaries. Typical also of today's usage is the 'sustainable yield ... that can be maintained indefinitely into the future' of the World Commission on Environment and Development. This is a splendid example of newspeak, because no species in nature has sufficiently invariant rates of reproduction, growth and natural mortality to produce an unchanging yield into any future. Indeed, advocates of sustainable development ignore the evident fact that the words sustainability and development when placed together create an oxymoron. Sidney Holt has suggested that 'sustainable' has now lost any meaning except as propaganda by those who claim that their activities have a light footprint on the natural world.[2] Because I prefer not to abandon perfectly good words, I use 'sustainable' to mean something very like the OED's 'to sustain a cavalry charge' – that is to say, that a sufficient number of infantry should be left on their feet afterwards to continue to operate as a viable unit. So, until the target population of fish is reduced to a condition where recruitment fails, we can consider the fishery to be sustainable, while continuing to try to manage it for higher levels of catch. But we can never expect present-day yields to be maintained indefinitely into the future!

It is perhaps appropriate now to examine the concept more directly by asking if any modern fisheries are really sustainable (never mind how their products are labelled) and so may serve as examples of what ought to be possible in other seas. However, the task of identifying such fisheries is not easy because, first of all, we have to distinguish those in which the population has collapsed due to natural environmental changes from those which have simply been fished out: this distinction has not generally been made by those who have compiled statistical data on stock collapses, yet it is critical for an understanding of what has occurred. What may help us in answering the question posed in the chapter heading is to locate some fisheries that do not appear to have been brought to ruin by over-exploitation, and then to suggest why collapse has not occurred and to what extent these examples may be extrapolated to other regions and other fisheries. Such a direct approach may be more useful than any amount of theorising about what ought or ought not to be possible.

It has been suggested that the management of fishing can be achieved provided only that its four intrinsic components of sustainability are treated comprehensively: these are said to be ecological, socio-economic, community and institutional sustainability.[3] However, I suggest that there are really only two dynamic components in any fishery system: the fish and the people involved in the fishery. So my closing thesis is that the performance of either of these components is capable of determining whether a fishery shall be sustainable or not. If this is correct, then it will be important to know if a fishery that we observe to be apparently sustainable is the result of (i) the quality of the management regime or (ii) of a peculiarly favourable biology of the fish population and of its environment.

The quality of any management regime can readily (given sufficient information!) be evaluated against the ideal regime, in which:

- the annual stock assessment data are accurate and comprehensive,
- an up-to-date and comprehensive review of the state of the marine ecosystem is routinely available for comparison with archived data,
- there is neither political nor industrial interference to modify the level of the annual permissible catch that is obtained by an appropriate scientific procedure,
- all regulations concerning total catches, permissible gears, closed areas and dates, and discarding are scrupulously respected by fishermen at sea and in their landings, and
- all required reports concerning landings and operations at sea are correctly maintained and deposited.

It goes without saying that this ideal has never been achieved, but where might we expect such conditions to be approximated? Unfortunately, only in a few unusually favoured regions off the coasts of a few highly developed nations, especially in those seas where:

- the control of fishing is in the hands of an organisation that lacks complex and conflicting interest centres, especially those where personal interests may be concerned,
- the social and economic environment is more evolved than evolving,
- communications are easy, and practical management measures such as policing can be adequately funded, and, finally,
- fisheries science and oceanographic research is active, independent and has a long history.

There are a just a few regions where we might conclude that these conditions are fulfilled, and I shall make a somewhat arbitrary choice among these below, although I shall defend my choice with observational evidence. That such regions comprise only a very small fraction of the oceans is obvious to anybody with any background in fishery management, and it would be relatively easy to list such areas; this subjective judgement has recently received some sort of confirmation as the result of a global expert opinion survey[4] in which information was obtained from almost 2000 fishery experts appropriate to 118 national EEZs. The results were entirely as one would expect: in only a very small number of high-latitude zones did overall management effectiveness score even as much as 50% of the optimal score that would be attributed if everything was done right. In the vast majority of zones, the score was abysmal. Tactfully, the authors did not list national scores, but their global map is largely covered in red or orange spots representing very low scores!

On the other hand, it is somewhat more difficult to define the special qualities of a fish that would be a suitable target species in a sustainable fishery, but I suggest that the ideal candidate would likely be:

- a teleost rather than a selachian,
- abundant, widely distributed and genetically homogeneous,
- very fecund, and have low between-year recruitment variability,
- have a wide and discretionary diet,
- be very well accepted in the marketplace and
- normally taken with gear that was highly selective.

Our knowledge of the comparative ecology and biology of teleosts should be sufficient to enable us to locate suitable candidates: they will be neither the clupeids of mass occurrence, nor will they be the demersal species that occupy continental shelves, because both of these groups exhibit a high level of sensitivity to changes in ocean conditions. Rather, I shall suggest that it is, paradoxically, to the large ocean predatory fish that we should look – even though it has been suggested by some that these have been very severely reduced by fishing (Chapter 7).

Using these three sets of criteria, I propose to identify two candidate fisheries for consideration as potentially sustainable: these are selected because sustainability in the first case depends on the unusual qualities of the ecology of the target species while, in the second case, it depends on the unusual qualities of the management system.

A FISHERY FOR AN IDEAL FISH

I propose that the fishery for tropical tunas in the eastern Pacific Ocean is an excellent example of the situation in which a fishery appears to be sustainable because of the qualities of the target species, rather than because of good management.

This fishery is managed under the 1950 Convention that established the Inter-American Tropical Tuna Commission (IATTC) between Costa Rica and the USA; there are now 15 adherent nations, with all the conflicts of national interest that such a number must inevitably raise. The degree of success achieved by IATTC management cannot, of course, be evaluated by reference to one species alone but also with reference to other target species (bigeye, skipjack and yellowfin) and to the broader ecological consequences on populations of non-target fish (*Coryphaena*, *Acanthocybium*, *Elegatis*, *Seriola* and the rest) and on turtles and dolphins. Long-lining is aimed at bigeye, which formed 65% of that catch (by number of fish taken) in 2007, while purse-seining targets yellowfin and skipjack. The by-catch of sharks, rays, billfish and smaller species is very high in this fishery, especially in purse-seine sets around floating objects.

The IATTC proposes action to the member governments that is intended to maintain fishing yellowfin tuna and other species at or below levels appropriate for F_{max} and, from 1966 to 1979, the Commission set annual yellowfin catch quotas (usually below 200,000 tons), which member nations implemented. The tension in the Commission between major fishing nations and what were termed resource-adjacent nations, or RANs, was the subject of a scholarly review by Joseph and Greenough in 1979.[5] These authors analysed the actual and potential problems arising from what they termed regional coalitions of RANs and non-RANs, and arising also from the primitive control by RANs of their own EEZs. They proposed the use of what they termed partial allocations and, eventually, the formation of a global umbrella organisation for tuna management, having both executive and research functions, but concluded that the outlook was rather bleak.

Unfortunately, the future rapidly became even bleaker than Joseph and Greenough anticipated, because in 1979 – the same year in which they wrote – the entire conservation programme was suspended because several member countries of IATTC, including Mexico, withdrew from the Commission. Obviously, the remaining members were reluctant to agree to total catch quotas if several fishing countries, previously members of the Commission, no longer agreed to be limited by overall quota. Nevertheless, IATTC continued informally to recommend an annual international yellowfin tuna catch quota 'as the basis for all participants in the fisheries to evaluate the conservation needs of the resource'. Even though Mexico later rejoined the Commission, the IATTC had by then lost its management function, although science and assessment work continued as before; it was only in 1998, after a 20-year gap, that restrictions were again imposed, especially to protect bigeye tuna: in recent years, these have taken the form of periodic closures of the fishery on floating objects for periods of up to 3 months and/or closure or reduction of the long-line fishery.

Since the annual allocated catches of individual member nations differ widely in quantity, this is an ideal situation for the trading of votes in the Commission for favours in other areas of mutual interest. There are two major groupings of participants: those flying flags from nations on the American continent (the RANs), which fish with purse-seines and mostly for yellowfin and skipjack, and the long-line fleets bearing national flags from the western Pacific (the non-RAN nations), which concentrate on deeper-swimming bigeye.

Recall that each of the national representatives at meetings of the Commission arrives with baggage, comprising rather precise instructions that limit their freedom of argument. Recall also that the national representatives are under pressure from 'a growing coalition, including international conservation organisations and recreational and commercial fishing interests ... repeatedly issued joint statements calling upon the members of IATTC to take urgent action and follow the advice of the IATTCs own highly qualified scientific staff. This advice continues to be disregarded.'[6]

So it is not surprising that, as I write, the fleets are at sea and are fishing without restrictions other than those imposed unilaterally and voluntarily by national governments on their own fleet; the 2008 meetings of IATTC failed to produce any agreement on the annual closure periods and other issues, because of a major confrontation in the Commission between the nations with the two largest fleets of purse-seiners: Mexico, whose fleet mostly sets on dolphins to take yellowfin,[7] and those of Ecuador and Spain, whose fleets use floating Fish Aggregation Devices (FADs) and preferentially catch skipjack. The quarrel between these fleets is based (amongst other matters!) on the fact that FADs[8] continue to attract tuna even during the annual closed period, so fattening the later catches of the Ecuador and Spanish fleets, while the Mexicans argue that they obtain no such windfall at the end of the closure. Further, the Mexicans point out, FADs aggregate tuna too small to retain but which do not survive the seining operation, while young tuna do not swim with dolphins. Further, the by-catch of sharks, rays, mantas, turtles and other species under FADs is not inconsequential and these devices have now been rather widely accused of having the potential of destroying tuna populations.

The representatives of these three nations have, over the past several years, essentially brought the annual IATTC meetings to a halt while they attempted to negotiate their differences behind closed doors; at the end of the June 2008 meeting, a last-minute common proposal was made by them concerning closures in 2008–9 for Class 5 and 6 seiners, together with a line along 5°N to separate Ecuadorian and Mexican fleets. This was not accepted, perhaps because Columbia (a new adherent) simultaneously submitted proposals that could not be previously discussed, since the text had been negotiated behind closed doors, and by just three parties. The representatives returned home without engaging to respect any regulations during the forthcoming season.

I give these details[9] only to illustrate the distance that separates the management regime of this fishery from the ideal discussed above; in fact, what has happened in recent years at IATTC is not so very different from habitual levels of confrontation in other RFMOs.

Nevertheless, the biology of tropical tunas is so close to the ideal target species discussed above that the fishery has been sustained, in a loose sense of that term, for many decades; further, of the four principal species taken in the fishery, those that most closely approach the ideal target species appear to have best sustained the weight of 50 years of fishing: these are yellowfin and skipjack, the focus of whose habitat is in the tropical surface water mass. Bigeye tuna, characteristic of cooler water below the tropical surface water mass, and the subtropical albacore that migrates only seasonally into the eastern tropical Pacific have done less well.

Consider the biology of yellowfin tuna.

- This species is distributed continuously across the entire ocean, coast to coast, equatorward of the 21°C isotherm, a region that is oceanographically rather stable, being perturbed principally by episodic changes in the value of the ENSO index, when rates of primary production are reduced in the eastern ocean.
- There is very little evidence of genetic isolation or regular migration across this enormous area, and individual fish are usually recaptured within 1000 km of where tagged, although some phenotypic distinction are observed between coastal and offshore fish.

- Although the warm tropical surface layer is their basic habitat, individual fish make deep feeding excursions into the thermocline; their diet is rather catholic, being based on fish, cephalopods and crustaceans, according to their local relative abundance in the nekton.
- Yellowfin reproduce almost continuously throughout the year, wherever and whenever they are in water >24°C; the batch-spawning females produce around 2.5 million ova at less than 2-day intervals, representing a total of more than twice their body weight each year.[10] Extra-tropical seasonal spawning in summer may, in some years, result in a cohort distinguishable from fish spawned in tropical water.
- Recruitment of fish at 30 cm varied only from half to double the average count of about 178 million individuals annually in the IATTC area from 1975 to 2005, with a somewhat lower recruitment regime (Figure 11.1a) dominating the first decade of this period.[11]
- Despite the effects of the fishery, which has reduced natural biomass by about 50%,[12] spawning biomass in the IATTC-regulated area has varied only narrowly from approximately 250×10^6 to 800×10^6 tons annually; the principal period of lower biomass around 1980 was perhaps a response to the exceptional El Niño of 1978–79.
- The catches of yellowfin tuna in the eastern Pacific progressively increased after the inception of the fishery in the 1950s and have stabilised at between 180×10^3 and 400×10^3 tons annually, which is close to the MSY-associated level (Figure 11.1b).
- Throughout this period, the age structure of the IATTC area population has remained almost constant, because the loss of the oldest year-classes has been minor compared to the levels of truncation of the age pyramid with which we are familiar in other fisheries. In this fishery, the weight of fishing mortality falls mostly on the middle-aged fish; in any case, the largest fish (say >175 cm) comprised <5% of the pristine populations.[13]
- Some consequences of natural changes in rates of recruitment can be observed in the stock biomass trajectory so that biomass in 2001 was perhaps twice what it was at the start of the fishery in the mid-twentieth century.
- Some trickle-down consequences in the structure of the pelagic ecosystem have been imputed to the change in biomass and size structure of the populations of large tunas, including

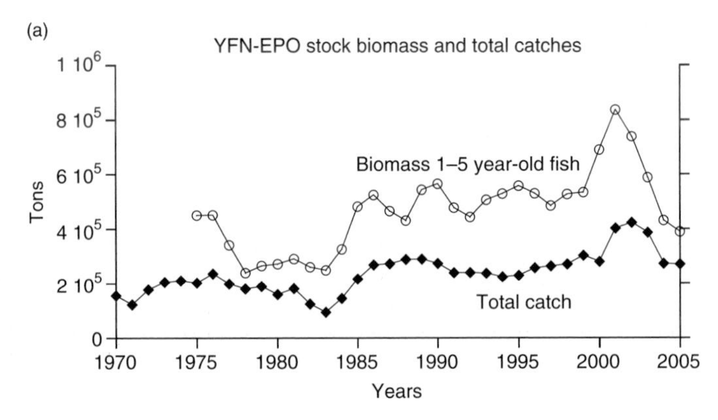

Figure 11.1a Relative stability of recruitment of yellowfin tuna (*Thunnus albacares*) in the eastern tropical Pacific, compared with annual rates of fishing mortality 1970–2005; the unusually strong recruitments of 1987 and 1998 are subsequently reflected in stock biomass and in fishing pressure.

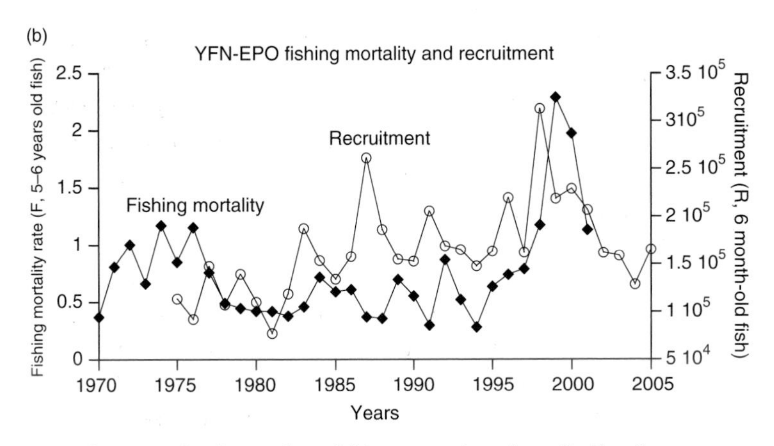

Figure 11.1b Assessed stock biomass and catches of yellowfin tuna (*Thunnus albacares*) in the eastern tropical Pacific monitored by the IATTC during a 35-year period after 1970.

yellowfin, but these seem unlikely to have fundamental consequences[14]: other pelagic predators may be advantaged, so that increased importance of skipjack and other small thunnids, blue sharks and dolphinfish (*Coryphaena*) may be expected.

- The markets for tuna products appear to be solid and have been so since companies like StarKist of California began producing canned tuna and the Japanese abandoned their post-war tuna sausage for more sophisticated canned and frozen products.

It would be very hard to argue, in the face of all these facts, that the fishery for yellowfin in the eastern tropical Pacific is not, in some sense of the word, sustainable, despite the lack of goodwill by at least some of the parties adhering to the IATTC; were individual national interests put aside in compromises, then the future of this fishery would seem to be assured by the obviously sensible management proposals of the IATTC scientific teams. This, then, is the fishery that I advance as my example of a sustainable fishery for an ideal species despite a less-than-ideal management climate: had yellowfin the same biological characteristics as southern bluefin, the fishery would have long ago been extinct.

A FISHERY WITH EXCEPTIONAL MANAGEMENT

Turning now to the other side of the coin, to a fishery that is well managed and probably sustainable despite far-from-ideal target species, let us examine the demersal fisheries off Alaska. Here, the ecology of the fish and the complexity and variability of their environment create much more difficult problems than are encountered in the tropical tuna fisheries yet, despite this, management is sufficiently well organised and sensitive that several large and profitable fisheries appear likely to be viable indefinitely, despite what some alarmists suggest[15]; these fisheries are dominated by catches of walleye pollock, Pacific cod, sablefish, Pacific Ocean perch and Pacific halibut. The biomass of the top nine species is computed to have been within the range $4.5–5.5 \times 10^6$ tons between 1984 and 2002; over a longer period, catches have fluctuated strongly with high catches during the mid 1960s and mid 1980s. The management of these fish is complicated not only by the nature of their ocean environment, but also by their co-occurrence with seven species of Pacific salmon and Steelhead trout, of great economic and social importance, but having much lower population biomass, in the range 250,000–300,000 tons during the same period. Fisheries for shrimp and snow crab are also very important.

The Gulf of Alaska and the Bering Sea offer the individual populations of these species with two very different environments, both of which are subject to major environmental regime shifts in response to decadal-scale changes in the atmospheric pressure field (see Chapter 5). The eastern shelf of the Bering Sea is sufficiently wide (>750 km) to support three different ecological domains, and is very productive although experiencing strong seasonality; the shelf along the fjord coastline of the Gulf of Alaska is narrower and extends to 200 km only in the region of Kodiak and the peninsula. Both of these shelf regions, together with large areas beyond the shelf edge, lie entirely within the EEZ of the United States; they are divided administratively into a 12-mile coastal zone of US waters, which is the responsibility of Alaska, while the rest of the EEZ is considered to be Federal waters, managed from Washington, DC.

The ecology and seasonal productivity of the Bering Sea responds to the duration of winter ice coverage, while the ice-free Gulf of Alaska is dominated by the rather variable coastal downwelling conditions associated with the cyclonic circulation of the Alaska Gyre. As was discussed in Chapter 5, the strength of this circulation is controlled by decadal-scale variability of atmospheric forcing; the positive phase of the PDO is associated with enhanced gyral circulation, nutrient pumping and enhanced production centrally in the gyre and, along the coast, a shallower mixed-layer depth and enhanced productivity. The biological consequences are significant and are timed according to the distance between environmental forcing and ecological effect; in general, the positive phase of the PDO is associated with positive (in the fisheries sense) effects in the Gulf of Alaska, and the negative phase with negative effects. The ecological consequences are sufficiently strong to cause regime shifts between states, and these have been identified as being associated with PDO changes in sign in at least 1925, 1947, 1977 and 1989.

The variability of the oceanographic regime is reflected in strong natural variability in recruitment success and population biomass of all the principal species, so that their biological characteristics are very far from those of the ideal species discussed above. Consider Pacific halibut (*Hippoglossus stenolepis*), one of the mainstays of the Alaskan demersal fisheries. This very large flatfish (<2.5 m) has great longevity (<55 years), a long pre-maturity period (*c.* 8 years), a long pelagic phase of westward drift (6–7 months) and a longer return migration. Both recruitment and growth rates are very variable, responding to environmental conditions; periods of slow growth occurred in the periods 1920–40, and 1990–99, while fish grew significantly faster and larger in the period 1970–90.

Over the period 1935–2000, Pacific halibut recruitment in the central Gulf of Alaska closely tracked the winter PDO value ranging from 8 to 12 million age-6 recruits at the beginning and end of this period, compared with only around 2–8 million in the central period 1960–80 when PDO values were low.[16] In the same region, the recruitment of several other species responded similarly to the value of the PDO during winter and hence to its shift towards higher values in 1979–80 (Pacific Ocean perch, arrowtooth flounder and Sitka herring), as does the catch rate of populations of chinook, chum, coho, pink and sockeye salmon in all three regions (west, central and east) of the Gulf of Alaska.[17] In the Bering Sea, seasonal ice cover, and hence planktonic productivity, shifted to values above the long-term mean in the same period.

Management of fishing in this region is thus done against a background of shifting ecological regimes; here, the consequences of regime shift (see Chapter 5) are sufficiently strong that they cannot be ignored or overlooked as they have been elsewhere. However, Alaskan management must also deal with a wide range of species having very different ecology from that of iconic Pacific halibut noted briefly above. The largest landings from this region have been habitually taken from walleye pollock (*Theragra chaligramma*), whose ecology strongly contrasts with that of halibut. Smaller (<1.0 m), lower longevity (<17 years), these bentho-pelagic gadoids migrate seasonally between feeding and spawning areas and are fished with mid-water trawls that also take pelagic salmon as a by-catch. There are also important fisheries for benthic invertebrates (snow crab, Alaska red crab and shrimp).

The resource base of the Alaskan fisheries is therefore greatly more complex both ecologically and spatially, and has significantly lower renewal rates, than the tuna resources of the eastern topical Pacific, yet the populations appear able to sustain the fishing levels that have been imposed. This suggests that the regional management must be unusually sensitive. Management decisions should be much simpler to take here than for the tropical tuna fisheries: instead of debate and quota bargaining among the representatives of 17 nations at the annual meetings of the IATTC, only 3 entities are involved: the federal North Pacific Fishery Management Council (NPFMC) which manages fishing in Federal waters (90% of groundfish landings), the State of Alaska Department of Fish and Game (ADFG) which manages those in State waters (10%), and the International Pacific Halibut Commission (IPHC) which manages halibut. The NPFMC comprises 10 voting members representing various interests in Alaska, Washington, Oregon and a federal government representative, the regional director of National Marine Fisheries Service (NMFS).

Management advice for the NPFMC is derived from the NMFS and from international scientific commissions established to study transboundary species with Canada. This relationship is the only foreign intervention in the process, because all foreign vessels have been excluded from the US EEZ since the 1990s, when the capacity of the US fleets was deemed sufficient to take any foreseeable TACs. There are conflicting interests within both the Gulf of Alaska and the Bering Sea, the principal problem being by-catch of salmon in the mid-water trawl fisheries; salmon are at once a valuable industrial resource, the target of a very vocal sports fishing lobby and the principal sustenance of isolated indigenous communities, especially in western Alaska: tensions are, therefore, sometimes rather intense.

Other conflicts arise over the protection of the important populations of marine mammals and birds, whose reproductive habitats are very sensitive to disturbance. Management decisions taken by the NPFMC are, of course, confirmed by the US Secretary of Commerce in Washington, to whom the Governor of the State of Alaska also has access in relation to fishery management decisions. It is important to note that the Secretary of Commerce is an appointed official and not an elected politician, so he is not subject to the same intense pressure from an electorate as fisheries ministers elsewhere: a Canadian fisheries minister having his electorate in, say, Newfoundland does not have the same freedom of action as the US Secretary of Commerce.

The NPFMC has taken a very proactive and sensitive stance to conflicting interests and has adopted very firmly the principle of EBFM and all that that involves[18]: in fact, the application of this principle by the NPFMC meets or exceeds procedures and techniques established by the US National Academy of Sciences as being essential to 'rebuild and sustain' fisheries resources and their habitat. In particular, the NPFMC claims that it routinely incorporates ecosystem-based goals into management and

- that it has adopted conservative harvest levels in setting annual catch limits and that for at least 25 years annual catch limits have been set at or below the Acceptable Biological Catch (ABC) level recommended by scientists,
- that ABC levels are set individually for each species and fisheries close when limit is reached, all by-catch being counted against limits,
- that federal observers are carried (at the expense of industry) on vessels >40 m LOA for 100%, and on smaller vessels 30% of the time; processing vessels in the pollock fishery carry 2 observers on each trip,

- that these observers produce data on by-catch that are monitored in real time ashore, so that areas with high by-catch levels are quickly identified and other vessels are warned to avoid them,
- that data from federal stock assessment surveys are available for several decades into the past, thus enhancing the value of annual surveys,
- that a comprehensive habitat protection policy allows the closure of very large areas of productive fishing grounds either seasonally or permanently, and, finally,
- all communities, including Alaska natives, are represented on the NPFMC; all meetings are open to the public, and public input is solicited on all proposed management measures.

Annual Fishery Management Plans (FMP) are established in each fishery in Federal waters and for each species having, as one objective, the creation of a stable management environment within which industry can plan its activities consistently. For this purpose, an Optimum Yield (OY) range is specified, within which annual catch limits are expected to be set. For the Gulf of Alaska groundfish complex, OY limits were set in January 2009 between 116,000 and 800,000 mt. The lower limit was equal to the lowest historical catch 1965–85 and the upper limit was established from MSY computations for the period 1983–87, recognising the volatility of two species: flounder and pollock. Within this fishery, six species are categorised as prohibited (and must be immediately returned to the sea), seven as target species (for each of which a TAC will be set), four as other species (for which a group TAC will be set) and nine families are listed as forage species (for which regulations are set in each FMP). For each species, in each fishery, account is kept of the levels of information on Essential Fish Habitat that is available to managers, specifying whether information levels concerning habitat status of eggs, larvae, early and late juveniles and adults are sufficient for management or not.

A Prohibited Species Catch (PSC) limit is established, as in the following example.

(i) When the number of mature female crabs is equal to or below the threshold of 8.4 million, or spawning biomass is less than 14.5 million lbs, then the Zone 1 PSC limit will be 32,000 crabs.

(ii) When the number of mature female crabs is above the threshold of 8.4 million mature crabs and the effective spawning biomass is equal to or greater than 14.5 but less than 55 million lbs, the Zone 1 PSC limit will be 97,000 crabs.

(iii) When the number of mature female crabs is above the threshold of 8.4 million mature crabs and the effective spawning biomass is equal to or greater than 14.5 but greater than 55 million lbs, the Zone 1 PSC limit will be 197,000 crabs.

Although such specifications are not unusual and resemble those employed in other fisheries administration, the unusual tension created by salmon by-catch requires unusually complex regulation. By-catch in the Bering Sea pollock fishery has been implicated in the decline of chinook salmon populations and this problem is being resolved by agreement between pollock cooperatives and community development groups on a rolling hotspot closure system so that small areas with very high chinook by-catch rates are identified and avoided. Vessels using benthic trawls for flatfish and cod are exempt, because they take few chinook.

In addition to such rolling closures, a very complex and extensive suite of area closures and restrictions (Figure 11.2) covers well in excess of 200,000 km^2. For instance, State waters are entirely closed to bottom trawling along the entire coast of Alaska west of Prince William Inlet, including Kodiak and other islands and the Alaska Peninsula, while the central part of Cook Inlet is closed to all trawling, as are the entire State and Federal waters along the Alaska panhandle down to the Canadian border. There are many other closed regions, especially in the east Bering Sea, and around the Aleutian and Pribilof Islands, for a variety of reasons: marine reserves and marine mammal habitat protection, crab protection areas, coldwater coral habitat protection, and so on. Each, of course, is established with a formal specification of restricted practices and periods.

Individual quotas and ITQs have been introduced extensively into these fisheries, although not so universally as in the New Zealand fisheries, and the sale and replacement of boats in such fisheries is strictly regulated. ITQs in the crab fishery got off to a rocky start because the new arrangements put very many fishermen out of work, as noted in the previous chapter. However, after some years, the boats remaining in the fishery have been enthusiastic concerning the effect on resource conservation and stability of the fishery; the ITQ licences require that 90% of their crabs on a set schedule to designated

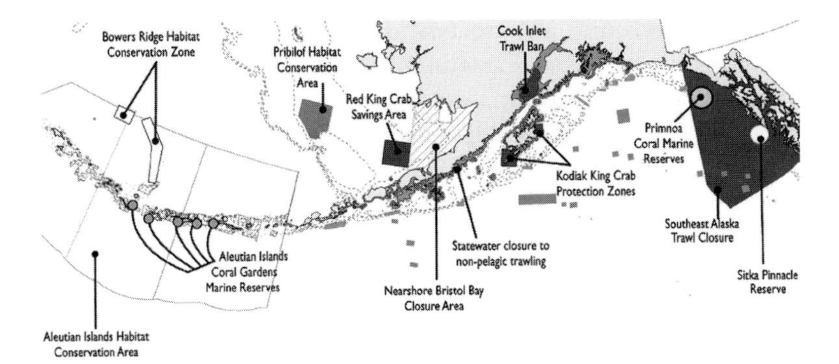

Figure 11.2 The closed areas and special protection regimes imposed over the Alaskan continental shelf and the national EEZ regions of the Gulf of Alaska and eastern Bering Sea. Only the larger special-status regions are labelled. (Courtesy of North Pacific Fisheries Management Council.)

processors located in each of eight crab-fishing districts of the Bering Sea and Aleutian chain. Nine coastal communities that are dependent on crab fishing received guarantees that boats and processors located in their community would receive equitable quota allocations. Prices are determined by bargaining between processors and fishermen's organisations at the start of each season and are binding for that season.

Management by the State of Alaska within coastal waters mirrors federal management in the EEZ, but with some important differences, which the case of the Pacific cod fishery in the central Gulf of Alaska will illustrate. When the NPFMC-managed fishery is open, a parallel fishery opens in State waters alongside, with similar gear restrictions, although with only a voluntary catch-reporting system. The State management plan for 2009 makes provision for a fishery in State waters to open seven days after the federal fishery closes; this is restricted to pot and jigging/trolling methods, defines complex individually designated areas, and specifies the rate at which the total quota is to be filled during the open season.

So much for this remarkable State/Federal management system, but what about the results? That question is not easy to answer for two reasons: first, of course, because of the naturally high variability of recruitment of most species and, second, because of the relatively short period during which the full management system has been in place: it cannot be held responsible for whatever occurred as the modern fisheries developed off Alaska.

The domination of present-day catches by pollock also somewhat confuses the issue, because it is a relatively recent development. In the 1960s, the Gulf of Alaska attracted a fleet of 70+ Soviet factory trawlers together with their supply ships and this severely reduced the populations of Pacific ocean perch, until then very abundant on the outer shelf and slope; a larger Japanese fleet was not long behind, attracted by the same species and sablefish. These fleets dominated the landings from the Alaska region for many years, and it was not until 1976 that the requirements of the Magnuson Act began to cause management to evolve, first towards joint ventures with foreign enterprises until, in 1991 management became an entirely domestic responsibility when foreign ships were excluded.

This history is revealed very clearly in the multispecies catch statistics showing a progressive increase in landings towards a plateau at around 2.0–2.5 million tons annually in the early 1990s to which the Bering Sea fisheries make by far the greatest contribution, particularly because the overall pattern of landings came to be dominated by Bering Sea pollock (Figure 11.3) during a period when the biomass of this species was unusually high. The overall catch statistics from this region show catches of ocean perch and other slope rockfish reaching a peak of around 250,000 tons in the mid 1960s, and sustained but much lower catch rates until the present day. Gadoids, principally *Merluccius productus* and *Gadus macrocephalus* catches (Figure 11.4), appear to be maintained, although the populations are naturally variable, at levels reached in the 1990s, as are those of *Limanda* and *Hippoglossus* in the Bering Sea, although other flatfish in the same region have reached rather low catch levels (Figure 11.5).

Intense fishing on pollock spawning concentrations during the 1980s were associated with serial population declines and closure of some areas although, since the start of US administration, overall catches have remained relatively high after recovering from the first wave of foreign trawling in the 1970s. Background material supplied with the January 2009 groundfish management plan for the eastern Bering Sea noted, retrospectively, that the population biomass had peaked in 1985, then declined by 1991 to about 6 million tons, but after 1995 had fluctuated around 12 million tons. The 2004 spawning biomass had been estimated at 3.5 million tons or well above that required to sustain an MSY of 2.5 million tons. However, as we saw in Figure 11.3, a serious decline in landings has occurred in the last few years and is creating some alarm.

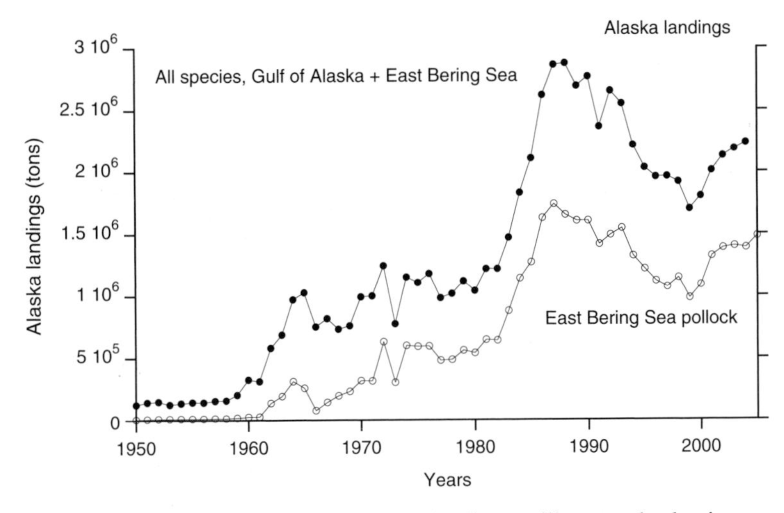

Figure 11.3 Historical Alaskan bulk landings to illustrate the dominance of walleye pollock (*Theragra chalogramma*) from the East Bering Sea since the earliest days of the mechanised fishery.

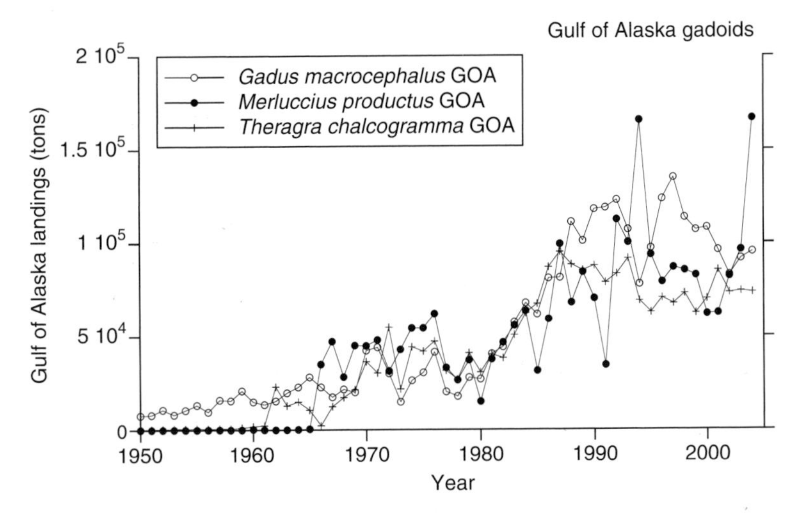

Figure 11.4 Progression of the landings of the principal gadoid species from the Gulf of Alaska demersal fishery, 1950–2004.

This is not the place for a detailed performance evaluation of the management practices of the NPFMC, but a glance at Figure 11.6 will convince the reader of the inherent difficulty of management of populations in this region: over the 25-year period prior to 1975, the catch rate

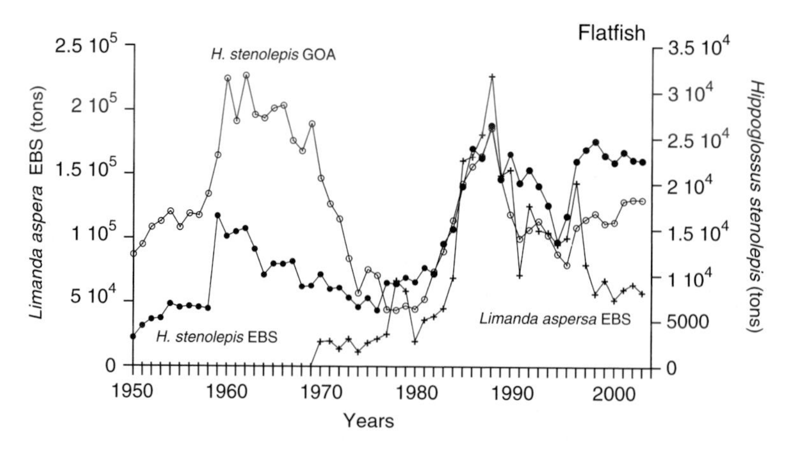

Figure 11.5 Progression of landings of *Limanda* from the eastern Bering Sea and *Hippoglossus* from there and also from the Gulf of Alaska.

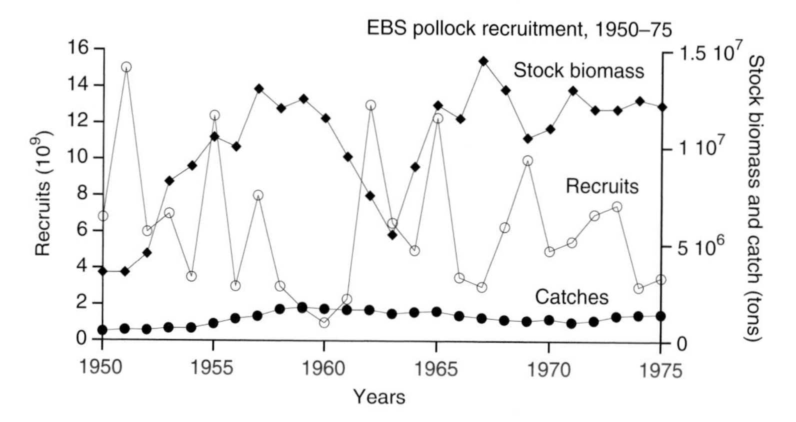

Figure 11.6 The highly variable recruitment of walleye pollock in the eastern Bering Sea, to suggest a relationship between change in size of stock biomass and episodes of poor and good recruitment.

of eastern Bering Sea pollock remained stable within limits acceptable to industry, while the total population biomass generally showed an increase: both of these trends occurred despite a background of highly variable recruitment of 1-year-old fish, the levels of which appear to be totally unconnected to population biomass of the spawning year. As the figure shows, the population biomass responded strongly to depressed recruitment in 1963–65 and to strong recruitment in 1951 and 1963. A similar plot of sablefish biomass in the Bering Sea reveals a recruitment pattern even more nightmarish for rational stock management.

The fish populations of this region (despite the cries of some alarmists) are far from general collapse, although if they had been managed during the last 20 years by a conglomeration of entities, and conflicting interest centres, such as those currently responsible for managing the comparable region of the NE Atlantic, then the Alaskan fisheries (based on an ecosystem as difficult to predict and manage as one could imagine) would not be in the relatively satisfactory state that we see today.

From the comparison between the performance of the management of tropical tuna and of Alaskan groundfish a simple conclusion may be drawn that can be widely generalised: it is not the quality of fishery science in each region that determines the quality of management there, but rather the organisation and administration of management decisions based on excellent science in each region. The fisheries literature has been replete in recent years with proposals that the fishery crisis can be resolved if only we were more sensitive to environmental variability, or that we must manage ecosystems and not fish populations, or that open access is disastrous (and socially incorrect); we are told that quotas must be personalised and transferable, that discarding at sea must be prevented, that novel stock evaluation methods must be adopted, and so on and so forth. At the same time, one subject has attracted too little attention, perhaps because it is intractable: that of political involvement in the implementation of scientific advice.

Unfortunately for the fish, fishing is an economic activity like any other and cannot be ignored in the political administration of any state; it is, perforce, subject to the changing wills of parties holding power in the central government. The relative economic importance and political sensitivity of fisheries will determine the response of any administration to the crisis of the day, and long-term solutions are not to be looked for, in general. Ray Beverton discussed an important consequence of the relationship between the state and fisheries management in the text of his posthumous Larkin Lecture: he related that British politicians were horrified when it was explained to them that the inevitable consequence of control of fishing mortality rate by TAC had been the build up of excess fishing capacity (and of IUU fishing) that would now have to be reduced if other management measures were to have any chance of success. The notion of having to undertake a fleet reduction programme, which would put many men out of work, was not appealing to them.

To this comparison of two very different management regimes and sets of target populations might be added a third, represented by the management of the EU Common Fisheries Policy, where both the management system and the target populations depart very far from the ideals that I discussed at the start of this chapter. Complex, bureaucratised and politicised management of fish stocks that are not inherently simple to manage in the first place represents a very unsatisfactory situation. It cannot help the stability of European fisheries that this is one of the more densely populated regions of the world and is also one of the more complex politically: it is inconceivable, for instance, that this region could accept such a sweeping protection of habitat as the 12-mile 'no bottom-trawling zone' that exists around the entire coastline of the Gulf of Alaska. It is really not surprising that the EU Fisheries Commission, speaking on behalf of 27 nations, all with their political fingers in the same pie and busily engaged in vote-swapping, has noted that European fish populations are in a significantly worse state than global populations, according to the evaluations made by FAO. Yet the quality of the fishery science that is performed in European seas is as high as anywhere.

It is self-evident that, to the extent that day-to-day management can be isolated from the mechanisms of the state and the interests of politicians, the more likely it is that fisheries will be successfully managed. I believe that political interest is at least as important in the dynamics of fisheries management as often-quoted problems such as those of open access and poaching, although, curiously, the influence of politicians and politics on natural resource management appears to have been an almost wholly neglected dynamic in comprehensive studies of fisheries and their management. Criticism of political interference in science-based management of populations has come largely from journalists and, more informally, has been discussed on the Internet. There have been some formal reactions, however, such as that of the Royal Society of London, which issued a warning in 2003 to the then Prime Minister's Strategy Unit on UK Fisheries: 'Current fishing practice is unsustainable ... the wrangling over quotas by EU fisheries ministers risks making the situation worse. The level of reduction (of quotas) that is adopted is often less than required, due to lobbying by the fishing industry and by disputes between different countries.' One has to wonder if anybody was listening ...

* * * * * * * *

I have placed the account of Plutarch's frugal traveller at the head of this chapter to remind us that, in the end, we are now so numerous on our little planet that even if we were all as spartan and rational as he was, our demand for food from the sea would still be sufficient to ensure that sea fish populations should become serially depleted in all but a few favoured and well-managed regions of our oceans. It is unreasonable to pretend that the future of industrialised and globalised sea fishing is very bright in a world as crowded as ours has become.

ENDNOTES

1. See Preface.
2. Holt, S.J. (2006) The notion of sustainability. In: *Gaining Ground: In Pursuit of Ecological Sustainability*, Ed. D.M. Lavigne. Guelph, Ontario: International Fund for Animal Welfare), pp. 43–81.
3. Charles, A. (2001) *Sustainable Fishery Systems*. Oxford: Blackwell Science.
4. Mora, C., *et al.* (2009) Management effectiveness of the world's marine fisheries. *PLoS Biology* **7**, 1–11.
5. Joseph, J. and J.W. Greenough. (1979) *International Management of Tuna, Porpoise and Billfish*. Seattle, WA: University of Washington Press, xvi + 254 pp., and see also Longhurst, A.R. (1979) Options for fisheries management. *Science* **205**, 779–80.
6. Erika Viltz, World Wildlife Fund, 1 July 2008.
7. 'Dolphin fishing' involves setting on a group of dolphins, together with the tuna that are habitually associated with them, using techniques – such as the Medina panel – to release the dolphins after the set is complete. In the 1960s, when dolphin sets were first introduced by Californian seiners, the number of dolphins killed became an international scandal, but the numbers killed today are very much smaller and probably sustainable.
8. FADs can range in sophistication from flotsam, encountered by accident, to smart objects that are moored subsurface and equipped with sonar and GPS for distant interrogation as to quantities of fish currently aggregated.
9. Letter of 6 August 2008 from Luis Ricardo Paredes M (Columbian representative to IATTC) to Erika Viltz (vide supra).
10. Schaeffer, K.M. (1998) Reproductive biology of yellowfin tuna in the eastern Pacific Ocean. *Bull. IATTC* **21**, 205–72.
11. IATTC (2008) Tunas and billfish in the ETP. Fishery status report, #6.
12. Far from the 90% suggested by Myers and Worm, a problem discussed at some length in Chapter 7.
13. Schaeffer, K.M. (1998), op. cit.
14. Sibert, J., *et al.* (2006) Biomass, size and trophic status of top predators in the Pacific Ocean. *Science* **314**, 1773–6.
15. See, for example, the Ocean Foundation post by John Hocevar and Jeremy Jackson, 16.4.09, entitled 'The fishery that's too big to fail', and positing the imminent destruction of the pollock populations.

16. Perry, R.I. (2004) *Marine Ecosystems of the North Pacific.* PICES Spec. Pub. 1, pp. 222–5.
17. Hare, S.R. and N.J. Mantua. (2000) Empirical evidence for the North Pacific regime shifts in 1977 and 1989. *Progr. Oceanogr.* **42**, 103–45.
18. Witherell, D., *et al.* (2000) An ecosystem-based approach for the Alaska groundfish fisheries. *ICES J. Mar. Sci.* **57**, 771–7.

Index